THE CREATION OF QUANTUM MECHANICS AND THE BOHR-PAULI DIALOGUE

STUDIES IN THE HISTORY
OF MODERN SCIENCE

VOLUME 14

JOHN HENDRY

THE CREATION OF QUANTUM MECHANICS AND THE BOHR-PAULI DIALOGUE

D. REIDEL PUBLISHING COMPANY

A MEMBER OF THE KLUWER ACADEMIC PUBLISHERS GROUP

DORDRECHT / BOSTON / LANCASTER

Library of Congress Cataloging in Publication Data

CIP

Hendry, John, 1952-
The creation of quantum mechanics and the Bohr-Pauli dialogue.
(Studies in the history of modern science; v. 14)
Bibliography; p.
Includes index.
1. Quantum theory - History. 2. Bohr, Niels Henrik David, 1885–1962.
3. Pauli, Wolfgang, 1900–1958. I. Title. II. Series.
QC173.98.H46 1984 530.1'2'09 83–27251
ISBN 90–277–1648–X

Published by D. Reidel Publishing Company,
P.O. Box 17, 3300 AA Dordrecht, Holland.

Sold and distributed in the U.S.A. and Canada
by Kluwer Academic Publishers,
190 Old Derby Street, Hingham, MA 02043, U.S.A.

In all other countries, sold and distributed
by Kluwer Academic Publishers Group,
P.O. Box 322, 3300 AH Dordrecht, Holland.

Printed in The Netherlands.

To Dee

TABLE OF CONTENTS

PREFACE

Many books have been written on the history of quantum mechanics. So far as I am aware, however, this is the first to incorporate the results of the large amount of detailed scholarly research completed by professional historians of physics over the past fifteen years. It is also, I believe, the first since Max Jammer's pioneering study of fifteen years ago to attempt a genuine 'history' as opposed to a mere technical report or popular or semi-popular account. My aims in making this attempt have been to satisfy the needs of historians of science and, more especially, to promote a serious interest in the history of science among physicists and physics students. Since the creation of quantum mechanics was inevitably a technical process conducted through the medium of technical language it has been impossible to avoid the introduction of a large amount of such language. Some acquaintance with quantum mechanics, corresponding to that obtained through an undergraduate physics course, has accordingly been assumed. I have tried to ensure, however, that such an acquaintance should be sufficient as well as necessary, and even someone with only the most basic grounding in physics should be able with judicious skipping, to get through the book. The technical details are essential to the dialogue, but the plot proceeds and can, I hope, be understood on a non-technical level.

The research for this book has extended over a number of years, and although the bulk of it is published here for the first time some aspects have appeared in my Ph.D. thesis and in a number of published papers. In the course of my work I have received assistance of varying kinds from many teachers and colleagues, and I should like to offer my thanks to them and also to those other colleagues on whose published research I have drawn. The latter, very considerable, debt is recorded in the bibliography. In the former class I should like to mention particularly Gerald Whitrow and Jon Dorling, who oversaw my doctoral research, and David Cassidy, John Heilbron, Erwin Hiebert, Karl von Meyenn and B. L. van der Waerden, who have provided much-needed criticism and valuable encouragement since. I should also like to thank all owners of copyright for permission to reproduce private correspondence, the Royal Society and the British Academy for invaluable financial assistance, and the staff of the old History of Science and Technology

Department at Imperial College, London, of the Niels Bohr Institutet, Copenhagen, and of the Historisches Institut, University of Stuttgart, for hospitality at various times.

Spring 1983 JOHN HENDRY

SCIENTIFIC NOTATION

In most cases, scientific symbols used are defined upon their first appearance. Exceptions are standard notations and the symbols for quantum numbers in the old quantum atomic theory, which are given below.

Standard Notations

c	Speed of light
ν	Frequency
e	Electron charge
m	Particle mass
h	Planck's constant

Quantum Numbers (for alkali atoms with single valence electron)

n	Radial quantum number
l, $k - \frac{1}{2}$, $K - \frac{1}{2}$	Azimuthal quantum number (orbital angular momentum of valence electron)
$j, J - \frac{1}{2}$	Inner quantum number (total angular momentum)
s, R, r	Spin, or angular momentum of atom minus valence electron
m	Magnetic quantum number (total angular momentum in direction of applied magnetic field)

Every sentence I utter is to be understood as a question and not as an affirmative.

NIELS BOHR

In my youth I believed myself to be a revolutionary; now I see that I was a classicist.

WOLFGANG PAULI

CHAPTER 1

INTRODUCTION

Quantum mechanics has evolved considerably since it first received a definitive formulation and interpretation in 1927. The quantum theory of fields grew up quickly over the next few years and after the discovery of a new range of elementary particles in the 1930s the formulation of field and particle theories has developed apace since the war. But in all this, two things have remained more or less unchanged. On one hand the quantum theory has continued in all its formulations to show a remarkable predictive power in respect of experimental observations. In this respect it must rank as an extraordinarily successful physical theory, and as one that will not easily be displaced. On the other hand, however, dissatisfaction with the conceptual foundations of the theory has also apparently endured. Many working physicists are seemingly content to accept what Einstein referred to as the "gentle pillow" of the Copenhagen interpretation without asking any further questions, and this has long been accepted as an orthodox position. But if we restrict our attention to physicists (or indeed philosophers) of the first rank, then we see immediately that such an orthodoxy is illusory. It was created in the late 1920s when many of the leading quantum physicists, among them Bohr, Born, Heisenberg, Pauli, Dirac, Jordan and von Neumann, sunk their more philosophical differences in an effort to repel the challenge of the semi-classical interpretations and get on with the job of developing quantum electrodynamics. But those differences remained. Copenhagenism was and is a generic term covering a whole range of related interpretations.

Even when these interpretations are taken together, they cannot be considered as an entirely dominant orthodoxy. Among their early opponents some physicists might arguably be dismissed as narrow-sighted conservatives. But such outright dismissal is very difficult to uphold in Einstein's case, and still more so in those of Schrödinger and de Broglie, neither of whose preferred interpretations could reasonably be labelled classical. More recently attention has shifted from the physical interpretation of quantum mechanics towards the logical and mathematical consistency of quantum field theory, but the issues remain closely connected and opposition to Copenhagenism remains strong. However, and here lies the crux of the matter, the opponents seem to be no nearer to providing a valid alternative than were their predecessors of the late 1920s. Beyond the limited compromise of Copenhagenism there is still no such thing as a consistent and generally acceptable interpretation of quantum mechanics, and the evidence of the last fifty years points unerringly to the conclusion that there will not be one until either the structure of our physical conceptions, or our expectations of physical theory, or the quantum theory itself should undergo radical changes more far-reaching than any yet seen.

Faced with this dilemma it is tempting to react as did Peter Debye to the problem of electrons in the nucleus, a problem that arose in the immediate wake of quantum mechanics, by treating it as something best ignored, "like the new taxes". And many physicists have indeed taken this course, either ignoring the interpretative problem altogether (paying the taxes without question) or proceeding stubbornly to seek fundamentally classical interpretations that are demonstrably not there (stalling the taxman). But whereas such attitudes may be expedient in the short term they are ultimately inconsistent with the very spirit of the scientific enterprise. Wolfgang Pauli, who will turn out to be one of the main actors in the story that follows, responded to Debye's suggestion by announcing publicly his hitherto private speculations on the existence of the neutrino. He was convinced that any positive line of exploration was better than none at all, and this conviction quickly bore fruit as the existence of the neutrino was confirmed. The interpretative problem of quantum theory is several orders more fundamental than that of nuclear electrons, and has proved immensely more resistant to attempts at a solution. But a theory with innate inconsistencies, whatever its present predictive success, cannot be expected to serve for ever. If the problem, like the tax, does not bear thinking about, then that is the strongest indication we can possibly have that it needs thinking about. And while it may not be so easily solved we can at least try to understand how such an extreme situation arose in the first place.

One aim of this study, then, is to approach the history of the theory of quantum mechanics as a means of exploring its philosophy. We shall not address ourselves explicitly to the various attempts to impose philosophical positions upon the completed theory. But by looking at the conceptual features surrounding its genesis, and at the philosophical and other preconceptions built into its structure and foundations, we shall try to come to a clearer understanding of these attempts, of the chronic differences of opinion they reflect, and of the essential nature of the problem they address.

The second aim is to present the history of quantum mechanics in such a way as to enable it to be related to the mainstream of intellectual history, not through the study of surface parallels reflecting the prevailing themata of the milieu, but through that of the essential conceptual changes underlying these.

With these aims in mind, the history will necessarily be a selective one, and the selection may sometimes appear arbitrary. The concentration will be upon the course of matrix mechanics, for example, at the expense of that of wave mechanics. This is not to deny the historical importance of the latter theory, but it is the former, the work of an interacting group rather than of a succession of individuals, in which the conceptual tensions can most easily be seen. Again, though the listed contents of the history will inevitably have much in common with those of the existing accounts, the emphasis will be very different. In particular there will be no attempt made to give anything like a complete account of the more technical developments within quantum theory, which have been well treated elsewhere and are covered by the bibliography. The same holds also for the prevailing philosophical milieu, so long as its notions were not applied directly and consequentially to physical problems. Instead the concern here will be with the middle area, with fundamental but specific epistemological and methodological beliefs, and with the manifestation of these beliefs both in the general debate on the quantum problems and, especially, in the major technical advances.

Straightforward as it is, this perspective requires some comment, for the concern of historians to date has tended to be with either the technical development of quantum mechanics or its conceptual background, as expressed through largely undefined philosophical debates, such as that on the causality issue. Even where the two aspects have been treated by the same historian, as by Paul Forman in his wide range of scholarly papers, they have remained imperfectly connected. But whatever the immediate concern of the historian it is this very connection that is most crucial to the dynamics of the events being considered.

The language of modern theoretical physics is that of symbolic models and mathematical equations, and although a physicist may think in terms of physical or metaphysical concepts he must continually translate his thoughts into this working language. It provides not only the means of communication within the physics community but also those between the underlying principles of physics on one hand and the highly artificial experimental results on the other. It is thus an omnipresent and irreducible part of the development of physical science. And, this being the case, the innovations of the 1920s were inevitably of a highly technical, symbolic and mathematical nature. Any attempt to reduce them completely to abstract conceptual developments automatically distorts the historical process and introduces a gulf between story and evidence that is historically unacceptable. But in seeking to understand why as opposed to how the innovations took place, and in seeking to extract the true essence of these innovations, it is the development of abstract concepts that is of the most relevance and interest. This poses a problem for the historian. The physicists concerned with a particular set of problems did not all share the same influences and concerns; each one approached the current issues, both technical and philosophical, from a different conceptual framework. Some no doubt did see the technical problems of quantum theory as just that, and others no doubt saw the historically elevated causality issue as the most fundamental imaginable. But this was patently not the case for the most innovative of the quantum physicists, who derived their views on both subjects from more fundamental concerns.

It is usually impossible to connect a scientist's deepest, usually religious, concerns with his scientific work, and no attempt will be made to do this here. But it is possible to relate both the technical development of quantum mechanics and the surrounding philosophical debate to more fundamental issues concerning the concepts used to describe the physical world. The aim here will therefore be to trace the development of these issues, and of their manifestation in the evolution of quantum mechanics. This done it should be possible to place the interpretative problem of quantum mechanics in a somewhat clearer historical light than hitherto, to relate the interpretations offered more closely and accurately to currents of thought prevailing in other fields, and even to get a clear idea of what type of interpretative problems remain.

The plan of the study is to begin with an analysis of the conceptual background to quantum mechanics in terms both of the commonly perceived issues (causality, energy conservation, the wave-particle duality) and of the fundamental epistemological and methodological problems manifest at about the same time in the context of general relativity theory and the search for a unified field theory. Having shown how these latter problems were carried over into the quantum context we shall then relate the conceptual background first to the technical problems encountered in the old quantum theory, then to the genesis and reception of the ill-fated virtual oscillator theory, and then to the genesis in 1925 of Heisenberg's "new kinematics", the foundation stone of quantum mechanics. Through this first part of the study the evolving theme will be one of a debate, its origins in the quest for a unified general relativistic field theory, between the established master,

Niels Bohr, and the up-and-coming Wolfgang Pauli. And in the second part we shall take this theme as a guide through the evolution of quantum mechanics, from Heisenberg's breakthrough to the establishment of a definitive formulation and interpretation in 1927. We shall look first at the early exploration of the meaning, significance and implications of the new kinematics, then at the early evolution of wave mechanics and at the attempts of Heisenberg, Pauli and Dirac to integrate the technical advances of this theory with the fundamental concepts of Heisenberg's. A predominant theme in this part of the story will be an ongoing attempt to base quantum mechanics upon a foundational framework consistent with these characteristic concepts; and in the ensuing sections we shall show how this attempt, building upon the insights achieved in the wake of wave mechanics, led first to the definitive formalism of the statistical transformation theory in Hilbert space, and then to Heisenberg's uncertainty principle. All these developments took place largely within the framework of Pauli's ideas, but in the last section of the main story we shall turn back to the other side of the fundamental debate and trace the evolution of Bohr's principle of complementarity and the compromise of the Copenhagen group of interpretations.

The characteristic of quantum mechanics that made it a truly revolutionary theory of physics was its explicit rejection of the classical criterion of a consistent visualisation of physical events. For centuries natural philosophers had argued whether a consistent picture in space and time was the ideal to which all theories should aspire or an idol imposed by the tyranny of the senses, the worship of which could only stand in the way of truth. By the time the issue was put to the test in the context of the quantum paradoxes, however, the old debate had apparently been long forgotten, and it was not immediately obvious just what was at stake. The role of visualisation was subsumed under more general questions as to the operational foundations of physical concepts, and indeed remained so subsumed throughout the creation of the theory. Throughout our study too, then, the central theme will be that of the status and applicability of physical concepts. As we reach the end of the story, this theme will merge into that of *Anschaulichkeit*, sometimes translatable as visualisability, sometimes as something less specific to the visual sense. Broadly speaking the path will be from the competing demands of visualisation on one hand and physical operationalism on the other towards a common recognition that neither ideal was obtainable.

CHAPTER 2

WOLFGANG PAULI AND THE SEARCH FOR A UNIFIED THEORY

INTRODUCTION: MANIFESTATIONS OF DUALITY

It has been customary for historians investigating the background to quantum mechanics to concentrate on one or other of the particular branches of quantum physics. In some cases this approach may be vindicated by a similar specialisation on the part of physicists whose work is being discussed. But even though they may have published only within restricted areas, most physicists took their perception of the problem complex from the wider field of their reading, correspondence and conversation. And most of those involved in the actual development of quantum mechanics had already contributed important work in several distinct areas. In general, the major conceptual issues ran through all such areas, and their most obvious and widely discussed manifestations were not necessarily their most significant. The wave-particle duality, which was almost certainly the most characteristic conceptual feature of quantum theory, was most commonly discussed in the contexts of experimental X-ray physics and individual absorption and emission phenomena. This discussion reached its peak, in the years immediately before the development of the new quantum mechanics, in the context of the Compton effect, viewed by at least one historian as a turning point in physics.[1] But by the time of Compton's experiments this whole area had long been one of only secondary importance for the active development of ideas pertaining to the duality problem, and dramatic though his results were to a wider audience they had relatively little impact on the physicists whose work was to be of importance in the creation of quantum mechanics.[2]

Ever since the turn of the century, and especially since the demonstration of X-ray diffraction in 1912, discrete X-ray phenomena had offered the clearest and best-known demonstration of the localised particulate properties of light, just as the continuous absorption of interference phenomena provided the clearest portrayal of its non-localised wave-like nature. But following the work of Poincaré, Ehrenfest and Jeans it was clearly recognised by about 1913 that any proof of this dual nature of light rested not on the individual wave and particle phenomena but on the statistical behaviour of black-body radiation, according to Planck's law. Although the fact had not been recognised at the time, it was in 1900, in his very first presentation of this law, that Planck had, albeit unwittingly, introduced the duality to physics; and it was in the details of the law's derivation that the duality manifested itself as a logical necessity. Even with the demonstration of this necessity and the full recognition of the wave-particle duality, this duality was not universally accepted, and for the best of reasons: in the face of a long-standing tradition requiring a physical description to be above all structurally visualisable and as such self-consistent, it seemed better to have an intelligible classical wave theory with flaws than a totally unintelligible "dual" theory.[3] The issue therefore remained a live one and a new wave of X-ray experiments in 1921 and the ensuing Compton effect results were the subject of heated discussion. But to physicists such as de Broglie and Schrödinger, in their struggles towards the wave mechanics, as Jordan and Dirac, who played a large part in the development of matrix mechanics, and as the old masters Einstein, Lorentz, and Planck himself, the phenomena of black-body radiation and quantum statistics were of far greater importance.[4] And meanwhile the development of theoretical spectroscopy through the Bohr theory of the atom had also encountered the wave-particle duality as a force every bit as potent as in the earlier contexts.

The problems of quantum spectroscopy were less explicitly related to the wave-particle paradox than either those of X-ray phenomena or those of quantum statistics. But partly because of this, because the paradox could be approached obliquely and because it did not stand in the way of a host of new results and predictions, many of the most brilliant physicists found this field the more suitable context for their endeavours to solve the quantum problem. In particular Bohr himself developed his ideas very largely within this context, and the problem complex of the Bohr theory was, as has indeed been recognised, the central source of quantum mechanics in the form of Heisenberg's new kinematics.[5] It was not, however, the only such source. For once we begin to look at just how the new kinematics developed out of the Bohr theory, the full extent to which the fundamental problems transcended the individual branches of physics becomes clear. And we find that Heisenberg's work is also rooted in yet another problem complex, that of the search for a unified general relativistic field theory.

The main debate over unified general relativistic theories, following Weyl's attempt at such a theory in 1918, does not in fact feature in the existing accounts of the development of quantum mechanics. Since neither Bohr nor Heisenberg were at all involved, and since the theories bore little direct relationship to the quantum postulate, this is not perhaps surprising. But the debate did occur in the years immediately preceding the genesis of quantum mechanics; and apart from Weyl it did involve Einstein and especially Pauli, who carried over what he had learnt into the quantum context and into his creative interaction of the early 1920s with Heisenberg and Bohr. The basic issues concerned a duality between field and particle theories that was in fact closely akin to the wave-particle duality of quantum theory. And the debate was so fundamental and important as to form an integral part of the perceived problem complex for any physicist who was actively involved in general relativity theory – including such important teachers and contributors to the quantum theory as Sommerfeld, Born, Schrödinger, Dirac and Hilbert. In this chapter we shall therefore look at the search for a unified field theory, and in particular at Pauli's response to this search, and show how it came to interact in general terms with the conceptual debate surrounding quantum theory.

WEYL'S THEORY AND ITS RECEPTION

The general theory of relativity had evolved gradually over a period of years, but had finally appeared in fully developed form in 1916.[6] From the observed equivalence between inertial and gravitational masses, Einstein had deduced that the gravitational field imparted the same acceleration to all bodies, and had postulated that it could therefore be transformed away completely and replaced for each infinitesimal region, and for its effects on all physical processes there, by a suitable choice of the space-time coordinate system. This "principle of equivalence", combined with the assumption of the special theory of relativity for each transformed infinitesimal region, led directly to an identification of the gravitational field with a geometry, or coordinate system, of space-time. Drawing on the postulate of "general covariance", according to which physical laws were required to be independent of the choice of coordinate system, Einstein had identified this geometry as Riemannian: whereas in the traditional, 'flat', Euclidean geometry the directions of vectors at different points could be directly compared, their relative direction in the 'curved' Riemannian geometry was dependent upon the choice of path joining the point. The parameters defining this choice could be identified with those defining the gravitational field.[7]

The development of Einstein's theory was closely monitored by David Hilbert and Max Born in the mathematics department at Göttingen, and also by Hermann Weyl, who was Hilbert's assistant and colleague until the end of

1913, when he moved to Zürich.[8] In Zürich for a year, Weyl worked mainly on mathematics; but he kept up his contacts with Göttingen, and also made direct contact with Einstein, then a professor in the same city. Weyl was called up for service in the First World War, but returned to Zürich late in 1916. Einstein, meanwhile, had moved to Berlin, but Weyl had become sufficiently interested in his work, and was sufficiently impressed by his newly published general theory of relativity, to drop his pre-war mathematical investigations and make Einstein's work the basis of a new programme of his own.[9] Early in 1918 he published a mathematical generalisation of Einstein's theory, and suggested that this generalisation might encompass not only gravitational but also electromagnetic phenomena.[10]

Weyl argued that the essence of the change from Euclidean to Riemannian geometry in the general theory of relativity lay in the change from a finite geometry, in which a uniform metric could be applied, to an infinitesimal one, in which a metric could only be applied directly to infinitesimally small regions. One consequence of this change was the dependence in Riemannian geometry already noted, of the relative direction of two vectors acting at two distant points on the choice of path joining them. But, argued Weyl, the Riemannian geometry still contained an element of finite geometry that made it mathematically something of a compromise, for the specific choice of which there was no sound *a priori* reason. Whereas the relative direction of two vectors acting at distant points could not be directly compared, their relative length could. Weyl therefore proposed a generalised geometry in which such relative lengths were also dependent upon the choice of path joining the two points, or, in other words, in which elements of length were specified only to within an arbitrary function of position, denoted by the term 'gauge'. In order to make this generalised geometry the foundation of a physical theory, as the Riemannian geometry was the foundation of Einstein's theory of gravitation, Weyl supplemented Einstein's principle of general covariance by the requirement of gauge invariance, namely that physical laws should be independent of the choice of gauge. He then proceeded to build a generalised analogue of Einstein's theory.

In Einstein's Riemannian theory, the connection between the coordinates of a point in space-time and another point infinitely close was given by the quadratic form corresponding to the 'distance', ds, between them:

$$\mathrm{d}s^2 = \sum_{\mu\nu} g_{\mu\nu}\,\mathrm{d}x_\mu\,\mathrm{d}x_\nu,$$

with $g_{\mu\nu}$ the coordinates of the gravitational potential. In Weyl's generalised theory this relationship still held. But the relationship between the two points, and the general metric structure of space, also depended upon an additional, linear, form,

$$d\phi = \sum_{\mu} \phi_\mu \, dx_\mu .$$

The interest of this for Weyl lay in the fact that just as the gravitational potential could be equated with the coordinates appearing in the quadratic form, so the electromagnetic potential could be expressed by linear coordinates of the form ϕ_μ. Pursuing this analogy he sought ways in which the classical electromagnetic theory could be definitely connected with the new geometry. Starting from the basic assumption, which he attributed to Gustav Mie, that all laws of a generalised physical theory should rest upon a generalised action invariant, or principle of least action, he suggested a function that might fill this role. And, leaving the actual identity of such a generalised action function undetermined, he showed that just as the laws of energy and momentum conservation could be identified as by Einstein and others with the general relativistic invariance of the normal action function, so the conservation of electricity could in principle be identified with the gauge invariance of this new, as yet arbitrary, function. This achieved, he interpreted his theory as a unification of gravity and electromagnetism, and, while stressing the mathematical difficulties in the way of further developments, he also held out hope for a deduction of the existence of the electron and of the quantum behaviour of the atom from the theory.[11]

Extraordinarily difficult though Weyl's work was, it was enthusiastically received. Mie wrote to him that it was the most fascinating mathematical work he had ever read,[12] Sommerfeld that he found it "truly wonderful",[13] Eddington that it was his "constant companion",[14] and Einstein that it was a "master-symphony".[15] The young Pauli started his research career by developing aspects of the theory,[16] and E. T. Whittaker's recollections confirm that it made an enormous impression upon many theoretical physicists of the period.[17] But with the praise came criticism.

Einstein had originally welcomed Weyl's interest in general relativity theory, but he had already recognised from the first exchange of ideas that they were unlikely to agree, and had expressed himself pleased in particular because "he who has the most powerful opponent excels".[18] When in March 1918 Weyl sent him proofs of his book, *Raum-Zeit-Materie*, incorporating his theory, Einstein responded with extravagant praise and agreed immediately to communicate a paper on the new theory to the Berlin Academy.[19] On receiving the paper he was again deeply impressed,[20] but a week later he wrote to Weyl that, despite the beauty of his theory, "I must say frankly that it cannot possibly correspond in my opinion to the theory of Nature, that in itself it has no real meaning."[21] The problem, Einstein explained, was that since the relative lengths of two space-time vectors depended in Weyl's theory upon the path joining them, the lengths of two measuring rods or the periods of two clocks, separated and then brought together again, would depend upon the paths each had taken in the meantime. This discrepancy would increase with the time of separation, and Einstein argued that if it really existed it would show up in the measurements of atomic spectra, the observed frequencies corresponding to the periods of the atomic clocks. Since no such discrepancy was observed, he could not accept that Weyl's theory had any real physical significance.[22] A few days later Einstein reported with apologies that Nernst and Planck were even less happy with Weyl's paper than he was, and that they were demanding on behalf of the Berlin Academy that his opinion be appended to Weyl's paper as a postscript. Together with a reply by Weyl, it was.[23]

In his reply, Weyl argued that he did not in fact see his theory as counter to experience, for the formal mathematical process of vector displacements that was the foundation for his geometry could not be identified with the physical motions of clocks and measuring rods, the behaviour of which had still to be worked out. He conceded that Einstein's own theory maintained the lengths of measuring rods and periods of clocks. But he insisted this theory had only been developed for the case of a static gravitational field, with no electromagnetic field present. In strongly varying gravitational and electromagnetic fields such as would be needed to effect the separation and return proposed, the constancy of the period of a clock could be predicted according to Weyl by neither Einstein's theory nor his own, as they had so far been developed. This being the case he returned to his claim that the geometry of his theory was the only one that could be mathematically justified, and that it would be strange indeed were Nature to have chosen a compromise geometry such as Einstein's, arbitrarily pasted onto the electromagnetic field.[24]

Weyl had already expressed the opinion in his book that *a priori* considerations could only specify what relations were possible in Nature, and that the choice of actual physical laws had to be based upon empirical considerations.[25] And he conceded that in this respect his own theory remained a speculative one to be proved or disproved.[26] But he had also set out his view that *a priori* considerations were fundamental to theoretical physics. Writing to Einstein in May 1918 he repeated his reluctance to "accuse God of mathematical inconsistency", and insisted that he still believed in his theory and in the importance of carrying it through to testable physical conclusions.[27] After initially retorting that it seemed to him just as bad to accuse God of a theoretical physics that did not do justice to human observations,[28] Einstein for his part accepted the validity of Weyl's defence – though not of his philosophy of physics – and while stating his belief that Weyl's theory would be proved false accepted that it had not yet been so proved.[29]

Although Einstein remained convinced that Weyl's theory would prove to be irreconcilable with physical reality, he was prepared to accept Weyl's defence and wait for further research to establish for certain which of their theories was correct.[30] But Wolfgang Pauli, in his famous encyclopedia article of 1921 surveying relativity theory, criticised Weyl's defence. According to Pauli, Weyl's argument implied that the coefficients $g_{\mu\nu}$ and ϕ_μ of his theory, in contrast to the $g_{\mu\nu}$ of Einstein's theory, were not directly observable; and this in itself meant in Pauli's view that the field of Weyl's theory could not have any physical significance.[31] The same year Eddington noted that although Weyl's geometry was indeed more general than Einstein's, it was itself still subject to one apparently arbitrary restriction, namely that zero lengths were preserved unchanged under all its allowed transformations. If this restriction were removed, the resulting theory would contain not only the 10 coefficients $g_{\mu\nu}$ of Einstein's theory and the 4 more ϕ_μ added by Weyl, but also another 26 coefficients, bringing the total to 40.[32] Weyl rejected this further generalisation on the grounds that non-zero values of the additional coefficients could not be reconciled with a generalised principle of least action such as his physical theory required,[33] but this defence was difficult to reconcile with that he had given against Einstein's objection. While accepting that only the 14 coefficients might be important in practice (no physical quantities could yet be associated with the other 26), Eddington argued in principle that Weyl's geometry was neither one with an arbitrary metric (as he claimed was his own version, with the 40 coefficients) nor one with a unique metric corresponding to observed physical measurements (as was Einstein's), and that it could therefore be defended on neither mathematical nor physical grounds.[34] Eddington agreed with Weyl in seeing the *a priori* mathematical theory as defining the possible laws of the world, and as providing a genuine if metaphysical connection between these laws. But he also agreed with Pauli that any account of actual material behaviour, defined in this case by the natural gauge corresponding to Einstein's theory, had to be based directly upon observation.[35]

Weyl's theory provided the context for the expression and exploration of fundamental differences of opinion as to what a physical theory could, and should, achieve. The issues raised were not peculiar to the theory but, as the disputants recognised, transcended the whole of physics. As we shall see, they were to impinge in several respects upon the debate surrounding quantum theory and its problems. And in one particular point, which was taken up by the young Pauli, they bore directly on the quantum problem of wave-particle duality.

THE CONTINUUM PROBLEM AND PAULI'S CRITICISMS

Although Weyl's work was stimulated directly by Einstein's, the prime concern of the Göttingen mathematicians had not been with Einstein's theory of gravitation but with a theory proposed by Gustav Mie in 1912.[36] Mie had attempted to explain both gravitational phenomena and the existence of the finite electron on the basis of special relativity and a purely electromagnetic field theory.[37] And in 1915 Hilbert, combining Mie's approach to a unified theory with the insights of Einstein and with his own mathematical facility, had independently derived the general relativistic gravitational field equations.[38] Born had also worked on Mie's theory while at Göttingen, and so had Weyl, whose basic concern with general relativity was to develop Hilbert's work into a fully-fledged unified theory, including an explanation of electron structure.[39]

The reconciliation of the corpuscular properties of the electron with the continuous electromagnetic field theory was one of the most persistent and difficult problems of classical physics, and had attracted considerable attention during the early years of the century. Mie's approach had been to seek additional electromagnetic field terms that would be significant only inside elementary particles, where they would combine with the traditional terms to allow a stable distribution of electricity.[40] Weyl, though incorporating gravitational as well as electromagnetic terms, essentially followed this approach and, as we shall see, based his physics upon an action principle derived from Mie's work.[41]

The terms required were however too complex to be realistically incorporated into the field equations. The approach failed in practice, and in 1919 Pauli attacked it strongly on principle.[42] Since the theory was symmetric with respect to positive and negative charges, he argued that it was contrary to experience.[43] Since it incorporated no direct relationship between its two basic entities, charge and mass, he argued that it was incomplete, in that it could not account for the physically observed elementary particles. And most fundamentally of all he argued that as a pure field theory it could not in principle offer any explanation of the cohesive structure of matter. As he put it two years later, in his survey of relativity theory,[44]

> The continuum theories make direct use of the ordinary concept of electric field strength, even for the field in the interior of the electron. This field strength is however defined as the force acting on a test particle, and since there are no test particles smaller than an electron or hydrogen nucleus, the field strength at a given point in the interior of such a particle would seem to be unobservable by definition, and thus be fictitious and without physical meaning.

According to Pauli, any attempt at a complete unified theory would have to account for the internal structure of particles, and would therefore have to incorporate complex but well defined properties of matter independent of and prior to those of the electromagnetic-gravitational field, which were therefore operationally meaningful only on a scale large compared with that of the elementary particles. But more important than this conclusion was the philosophy behind it. For Pauli's requirement that a quantity, to be physically meaningful, must be in principle capable of measurement had implications beyond the narrow realm of Weyl's theory. In a wider sense too Pauli's critique drew attention once more to the fundamental problem of the structure of matter (electrons) in a pure field theory and acted as a catalyst in the fusion of this problem with the related one of the structure of light in the electromagnetic field theory – the wave-particle problem of quantum theory.

DUALITY AND CAUSALITY

The strongest connection between the debate surrounding Weyl's theory and the genesis of quantum mechanics was to be through the development of Pauli's 'operationalist' ideas on the requirements of a physical theory.[45] But the two fields also interacted in other ways and it is instructive to look at these, if only to establish the general potential for cross-fertilisation between different branches of physics.

The greatest exponent of such cross-fertilisations was perhaps Einstein. He had himself been long searching for a unified pure field theory,[46] and he now treated Pauli's criticisms as applicable equally in respect of material particles and light-quanta. Writing to Born in January 1920 he recognised that "Pauli's objections apply not only to Weyl's but to any continuum theory, including one which treats the light-quanta as singularities."[47] Einstein had earlier reacted to Mie's theory of electron structure by suggesting how the additional terms of that theory might be replaced by purely gravitational terms connected with the solution of the cosmological problem.[48] This suggestion had run into the problem that (as with Mie's theory) all spherically symmetric distributions of charge seemed to be equally possible: there was no way in which the particular size, charge and mass of the electron could be singled out.[49] But Einstein did not give up hope of a pure field theory. In the letter to Born he suggested that the extra determination necessary might be imposed on the field by boundary conditions, and that both material particles and light quanta might be derived from the continuous pure field by overdetermination of differential equations.[50] The idea was still subject to Pauli's criticisms, but Einstein made it clear that he could not accept these, and his continued attempts to pursue the idea through the 1920s and to derive the light-quanta from the continuous field may be seen both as rooted in the problem complex of general relativistic field theory and as an assertion of his philosophical stance of objective physical realism against Pauli's more subjective position.[51]

Another physicist to recognise an intimate connection between the light-quantum and electron structure problems in this context was Eddington,[52] who in 1921 took Pauli's operationalist criticism a step further by arguing that the shape and size of an electron were observationally indeterminable, since they could only be measured using another electron.[53] In practice, he argued, one adopted the convention that the electron was spherically symmetric, and the mathematical expression of this convention corresponded to the choice of natural guage, and to Einstein's gravitational field equations.[54] Despite this restriction on the observed world Eddington hoped that one might get some further information on the possibilities of electron structure from the 26 parameters of his theory that could not be related to the gravitational and electromagnetic fields.[55] But writing to Weyl in July 1921 he suggested that this programme was severely hampered by ignorance as to the structure of light. Though his ideas were as yet "too vague to be formulated", the development of relativistic matter theory would have to be closely related to that of the quantum theory.[56]

Weyl too saw a close connection between the material and light-quantum problems, and in response to Pauli's criticism of a pure field theory he suggested that there might be an element of reality prior to the field, corresponding to acausality on the quantum level. In May 1919 he wrote to Pauli that he thought the non-equivalence of positive and negative electricity would reduce to that of past and future, to the unidirectionality of time.[57] As to his advocacy of a pure field theory, Pauli should not accuse him of being a dogmatist: he did not think he'd found the philosophers' stone, and he was himself quite sure that there was something in matter independent of the field.[58] Writing again in December, he approached the problem of deriving the world of physical observation from that of his predominantly mathematical theory, and suggested that the past-future distinction was more basic than that between the two signs of electric charge, and that it completely determined the latter through the world geometry. The process of events represented by this geometry was however fundamentally statistical and acausal, and it was the origin of this acausality, in terms of "independent decisions", that formed the existent in matter independent of the field.[59] From Weyl's highly metaphysical perspective it was the independent decisions that were 'real', and not the physical field; but it was the field that constituted what might be called the physical reality, the reality of the observed world.[60] Weyl did not explain just how the independent decisions operated to produce the observed phenomena, but in the fourth edition of *Raum-Zeit-Materie*, completed in 1920, he linked them with the apparent acausality of quantum phenomena.[61] And in the letter to Pauli he suggested that the complex structure of the electron to which they would lead might account for the quantum phenomena of the atom:[62]

> I believe, for example, that the fact that the electron does not radiate in the stationary Bohr orbits is an indication of the acceleration produced through the internal changes of the electron, which enables it to retain its energy; then why should it also in non-quasistationary acceleration behave as a rigid body with fixed charge?

Weyl's response to Pauli's objections may be seen, as by Forman, as reflecting a hostility towards causality pervasive in the prevailing intellectual milieu.[63] It was certainly based on philosophical as much as physical considerations, and in a paper of 1920 Weyl expressed his views on the rejection of causality in a thoroughly philosophical framework.[64] But these views were not without their merits, even on more restricted, purely scientific, grounds. The problem of the asymmetry between the elementary particles was one that Weyl had already tackled without success in the context of Mie's theory.[65] And Pauli's analysis had shown that an asymmetric theory of the type he had then proposed would inevitably fail to be invariant under time reversal, an important conclusion that must have been partly responsible at least for his identification of the problem with that of the unidirectionality of time.[66]

Turning back to the quantum context we should note also that Einstein's attempt at a causal pure field theory was not only subject to Pauli's methodological criticisms but also had every appearance of being untenable on empirical grounds. Einstein himself found all his attempts to derive the light-quanta from a continuum theory foundering on the necessity in this case of abandoning strict energy conservation in order to allow for the discrete absorption of light, and on the resultant possibility of an infinite energy arising from the gradually increasing Brownian motion made possible by energy non-conservation.[67] He had indeed demonstrated in 1911 and 1916 that the only alternative to the light-quanta, within the existing conceptual framework, was the rejection of energy conservation.[68] And although he seems to have been hopeful that by deriving the light-quanta from a continuum theory he might somehow retain the energy-momentum conservation, he could draw on nothing whatsoever to support this hope.

Einstein's paper of 1916 was a critical one for the future of quantum mechanics, and will often be referred to. In it he introduced for the discrete quantum behaviour of the atom probabilities of the absorption, free emission and stimulated emission of radiation in the presence of a field. These corresponded to the transition intensities of the classical continuum theory and were defined so that the total absorption and emission over a period of time would be the same on the two theories. Using the new terminology, Einstein offered a new derivation of Planck's Law and, in a second section of the paper, he also put forward convincing arguments to the effect that, assuming the conservation of energy, the emission or absorption of light by an atom was always accompanied by a transfer of linear momentum as well as energy. This result provided strong confirmation of the particulate aspect of the nature of light.

Following this paper of Einstein's it became widely accepted that the mildest possible modification to classical electromagnetic theory if quantum effects were to be explained would be the abandonment of energy conservation;[69] and to extend this to the abandonment of causality, as did Weyl and later Schottky, Schrödinger and Von Mises, was not at all unreasonable. Given a preference for a field over a particle theory it was indeed the only path that seemed to offer any hope for advance, and even Weyl's metaphysically-based introduction of independent decisions must therefore be associated with the fundamental wave-particle and field-particle problems.[70]

It should by now be apparent that the issues of the search for a unified field theory and those of quantum theory, whether in the form of conservation, causality or duality, could not be taken in isolation. In particular both Weyl's rejection and Einstein's defence of causality, though expressed in the quantum context, were deeply rooted in the problems of a unified general relativistic field theory. The same is true also of Schottky's rejection of causality in 1921, which was closely related to Einstein's work.[71] So far as the actual development of quantum mechanics was concerned, Weyl and

Schottky played relatively minor parts (though Weyl did contribute on a technical level to both matrix mechanics and wave mechanics). But Einstein's ideas were to be of the greatest importance for the genesis of wave mechanics, and Schrödinger himself, the author of that theory, also appears to have drawn on the general relativistic context.

Schrödinger joined Weyl in Zürich in 1921,[72] and the following year he published two papers, both of which bear the stamp of Weyl's ideas. In 1920 Försterling had sought to reconcile Bohr's quantum condition, $E = h\nu$, with the special and general theories of relativity; but his treatment had been criticised, by Pauli once more, on the grounds that it contradicted Einstein's result on the linear momentum (particulate light-quantum nature) of emitted light.[73] In June 1922 Schrödinger rederived the reconciliation with special relativity, apparently upholding the light-quantum hypothesis.[74] But the following November, responding to further criticisms by Pauli, he explained that he now interpreted Einstein's result as applicable only to the recoil of the emitting atom, with which Einstein's calculations had in fact been concerned, and not to the light itself.[75] For he stressed that he saw no way out of the quantum paradox other than to abandon energy-momentum conservation on the atomic level. And he put forward his own suggestion that one might replace the quantum *identity*, $E = h\nu$, by an *equivalence* between the effect produced by a material particle of energy E and that produced by a (spherical) wave of frequency E/h, but without the energy E necessarily being attributed to the wave, and thus without the wave having to be localised as a light-quantum. This would not be possible for a conservative mechanical system of course; but "the devil knows", he argued, whether atomic systems are strictly causally determined, let alone whether they are conservative mechanical systems.[76] A few weeks later, in an inaugural lecture, he explicitly rejected the "rooted prejudice of causality".[77]

Writing to Pauli, Schrödinger insisted that their differences were purely philosophical,[78] and his rejection of causality was explicitly a philosophically based decision rather than one dictated by the requirements of physics.[79] In 1924–5 he changed his views on both light-quanta and causality and moved closer to Einstein's position,[80] but in contrast with Einstein he maintained his metaphysical bent, and he read de Broglie's *Thèse* in 1925 not as offering a straight-forward fusion of wave and particle theories (which was more or less how it had been put) but as revealing the "inner, spiritual" connection between the two theories.[81] This perspective must have brought him close to Weyl, with whom he worked closely in 1922, and it would seem likely that his rejection of causality was linked with Weyl's. His second paper of 1922, completed in October, was explicitly based upon some of Weyl's work.[82] In this paper Schrödinger again pursued the programme of reconciliation initiated by Försterling, and following up the lead given earlier by Weyl himself he attempted to derive the Bohr atomic orbits directly from Weyl's unified field theory.[83] His efforts bore no immediate fruit, but as Raman, Forman and Hanle have shown, this work appears to have played a crucial part in his later creation of wave mechanics.[84]

THE DEVELOPMENT OF PAULI'S VIEWS

Important as it was, we shall return to Schrödinger's work only briefly, for it was in the development of matrix mechanics that the flow of ideas found clearest expression, and in this context the link with Weyl's work arose through the developing ideas of Pauli.

We have referred to Pauli's criticisms of both Weyl and Schrödinger, and have described his philosophical position loosely as 'operationalist', a term that was coined by Bridgman in 1927 and is also used sometimes of Eddington. But what more precisely was Pauli's position and how did it translate into the quantum-theoretical context? One outcome of the debate over unified field theories was the explicit development of Eddington's natural philosophy, which found its first extended expression in the closing sections of his book *Mathematical Theory of Relativity*, in which both Weyl's theory and his own extension of that theory were presented.[85] And it was in response to this book, and in a letter to Eddington, that Pauli was prompted to give the most detailed account we have of his own position. It was already clear from his publications that this entailed some form of methodological or epistemological operationalism, summed up in the statement from his 1921 survey that "we should hold fast to the idea that in physics only quantities that are in principle observable should be introduced."[86] But in the letter to Eddington the status of this demand was clarified and the demand itself expanded and carried into the quantum context.

So clearly did Pauli express himself in his letter to Eddington, and so crucial were his views to be to the development of quantum mechanics, that it is worth quoting them directly and at length. Having first discussed the formal connection between electromagnetism and gravitation in Eddington's theory, Pauli restated the limitations of a pure field theory such as the classical electromagnetic theory:[87]

The most unsatisfactory thing about the electron theory of Lorentz and Larmor is indeed that it is not a theory of the electron, as one might somewhat jokingly say. Why are there only two kinds of elementary particle, negative and positive electrons (Hydrogen nuclei)? Why are their electric charges the same, when their masses are so completely different? Generally, how can an elementary particle stay together despite the Coulomb repulsion between its parts? This cannot happen according to the Maxwell-Lorentz theory, nor can either the energy or momentum of a uniformly perturbed electron be calculated in its entirety on the basis of this theory of the electromagnetic field. The reduction of the whole inertia to the electromagnetic field also does *not* succeed, so long as we hold to this theory.

He then reaffirmed his belief that the problem of the electric elementary particles could not be solved by any theory using the concept of continuously varying field equations, satisfying known differential equations, for the region *inside* the elementary particles.

Pauli next turned to quantum theory, and used the problems in that context to expand upon his viewpoint:

As is shown in quantum theory, classical electrodynamics also fails in the domain in which dimensions are large compared with those of the elementary particles, as soon as it treats swiftly varying fields. It holds in this domain only when the fields are static or quasistatic, as for example when an electron is moving with constant velocity. For the other cases it leads to the right answer only in respect of the statistical average over many individual processes within a certain range.

I might emphasise especially that the quantum theory in no way calls for only a modification of light theory, but generally calls for a new definition of the concept of electromagnetic field for non-static processes. Indeed the classical theory fails also in the description of electron-atom collisions, as well as for the behaviour of the electron in the atom in time-varying fields (dispersion theory). The famous contradiction between the interference capability of light emitted in various directions and its property of remaining constantly disposed in energy quanta (Einstein's "light-quanta"), for example in the photoelectric effect, comes only from the fact that we give up the *laws* of classical theory, but still always work with the *concepts* of that theory.

Thus in Pauli's view the wave-particle duality was neither to be dismissed (as many more conservative physicists still maintained) nor to be attributed to any amendment of the existing theory, such as the absence of causality or energy conservation. Its origins were rather more fundamental, lying in the fact that the concepts being used in physics were themselves inappropriate to the situations concerned. Just how they were inappropriate, Pauli went on to explain:

We now know both from experimental results and from quantum-theoretical considerations, which I shall not however go into here on account of brevity, that even an electron in a light wave does not in fact move at all as specified [by the classical theory]. The field of a light source is therefore at present undefined. But the path of an electron (in general in a swiftly varying field) is just as little so. Indeed one can only define this through the action of the electron on other elementary particles, and in a non-static field this again fails to satisfy the classical laws. Some physicists have fallen upon the idea, considering the above-mentioned paradox of light, of completely abandoning the concept of the field of a light wave and only describing the motion of the electron. This we wholly reject; for on the one hand the idea scarcely helps us over the difficulty, and on the other hand it makes a connection with the classical theory, which has proved itself so brilliant for large scale phenomena, quite impossible. We can proceed only through a modification, not through an abandonment of the field concept. So much would the return to the pre-Maxwellian standpoint be considered a retrogression, so much must we hold before our eyes the fact that the field concept only has a meaning when we specify a reaction, which is in principle possible, in which we can if we want to measure the field strength at each point of space-time. We can attribute a reality to the field intellectually, even if we do not exactly execute the reaction. It is only essential that we *can* always execute it in principle, if we *want* to. But as soon as the reaction ceases to be always specifiable or in principle executable the respective field concept is no longer defined.*

In a footnote, Pauli illustrated this point by an example:

* It has often been said that one could retain the Maxwell equations in the vacuum (charge-free space). Only the interaction between light and material systems could not be treated with the help of the classical theory. In fact the two are inseparable. Only by means of this interaction can the electromagnetic field of a light wave be defined. The Maxwell equations in a vacuum are neither correct nor incorrect; they are *meaningless* in every application where the classical theory fails and where it is not a question of statistical averages.

If we replace "reaction" by "operation", then Pauli's position is very close to that later adopted by Bridgman. Only whereas Bridgman was concerned clearly with epistemology, Pauli's concern oscillates between a form of effective epistemology – what is "meaningful" – and methodology. Perhaps under the influence of Bohr, under whom he was working at the time, Pauli was concerned not only with building up a system of concepts from scratch, but also with a correspondence between the new concepts and those that had proved themselves valuable in the classical context. He was certainly deeply concerned that one should seek, within the limits set by the operational definition of concepts, to go behind the phenomena, to provide an explanation not just an account. And he had some ideas too on the direction in which the new concepts required should be sought:

In the definition of the electromagnetic field strength in the classical theory the reaction concerned at present is the force on a charged test particle. This means that already in the definition of the electro-magnetic field strength we use mechanics (force and mass concepts). Already therefore the reduction of mechanics to electrodynamics is merely apparent. It would be much more satisfying if the measurement of the field strength could be based exclusively on a counting process, without making use of mechanical concepts. Instead of introducing discontinuous functions one would perhaps associate suitable of these with continuous functions; and as the laws of relativity theory are invariant with respect to coordinate transformations, so would the laws of the future quantum physics perhaps also be invariant with respect to alterations of the field functions, so long as these left unchanged only certain counting process results.

To summarise, may I say this: we do not yet know today which fast-varying electromagnetic processes we can in principle observe, and which not. As soon as we know this, we shall have solved the quantum riddle. Because of the relationship between the definition of electromagnetic field strength in the classical theory and mechanics, I believe that the *same* dissociation of the field concepts from mechanics, with the simultaneous introduction of some kind of atomism (countable quantities instead of continuous ones) will bring the understanding of the quantum problem and of the elementary electric particles.

In the last passage, Pauli made two specific recommendations that were indeed to be relevant in the development of quantum mechanics. Insisting on an operationally based definition of any concepts used in the new theory, he suggested that this would entail the removal of the existing mechanical content of the field concepts, and the replacement of continuous concepts by discrete ones. This last point had obvious implications for the quantum problem, and it was closely linked with Pauli's earlier rejection of pure continuum theories and his repeated defence of the particle aspect of the wave-particle duality as being if anything the more fundamental. Since our observations of the physical world consist of instantaneous and localised, effectively discrete measurements, then the concepts used to describe that world had, according to Pauli's principles, to reflect that discreteness.

In his long letter to Eddington, Pauli next returned to the original context of Eddington's theory, and to his conviction that the connection therein between the electromagnetic and gravitational fields had no physical significance. In the course of discussing this very difficult problem, which he saw as "rooted in the quantum theory", he extended his discussion to suggest that, in the absence of any solution to the quantum problem, one could legitimately proceed "phenomenologically", that is without any insight to the fundamental nature of the elementary particles:

We do not know a priori whether it will be possible to answer this question without regard for quanta. It *can* indeed be the case. But then the question must in my opinion be answered purely phenomenologically, without regard to the nature of the elementary electric particles. Further, I stand by my position (naturally not demonstrable) that this question must begin with a definition of the field quantities used, which specifies how the quantities can be measured. It must further disclose relations between electromagnetic and otherwise measured effects. (The greatest achievement of relativity theory was indeed to have brought the measurements of clocks and measuring rods, the orbits of freely falling mass points, and those of light rays into a firm and profound union.) Logically or epistemologically, this postulate cannot be demonstrated. But I am convinced that it is correct.

Pauli had already criticised the adoption of what might be called a negative or passive phenomenological approach, but he now indicated how such an approach might be positively used, in the absence of any sound foundation for a theory, to explore the relationship between parts of that theory. If the discrete and non-mechanical conceptual foundations he had advocated could not yet be found it was important above all to ensure that those concepts that were used, though they could not of course provide a unified base for the whole theory, were at least operationally well defined; and that the relationship between their operational definitions was explicit. Then one could at least proceed, albeit only on a phenomenological level, without danger of contradiction. Pauli's immediate concern was that neither Weyl's theory nor Eddington's satisfied these conditions, and they were indeed far from trivial. But once again the perspective was one that was to be of continuing use to Pauli himself in the course of the development of the new quantum mechanics.

CHAPTER 3

NIELS BOHR AND THE PROBLEMS OF ATOMIC THEORY

INTRODUCTION

Having completed his survey of relativity theory in 1921, Pauli switched his concern to quantum theory.[1] He worked first as Max Born's assistant for the academic year 1921–1922, sharing in Born's own first attack on the problems of the atom. By the time he wrote the letter to Eddington in September 1923 he had also worked for a year in Copenhagen with Bohr.[2] Neither of the physicists with whom he worked most closely in this period, Bohr or Heisenberg, had played any active part in relativity theory or the search for a unified theory. Indeed Bohr, uniquely among the great theoreticians of the period, appears to have displayed little interest in unification and little if any sympathy with relativistic thinking and analysis. But despite this difference, the new environment was far from unconnected with the old. Born started work on quantum theory at about the same time as Pauli, at the urging of the experimentalist James Franck, who was his colleague at Göttingen and a close ally of Bohr's.[3] Before then his orbit had been much closer to the problems of relativity theory than to those of quanta, and he had, moreover, shown some sympathy with Pauli's ideas.[4] Heisenberg had been Pauli's colleague as a student in Munich under Sommerfeld, and had already then discussed the fundamental problems of physics with him. Sommerfeld himself was part author and leading exponent of the quantum theory, but equally at home with relativity theory, and again of course familiar with Pauli's ideas. But of all his new colleagues the one who provided the greatest stimulus to Pauli's thought, and whose ideas had most relevance to his own, was the one whose work had been most removed from his, the father of quantum spectral theory, Niels Bohr. Though developed in a completely different context Bohr's ideas applied to the same fundamental problems as did Pauli's – and were completely opposed to them. As in the earlier discussions of unified field theories the argument was conducted with friendship and respect. Pauli never ceased to admire Bohr as a truly great physicist and as the teacher who made the greatest impression upon him.[5] Bohr was delighted to find in the young Pauli a worthy if troublesome opponent, the soundness of whose judgement he could trust completely.[6]

But argument there was, for the division between their views was a deep one, and if Pauli's position was based on a strong conviction Bohr's, developed over a longer period and greater experience, was even more so.

THE CONFLICT OF BOHR'S ATOMIC MODEL

Like the quantum mechanics it preceded, Bohr's theory of the atom appeared, and by any reasonable standards was, a success. But from its first enunciation in 1913 the theory's foundations were problematic. Bohr had early become convinced that in the solution of the quantum problem classical mechanics would have to be superceded, and his model of the atom was explicitly at variance with this theory, which could not accommodate the discrete transitions between stationary states. It was also at variance with classical electrodynamics, which could not account for the absence of radiation in these states. But at the same time it was framed entirely in the language of, and rested entirely upon the foundations of, these classical theories. The concept of electrons orbiting the atom was a classical mechanical concept, and as the Bohr theory developed the choice of possible orbits continued to be governed entirely by the requirements of the classical theory. Bohr's orbital model of the atom was explicitly a model, of course, and Bohr himself was the first to insist that it could not be interpreted as in any sense structurally true.[7] Rather as in mediaeval planetary theory, the details of the atomic model were varied at leisure, within the basic form, so as to produce the required observed results. But still the model was in some sense intended, and taken, as real. It was not intended, as for example were J. J. Thomson's models, as a merely heuristic device directing the way to further investigation, but rather as a genuine if inadequate image of reality. Bohr sacrificed elements of the classical mechanics and electrodynamics, but he did not depart from the basic conceptual entities of those theories. Apart from the fundamental but conceptually undefined postulate governing the energy differences between stationary states, $E = h\nu$, he did not introduce anything to replace what had been lost. And so the Bohr theory entailed a fundamental conflict, as on the one hand it departed from the classical theories radically, while on the other hand it remained both technically and conceptually dependent upon them.

As the theory developed, this conflict heightened. Since the structure of the model took second place, as it had to in the absence of any consistent foundation, to the results derived from it, the details of this structure were varied at leisure. Just as astronomers of old had adjusted their epicycles to agree with any new data, so the quantum theorists, notably Sommerfeld, Landé and their colleagues, adjusted the choice and relative configuration of electron orbits at will in trying to apply the model to higher elements.[8] They did not concern themselves much with why this or that orbit should be preferred, let alone with why any such choice made sense, given the admitted failure of the foundations upon which the model was based. Their only criterion was whether it would give the observed spectra. With the introduction of quantum numbers to describe the stationary states this trend was exaggerated, for although the numbers were originally descriptive of physical properties they eventually came to precede these properties, with the choice of a new quantum number, i.e. a new degree of freedom, preceding the debate as to what this number represented physically. The positivistic attitude in the choice of orbits was thus continued, and strengthened, in the treatment of quantum numbers.

But still the theory rested upon the very classical concepts it spurned. The orbits were restricted to those that, given the absence of any continuous radiation from the orbiting electrons, were possible according to the classical theory. The quantum numbers, to be accepted, had ultimately to be interpretable in terms of the classical concepts. Even if it were recognised that the planetary model was too naive to be a genuine structural representation, there was simply no alternative visualisation. This situation became even more acute, moreover, when in the early 1920s the theory progressed from the mere prediction of transition *frequencies* to take into account also the *intensities* of the emitted and absorbed radiation of each frequency. The extension was made possible in 1916, when Einstein introduced the notion of transition *probabilities* as the quantum equivalent of the intensities of the corresponding radiation on the classical theory.[9] But it was possible only through such a connection, justified conceptually by Bohr's correspondence principle of 1919.[10]

Since the Bohr atom transitions were discrete, and since there was no known mechanism governing when they took place, Einstein had seen that the classical intensities of continuous emission and absorption could only be replaced on the quantum theory by probabilities. In order to provide a theoretical derivation of these probabilities Bohr then argued that there must be a "necessary connection" between the quantum and classical theories "in the limit of slow vibrations" ($h\nu$ small). He then assumed that such a connection also held in the slightly wider limiting range of high quantum numbers, and, more generally still, wherever the classical theory gave what seemed to be the right answer. The correspondence principle was based on this connection, which was made through the quantum theory of conditionally periodic systems in the limit in which these gave frequencies that could be associated with those of classical oscillators. Since, however, the only cases of physical importance tended to be those in which the frequencies could *not* be so associated, the principle tended to be used in isolation from its theoretical foundations, as a blanket excuse for carrying over empirical results from the classical to the quantum theory.[11]

Bohr was of course well aware of these problems, and also of the closely related problem of the wave-particle duality, clearly manifest in Einstein's 1916 paper. The dominant theme of that paper was that of light-quanta. Einstein's analysis of the momentum exchange in radiative processes supported the light-quantum hypothesis, and his introduction of transition probabilities opened the way for the description of such processes in terms of discrete quanta. At the same time, however, his accompanying derivation of Planck's law, though apparently based on the light-quantum hypothesis, in fact included a limiting process that was valid only for the wave theory of light. Moreover Einstein's hypothesis of stimulated emission in the presence of a radiation field, which was also necessary for his derivation of Planck's law, could be explained naturally only in terms of wave resonances and not in terms of particle mechanics.[12] Bringing the same considerations to bear on this complex of problems as on those of the atomic model, Bohr came to a series of fundamental and closely related conclusions.

THE DEVELOPMENT OF BOHR'S POSITION

As we have noted, Bohr's atomic model represented a departure from both classical electrodynamics and classical mechanics. But although the former departure appears to have exercised his mind the more by the 1920s, it was the latter that was most central to his theory, and with which he had originally been primarily concerned. Indeed, while he had from the beginning aimed to depart from classical mechanics, Bohr had not actually sought to depart from classical electrodynamics. He had, on the contrary, a very strong belief in the power of that theory,[13] and even after 1913 he argued persistently in its favour, trying to retain as much of it as possible. Thus, although his atomic model appeared to imply the necessary discrete absorption of light, and hence its localisation in light-quanta, Bohr could not accept this implication, but believed as did many others that the demonstrated success of the classical electrodynamical wave theory of light argued convincingly for its retention. Over the years he spoke out gradually more strongly in favour of the wave, and against the particle concept of light. In a survey of quantum theory written in 1918, he adopted Einstein's transition probabilities but pointedly ignored the arguments for light-quanta that had accompanied the derivation of these probabilities in Einstein's paper.[14] In 1920, he wrote that "I shall not here discuss the familiar difficulties to which the hypothesis of light-quanta leads in connection with the phenomenon of interference, for the explanation of which the classical theory has shown itself to be most remarkably suited."[15] And the same year he explained to his close colleague Darwin that since one could only define a frequency through a wavelength, and a wavelength through interference, and interference through the wave concept of radiation, this concept had to be retained.[16] In 1921, at the Solvay congress of that year, he did discuss the light-quantum concept for the first time in a published paper, and concluded that it "presents apparently insurmountable difficulties from the point of view of optical interference."[17] And the following year, in the course of the most clear and extensive account of quantum theory to date, he brought together his criticisms in a sharp attack on the light-quantum hypothesis:[18]

As is well known, this hypothesis introduces insuperable difficulties, when applied to the explanation of the phenomena of interference, which constitute our chief means of investigating the nature of radiation. We can even maintain that the picture, which lies at the foundation of the hypothesis of light-quanta, excludes in principle the possibility of a rational definition of the conception of a frequency ν, which plays a principal part in this theory.

Apart from Sommerfeld, who wrote to Bohr in 1918 of his conviction that[19]

The wave process occurs only in the aether, which obeys Maxwell's equations and acts quantum-theoretically as a linear oscillator with arbitrary eigenfrequency ν. The atom merely furnishes a definite amount of energy and angular momentum as material for the process.

and who continued to follow this line until about 1923, Bohr was the only one of the group working on the quantum theory of spectra to hold such firm views on the wave-particle duality. In this respect his position bore more resemblance to those of the physicists on the fringe of quantum theory than to those in its midst.[20] Unlike many of those physicists, however, Bohr did not underestimate the force of the arguments for light-quanta. On the contrary, his opinion was considered, well supported, and far from lacking in subtlety. In respect of both the wave-particle duality and the atomic theory he had sought hard for a more satisfactory solution. And although he had failed to find such a solution he had succeeded in tracing the problem back to the concepts used to describe the physical world.

Whereas Pauli was to call for a radical revision of the concepts of physics so as to ensure their consistency both with each other and with the operations used to define them, Bohr came to the conviction that such a replacement was impossible. The concepts of classical physics were demonstrably insufficient to describe the processes of physics, but their choice seemed to Bohr to be psychologically as well as physically determined, and there seemed to him to be no alternative to them. In his survey paper of 1922, which was written and may be read as his manifesto of the period, he admitted that "a description of atomic processes in terms of space and time cannot be carried through in a manner free from contradiction by the use of conceptions borrowed from classical electrodynamics." But in the opening paragraph of the very same paper he insisted that "from the present point of view of physics, however, every description of natural processes must be based on ideas which have been introduced and defined by classical theory."[21] And as is clear from his attitude to the wave-particle duality, Bohr saw the concepts of classical wave theory, of classical electrodynamics, as peculiarly fundamental in this respect.

In purely philosophical terms, Bohr was not troubled by this situation. As he explained in September 1923 to his philosophical friend and mentor Høffding, he saw no reason why atomic science, or anything else for that matter, should conform completely to the visual world of our perceptions and of the classical physical conceptions:[22]

> In general, however, and particularly in some new fields of investigation, one must remember the obvious or likely inadequacy of pictures: as long as the analogies show through strongly, one can be content if their usefulness – or rather fruitfulness – in the area in which they are being used is beyond doubt. Such a state of affairs holds not least from the standpoint of the present atomic theory. Here we find ourselves in the peculiar situation, that we have obtained certain information about the structure of the atom which may surely be regarded as just as certain as any one of the facts in natural science. On the other hand, we meet with problems of such a profound kind that they seem to defy solution; it is my personal opinion that these difficulties are of such a nature that they hardly allow us to hope that we shall be able, within the world of the atom, to carry through a description in space and time that corresponds to our ordinary sensory perceptions.

However, while this may not have been philosophically disturbing it did have very serious consequences for the development of physics. The key problem for Bohr was how to overcome the conflict he had identified, and in particular how to preserve the continuity both of and with the classical conceptions. Although even his own correspondence principle rested necessarily upon Einstein's considerations of 1916, he described these as providing "only a preliminary solution", and throughout his 1922 survey paper he treated the discreteness of quantum theory not only as its defining characteristic but also as its major problem. The task as he saw it was to reconcile the discontinuities of quantum phenomena with the continuous nature of our natural understanding, manifest in the classical continuum electrodynamics, and to do so by effecting a continuous transition from the old theory to the new.

Setting about this task, Bohr proceeded in two directions. On one hand he sought to *extend* the range of the classical concepts into non-classical phenomena, and this he did through his famous correspondence principle. At the time of its enunciation the status of this principle was still somewhat obscure. But Bohr stressed in his survey paper that it was intended as far more than a mere link between the quantum and classical theories and that it, and the classical theory of electrodynamics itself, must be treated logically as part of the foundations of quantum theory. This was consistent with the dependence of his quantum theory upon results that could not be obtained without reference to the classical theory, and it also fitted in with the status he afforded to the classical concepts. On the other hand, Bohr also sought to *limit* the range of these classical concepts so as to allow completely non-classical phenomena to enter into the theory without leading to serious contradictions. Here he pursued two related possibilities, the abandonment of strict energy-momentum conservation, and the abandoment of a causal description in space and time.

We have already noted in the context of Schrödinger's work that Einstein had mooted the possibility of energy non-conservation – a possibility that he himself had found totally unacceptable – in 1911. At the first Solvay congress he had argued that the only alternative to acceptance of the light-quantum hypothesis was to "resort to abandoning the law of conservation of energy in its present form, giving it for example only a statistical kind of validity, as one does already for the second principle of thermodynamics."[23] For the particular case he was examining, he stressed that "one can only choose between the structure of radiation and the negation of an absolute validity for the law of conservation of energy."[24] Einstein himself was not prepared to sacrifice energy conservation, a position he justified later by its implications for causality and by consideration of a possible infinite Brownian motion that might result in this case.[25] He preferred to keep both the light-quanta and "the indispensable Maxwell equations",[26] hoping to derive both in due course from a unified field theory. But as we have noted before, this position was very difficult to uphold, and the pressures in support of the alternative of energy non-conservation were therefore strong. Both Bohr's theory of the atom and Planck's 'standard', 1911, derivation of his black-body radiation law seemed to require discrete changes in the energy content of matter at the same time as their authors upheld a continuous theory of the transmission of energy, and the abandonment of strict conservation was the most natural response to this situation.[27] Such at least was the view of the British physicist Darwin, who wrote to Bohr at length on the subject in 1919,[28] and in apparent reply to this letter Bohr agreed that "on the quantum theory, conservation of energy seems to be quite out of the question."[29] His own feeling at that time was that something funny must go on inside the atom, triggered by the incident light.[30]

In the paper to the 1921 Solvay congress in which Bohr first came out strongly against the light-quantum concept, he nevertheless noted that this concept seemed "to offer the only possibility of accounting for the photo-electric effect [one of the phenomena discussed by Einstein in 1905 as characteristic of particulate light-quanta], if we stick to the unrestricted application of the ideas of energy and momentum conservation."[31] In a manuscript of the same year that did not reach publication, Bohr repeated the same statement verbatim, but followed it up with the comment that[32]

> At this stage of things it would appear that the interesting arguments brought forward recently by Einstein [those in support of the light-quantum concept in his 1916 paper] ... rather than supporting the theory of light-quanta will seem to bring the legitimacy of a direct application of the theorem of conservation of energy and momentum to the radiation processes into doubt.

In his survey paper of 1922 Bohr insisted simply that "a general description of the phenomena, in which the laws of the conservation of energy and momentum retain in detail their validity in their classical formulation, cannot be carried through."[33] In support of this contention he cited with approval Schrödinger's treatment of Einstein's analysis of momentum exchange during radiative processes as applying only to the recoil of the atom and not necessarily to the light, and he pointed out that because of the relative mass of the atom any such recoil would be of small velocity and effectively unobservable.

One problem with the abandonment of energy conservation was its implication for the causality principle. There was no *a priori* reason why a statistical energy law, or systematic changes in energy, should not have been incorporated in a deterministic physics much as was the statistical entropy law. But this would not have been possible without some refinement of either the energy concept or the structural conception of matter: it would have been necessary for example to keep track somehow of the deviation of energy from its statistical norm. The simplicity of the energy concept and the conservation principle gave classical physics a large measure of its security, and the abandonment of energy conservation was therefore likely to be seen by many as tantamount to, either logically (within the existing conceptions) or at least in the sense of being as bad as, an abandonment of causality. This appears, for example, to have been true in Einstein's case,[34] and the connection may also be seen in Schrödinger's work as discussed in the last chapter. Planck's use of an *a priori* emission probability in his 1911 derivation of the black-body law and Einstein's introduction of transition probabilities in 1916 both had to be accompanied by clarifying statements to the effect that causality was somehow, though how they could not tell, maintained.[35] The absence of any mechanism for the discrete changes in either Planck's theory or Bohr's cried out for the abandonment of causality and the resort to a purely statistical theory. Moreover, beyond the realms of the quantum theory itself there were also strong ideological pressures in the same direction.[36]

Bohr is unlikely to have been swayed by the anticausal pressures of the German Weimar milieu, strong as these seem to have been, but he was acutely conscious of those within quantum theory, and he also appears to have been influenced by aspects of Danish philosophy. He himself talked of such an influence in respect of his principle of complementarity,[37] and there are striking parallels between the exposition of his atomic model and that of some aspects of Kierkegaard's philosophy, to which he was strongly drawn.[38] His philosophical mentor, Høffding, had made the rejection of causality an explicit part of his system.[39] It was not surprising that Bohr himself should move in this direction. But whereas the abandonment of energy conservation seemed to Bohr to offer a freedom that was at once sufficient to allow the incorporation of non-classical behaviour in the theory and at the same time restricted enough, given for example statistical conservation, to allow the theory to retain some positive predictive potency, a mere abandonment of causality did not have this virtue. It gave up too much. And in these circumstances Bohr's thoughts seem to have run along lines, perhaps tied in with his philosophical background, that anticipated in some respects his later principle of complementarity. Some of his colleagues rejected the causality principle outright. Others talked of the impossibility of a consistent description in three-dimensional space.[40] But Bohr, recognising that these factors were intimately linked but that neither a spatial description nor a causal description could be sacrificed altogether, moved only tentatively toward the idea that they might be incompatible. He referred to the role of probability in determining the atomic transitions, but always qualified his reference as applying "in the present state of the theory." In his 1922 survey he piled caution upon caution, stating only that[41]

> In the present state of the theory, it is not possible to bring the occurrence of radiative processes, nor the choice between various possible transitions, into direct relation with any action which finds a place in our description of phenomena, as developed up to the present time.

It was only in a manuscript dating from the Winter of 1923–1924, by when the pressures had increased considerably, that he was even so bold as to suggest that, in relation to the wave-particle duality,[42]

> It is more probable that the chasm appearing between these so different conceptions of the nature of light is an evidence of the unavoidable difficulties of giving a detailed description of atomic processes without departing essentially from the causal description in space and time that is characteristic of the classical mechanical description of nature.

Bohr was cautious. But there can be little doubt that he was already heading towards the abandonment of a causal space-time description for atomic processes. Writing strongly under his influence, Darwin, in the letter of 1919, suggested the last resort of endowing electrons with free will.[43] Von Mises's rejection of causality in 1921 appears to have been directly related to Bohr's views as they were expressed in his paper to the Solvay congress that year.[44] And in 1923 H. A. Senftleben, who was also strongly influenced by Bohr, observed that "Planck's constant h limits in principle the possibility of describing a process in space and time with arbitrary accuracy."[45] Moreover, he drew the conclusion that in this case the principle of causality, expressed as "Given a situation A in space-time we can determine a later situation B", was simply inappropriate: we were not given such a situation A.

THE STAGE IS SET FOR A DIALOGUE

Bohr's views on the causality principle at the time he was joined by Pauli are unclear and, in the last analysis, unknown. For although we have tried to indicate a trend in the expression of his ideas the evidence is slight. Although the causality issue was the most emotionally potent and the most publicised and argued aspect of the quantum theoretical problem complex, it was not however the most fundamental. And on the most fundamental issue Bohr's views, and their relationship to Pauli's, are clear. At the very end of his 1922 survey paper Bohr expressed "a hope in the future of a consistent theory, which at the same time reproduces the characteristic features of the quantum theory, important for its applicability, and, nevertheless, can be regarded as a rational generalisation of classical electrodynamics."[46] The quantum concepts were "important for [their] applicability", but the concepts of classical electrodynamics were fundamental. The quantum theory was based upon these concepts, both through the atomic model itself and through the correspondence principle, and they were treated as basically immutable, so that any account of realms in which they were not directly applicable had to proceed by their limitation rather than their replacement. In Pauli's view, on the other hand, these concepts had to be replaced by new ones, to be derived in accordance with the criteria of internal and operational consistency. Bohr's fundamental concepts of understanding were those of the wave theory, and he rejected the light-quantum hypothesis in favour of energy conservation. Pauli's fundamental concepts were to be derived from discrete counting processes, and he insisted on the validity of the light-quantum hypothesis,[47] and on the retention of energy and momentum conservation.[48] The concerns of the two physicists were similar, but their convictions opposed.

CHAPTER 4

THE TECHNICAL PROBLEM COMPLEX

The dialogue between Bohr and Pauli was to be central to the development of the new quantum mechanics, but it could be so only once it had been incorporated into the technical problem complex of quantum theory, and this was no easy matter. Pauli's ideas, developed outside the quantum context, seemed strange to some of his colleagues and could have little impact until applied to detailed quantum problems. Even Bohr's views, though developed in the quantum context, lacked the precision conferred by concrete application. Moreover, there was a further complicating factor in the anti-causal pressures of the Weimar intellectual milieu. That such pressures existed, were strong, and were recognised and to some extent accommodated to by German physicists in the early 1920s has been clearly demonstrated by Forman.[1] And at first sight their existence would seem to offer support for Bohr's views on the absence of causality in quantum theory. But so far as the mainstream quantum physicists were concerned, the pressures do not seem to have been anything like so strong as has sometimes been suggested, and their existence tends historically to conceal as much as it reveals.[2] Not only for Bohr and Pauli but also for most of the main quantum atomic physics community, causality was an issue, but only a secondary one, a decision on which was to be derived from other more fundamental considerations. Within this community there was an awareness that a fully causal qauntum theory did not yet exist, and that the retention of causality in any future theory could not be assumed. But a discussion of causality *per se* did not seem a very useful way of going about things, and the subject was not really an issue.[3]

More popular among the quantum physicists was Bohr's rejection of strict energy-momentum conservation. Heisenberg has recalled that after the implications of Einstein's 1916 paper had been absorbed most members of the quantum physics communities in Munich and Göttingen were open to the possibility of statistical energy conservation.[4] And in 1921 Sommerfeld included in the latest edition of his famous and influential book *Atombau und Spektrallinien*, the "bible" of quantum atomic theory, the statement that[5]

The mildest modification that must be applied to the wave theory is, therefore, that of disavowing the energy theorem for the single radiation phenomenon and allowing it to be valid only on the average for many processes.

A little later, Einstein and Ehrenfest found themselves obliged to consider seriously, though they could not accept, the possibility of energy non-conservation. In the Winter of 1921–1922 Stern and Gerlach demonstrated that if a beam of silver atoms was passed along a strong magnetic field gradient the beam split into two well-defined beams, deflected from the original path in opposite directions, and with no atoms at all remaining on the original path. This result was in clear agreement with the 'space quantisation' prediction of quantum atomic theory, according to which the atoms could possess only certain discrete values of magnetic moment, producing the discrete deflections observed. It was in complete disagreement with the classical theory, according to which any change in magnetic moment had to be continuous and should have resulted in spreading of the beam of atoms with the peak of the resulting distribution remaining along the undeflected path. What concerned Einstein and Ehrenfest, however, was that while the behaviour did give the result predicted by quantum theory, they could see no way in which the process leading to that result could take place, no way in which in the time available and in a radiation-free vacuum the continuously varying field could impart sufficient energy to the atoms to make possible the discrete change in magnetic moment observed. Either one had to give up altogether any description of the process, and merely rest content with the result, or else one had to assume at the least that energy was not conserved.[6] Finally, in a study of 1923 Born and Heisenberg ran into some closely related problems when considering the behaviour of an atom in crossed electric and magnetic fields. A continuous change in the field specification, which change could be infinitely small, led according to the quantum theory to a discrete change in the state of the atom, but without apparently being able to provide the discrete amount of energy required for such a change. Their provisional conclusion had to be that energy was not strictly conserved.[7]

Bohr's ideas on the failure of mechanics were also reflected quite widely at about this time. In February 1923 Heisenberg admitted in correspondence that he was beginning to follow Bohr and Pauli in accepting the failure of mechanics,[8] and suggested that "either new quantum conditions, or proposals for the modification of mechanics" were needed.[9] And that Summer Landé and Born also wrote of this same "failure of mechanics".[10]

In general at this stage, Bohr's views dominated over Pauli's. He was after all the accepted and undisputed master of quantum theory. But to some extent, insofar as they talked of the problems rather than the remedies, their views did overlap; and even where they differed Pauli's more radical attitude was not without support. Thus Born, who had earlier shown sympathy with Pauli's views on the inapplicability of field theory to the inside of the atom, wrote at this time that "the whole system of concepts of physics must be reconstructed from the ground up."[11] Heisenberg was now becoming increasingly sympathetic to Pauli's viewpoint. And there was a growing feeling that while statistical energy conservation might indeed be the "mildest modification" needed to solve the immediate problems, it would not ultimately be enough; that more radical conceptual changes would be necessary. Only a few weeks after advocating statistical conservation in his textbook, and after arguing in its favour against Heisenberg, whose study of the anomalous Zeeman effect had led him to "place ourselves deliberately in opposition to classical radiation",[12] and in support of light-quanta, Sommerfeld admitted to Einstein that "inwardly I too no longer believe in the spherical waves."[13] Statistical conservation had been introduced as an alternative to the light-quantum hypothesis, and before the results of Geiger and Bothe in 1925 there was nothing that could experimentally distinguish between them.[14] But as the light-quantum gradually found wider and more convincing application so energy non-conservation appeared more and more as an inadequate substitute, avoiding rather than addressing the crucial problems involved. At the end of 1922 the discovery of what quickly became known as the Compton effect provided what is still the clearest and most natural application of the light-quantum hypothesis. The scattering of X-ray light by an electron was explained perfectly by the simple laws of particle collisions, and Compton drew the "obvious conclusion" that the light was composed of discrete and localised quanta.[15] A few months later Debye also came upon the same effect, and drew the same conclusion.[16] Soon after, the same light-quantum treatment was extended by Compton and Duane to Fraunhofer diffraction, and this new run of success for the light-quantum was consolidated by Pauli who, building on Einstein's 1916 paper and combining this with the insights of Compton and Debye, derived the first successful probabilistic treatment of the temperature equilibrium between radiation and free electrons.[17]

The Compton effect did not "prove" the existence of light-quanta. Ehrenfest and Epstein failed in their attempt to extend the light-quantum analysis to Fresnel diffraction; and Compton himself, between writing and submitting a paper on Compton scattering, published another important article on the wave-like total internal reflection of X-rays, a phenomenon which, as he admitted, was "not easy to reconcile" with the conclusions he had drawn from the scattering effect.[18] Moreover it was still the case that the admission of light-quanta entailed what appeared to be more fundamental inconsistencies than the absence of strict energy conservation. But the Compton effect and related results did help to redress the balance between the view of Pauli, that far more radical conceptual changes were needed than the mere acceptance of energy non-conservation, and the previously dominant view of Bohr, that the way lay through modification rather than replacement of the classical wave theory. To see how these two viewpoints interacted so as to lead to the creation of the new theory of quantum mechanics we must now turn away from general statements and issues to the specific and highly technical problems of the quantum theory of the atom.

PROBLEMS WITH STATIONARY STATES

Prior to the advent of quantum mechanics the quantum theory of the atom was divided into two distinct parts, one dealing with the determination of stationary states and allowed transition frequencies, the other with the transition probabilities or intensities. The theory of stationary states was effectively based upon a mechanical model of the atom as a conditionally periodic system, with the quantum-theoretically possible orbits or vibrations of the atomic electrons given by Sommerfeld's generalisation of Bohr's original quantum conditions,[19]

$$(1) \qquad J_k = \oint p_k \mathrm{d}q_k = n_k h$$

(J_k action, p_k momentum, q_k displacement, h Planck's constant and n_k integers).

The theory was able to predict correctly the spectral frequencies of the hydrogen atom, both with and without the application of external electric or magnetic fields.[20] But as soon as it encountered the more complex problem of mutli-electron atoms, and especially that of the anomalous Zeeman effect (splitting of spectral lines in a magnetic field) in alkali atoms, which on Bohr's shell theory of periodic structure had a single electron in the outer orbit or uppermost set of energy levels, it ran into serious problems. One did not in fact have to go very far up the periodic table to run into trouble and in 1921–1922 Langmuir, Epstein and Van Vleck in America, and Kramers in Denmark, all reported apparently insurmountable difficulties with the helium atom.[21]

Meanwhile in Göttingen, Born embarked on a programme of pushing the theory as far as it would go, trying with Pauli's assistance to find out its limitations and shed light upon the modifications it needed.[22] But they were able to make little progress, and Born could only conclude, writing to Einstein, that "the quanta really are a hopeless mess."[23] In 1922 Pauli applied the theory to the hydrogen molecular ion and showed that even for this structure, which possessed just one electron, it gave the wrong answer.[24] Indeed the only success obtained at this period was with a model of the atom introduced by Heisenberg, while still a student under Sommerfeld at Munich, and developed by Landé.[25] This "core model" of the atom, however, represented an obscure but essential departure from the Bohr theory. Abandoning any detailed decription of the electron orbits such as the theory required, Heisenberg restricted the description of the atom essentially to that given by the quantum numbers, and treated the multi-electron alkali atom as composed of just two parts: the single outer or "series" or "valence" electron, and a "core" composed of the nucleus and other electrons but treated as effectively a single particle.[25] And while this model was applied with some success it was a glaring feature of that success that it entailed hypotheses which, if interpreted physically, were quite contrary to the established theory. To obtain the observed splitting of the spectral lines in a magnetic field, half-integral quantum numbers had to be introduced for some components. The selection rules determining which transitions between stationary states were possible violated those established on theoretical grounds by Rubinowicz and appeared to be quite inconsistent, even arbitrary. And the failure of the Larmor theorem on the precession of an electron orbit in a field manifested itself in the fact that to obtain the correct atomic energy in a magnetic field, the core of the atom (i.e., all but the single outer electron) had to be counted twice in its contribution to the "g-factor" defining the atom's angular momentum and magnetic moment.[27]

Of these inconsistencies the anomalous g-factor and the half-integral quantum numbers were seen as the most problematic. Bohr objected of the former that the break with the classical theory of conditionally periodic systems that it involved "immediately removes any ground for the calculation of the energy of the atom in the field of the sort that Heisenberg undertakes."[28] And of the half-integral quantum numbers he wrote simply that "the entire method of quantisation . . . appears not to be reconcilable with the fundamental principles of quantum theory."[29]

But what was the alternative? By the Winter of 1922–1923 Pauli had moved from Göttingen to Copenhagen and Heisenberg had followed in his footsteps from Munich to Göttingen; and that Winter was devoted to a thorough investigation of the status of Bohr's atomic theory, in which the use of perturbation theory was treated as acceptable, but half-integral quantum numbers and their associated contradictions were not. Born and Heisenberg, pursuing the perturbation theory approach, published two papers, the first of which ran into the troubles with crossed fields already mentioned.[30] In the second paper, in which they also acknowledged the assistance of Pauli, they subjected the helium atom to its most thorough investigation yet, and their conclusions were quite clear:[31]

We have now set ourselves the problem of examining all possible orbital types in excited Helium atoms, of selecting the quantum-theoretically possible solutions, and of calculating the energy values so as to be able to establish whether or not orbits are present which give the empirical terms correctly. The result of our investigation is negative: one reaches through the consequent application of the known quantum rules no explanation of the Helium spectrum.

This paper was taken as decisive proof that the old quantum atomic theory, with integral quantum numbers, failed for the helium atom.[32]

Meanwhile, late in 1922, Bohr had already anticipated that Born's perturbation theory approach must fail, if only because the perturbations were themselves of the same order as the unperturbed results.[33] And in his major survey paper on the foundations of quantum theory he had expressed strongly his conviction that classical mechanics, and with it the existing quantum atomic theory, must fail.[34] Though differing in their beliefs as to how the problems of the theory might eventually be resolved, both Bohr and Pauli were convinced of the need for major innovations, and their approach that Winter was to push the existing theory as far as it would go, especially in the context of the anomalous Zeeman effect, and to look for a resolution through the exploration of the resulting paradoxes.[35] But sticking, as they felt they must, to the integral quantum numbers, they made little progress.[36] And the situation continued to deteriorate.

Further confirmation of the failure of the theory continued, moreover, to arise. Towards the end of 1922 Van Vleck had obtained additional evidence of the failure of integral quantum numbers in a situation in which the application of half-integral numbers gave results to within the experimental error.[37] And this had increased Heisenberg's (and Sommerfeld's) conviction that "the half-integral quantum numbers are right."[38] Then in early 1923 Landé made explicit yet another problem arising from the successful use of the core model.[39] In 1921 Bohr had attempted to construct a quantum periodic table by means of a "building-up principle".[40] For conditionally periodic systems, the quantum numbers and statistical weights of stationary states could be treated as adiabatic invariants, that is as invariants with respect to certain gradual "adiabatic changes" in the systems. This enabled Bohr to build up the theory to describe the higher elements by starting from hydrogen and adding electrons one by one such that the change was in each case adiabatic. Although the procedure could be applied rigorously only for true conditionally periodic systems, already recognised as insufficient representations of the higher elements, it gave qualitatively good results, and Bohr was sufficiently convinced of its general validity to stress the role of the "adiabatic principle" and to formulate a corresponding "principle of the existence and permanence of quantum numbers" in his 1922 survey paper.[41] No sooner had this been published, however, than Landé demonstrated that the core model not only necessitated half-integral quantum numbers but also involved a change in the quantum numbers and statistical weights during the building-up process. To make matters worse, it was the building-up process itself that generated the higher elements for which the core model was used.

Although the required change of quantum numbers in the building-up process was a new result, the general problem of statistical weights was not. In terms of the core theory it could be expressed by saying that the theory gave the core of the atom either one too few or one too many degrees of freedom, and the outer electron one too many; but it had already existed in essence before the advent of that model. As early as 1920 Bohr had introduced a *Zwang* ("constraint") to regulate the choice of possible orbits, despite the fact that such a *Zwang* (justified in 1922 on the basis of the correspondence principle) could not be explained in terms of the mechanical model of the atom.[42] In 1922 Landé had referred to the problems of the core model as involving an "unmechanical adjustment of the core",[43] and it was to a combination of these ideas that Bohr turned with Pauli in 1923. To preserve what consistency they could, they insisted on integral quantum numbers, and the invariance of quantum numbers, for all those numbers that were interpreted as characterising the individual electron motions in the atom. But for the "inner" quantum number, *j*, which was associated with the relative orientations of the electrons and with the coupling between the outer electron and the core, both conditions were finally sacrificed. And their absence was explained by an "unmechanischer Zwang":[44]

In the electron assemblage in an atom, we have to do with a coupling mechanism that does not permit a direct application of the quantum theory of mechanical periodic systems; in particular, there can obviously be no question of accounting for the complex structure in terms of the exclusion, based on the consideration of adiabatic transformation, of certain motions, compatible with this theory, as stationary states of the atom. Rather, we are led to the view that the interplay between series [outer] electron and atomic core, at least so far as the relative orientation of the orbit of the series electron and those of the core electrons is concerned, conceals a "*Zwang*" that cannot be described by our mechanical concepts and that has the effect that the stationary states of the atom, in essential respects, cannot be compared with those of a mechanical periodic system. According to our view it is just this Zwang that finds its expression in the regularity of the anomalous Zeeman effect, and, in particular, is responsible for the failure of the Larmor theorem.

The introduction of the *unmechanischer Zwang* did not solve the problems of atomic theory, and the paper was not indeed published; but it did mark an important step in their development. At this stage neither Bohr nor Pauli could accept Heisenberg's rather buccaneering approach to the quantum theory. As Pauli wrote to Bohr at the beginning of 1924,[45]

If I think about his ideas they seem monstrous and I curse to myself a lot about them. Because he is so unphilosophical, he pays no attention to clear presentation of the basic assumptions and their relationship to previous theories.

Bohr's feelings appear to have been rather stronger,[46] and the fundamental differences between Bohr and Pauli also remained. But the new development did represent a general recognition that while Heisenberg's methods might be appalling, and many of his suggestions unacceptable, his was the approach that was getting results. Moreover, it also provided a common language in terms of which the problems could be further explored.

When Pauli wrote up his work on the anomalous Zeeman effect in April 1923 he stuck hard by his criterion of internal consistency and adopted what he described as a "purely phenomenological" description which "abandoned all use of models".[47] In June, he wrote to Sommerfeld that the quantum theory supplied "no sufficient grounding" for the treatment of complex spectra, and that something "in principle new" was needed for the anomalous Zeeman effect.[48] The old quantum theory had failed:[49]

This failure can scarcely be doubted any longer, and it seems to me to be one of the most important results of the last few years that the difficulties with many-body problems lie in the physical atom, not in the mathematical treatment (when, e.g. the Helium term comes out wrong in Born and Heisenberg, this certainly does not lie in the fact that the approximation is insufficient.

But beyond the adoption of his phenomenological approach and a consequent rejection of the orbital model of the atom, Pauli had no idea what to do about this failure. In July he wrote to Bohr optimistically that "perhaps after all you may in the course of the Summer get a saving idea about complex structure and the anomalous Zeeman effect."[50] But it was not to be, and rather than pursue further the details of what appeared to be an innately inconsistent theory Pauli turned his attention elsewhere. In August he submitted his paper on the thermal equilibrium between radiation and free electrons.[51] In September he wrote the letter to Eddington based on the latter's mathematical theory of relativity.[52] Later in the year he returned to Copenhagen for a two months "holiday", and early in 1924 he explained to Bohr how he felt about the quantum atomic theory:[53]

> The atomic physicists in Germany today fall into two groups. The one calculate a given problem first with half-integral values of the quantum numbers, and if it doesn't agree with experiment they then do it with integral quantum numbers. The others calculate first with whole numbers and if it doesn't agree then they calculate with halves. But both groups of atomic physicists have the property in common, that their theories offer no *a priori* reasoning which quantum numbers and which atoms should be calculated with half-integral values of the quantum numbers and which should be calculated with integral values. Instead they decide this merely *a posteriori* by comparison with experiment. I myself have no taste for this sort of theoretical physics, and retire from it to my heat conduction of solid bodies.

HEISENBERG, BORN AND DISCRETE ATOMIC PHYSICS

As it turned out, Pauli's absence from atomic physics lasted only a few months. But in that time the theory saw two further striking developments, both of which were to play a large part in shaping its future development. One of these, to which we shall return, was in the branch of the theory devoted to transition probabilities or intensities, and was to lead to the infamous virtual oscillator theory of Bohr, Kramers and Slater. The other was a continuation of Heisenberg's work on the core model, and although it is far less well known, it was to be equally important for the development of a new quantum mechanics.

Early in 1923 Heisenberg's reaction to the failure of the existing atomic theory had been to continue developing this theory but to suggest that "either new quantum conditions or proposals for the modification of mechanics" were needed.[54] And returning to the anomalous Zeeman effect in October of that year he did in fact derive some new quantum conditons:[55]

$$(2) \qquad H_{\text{quantum}} = \int_{-1/2}^{+1/2} H_{\text{classical}}\, \mathrm{d}j = F(j + \tfrac{1}{2}) - F(j - \tfrac{1}{2}),$$

with H the Hamiltonian and F a new function to be determined. The immediate context of this formula was the problem of the anomalous g-factor (governing the magnetic moment) in Heisenberg's core model of the atom, and Serwer has suggested that the problem of statistical weights may also have played an important part in its origin.[56] In some of its manifestations this required that the number of atomic states should be one less than that given by the quantum numbers, and by associating each single state with a pair of quantum numbers instead of a single quantum number, Heisenberg's new formalism overcame this difficulty. From the letter to Pauli in which the formalism was first developed, however, its origins appear to have been more fundamental.[57] Heisenberg wrote that so far one had obtained the frequency of a transition simply by taking the difference between two energies as in the Bohr quantisation conditions,

$$(3) \qquad h\nu = \Delta H,$$

but that this was only appropriate for the simple case of hydrogen. In the general case, one would also have to derive the energies themselves from a difference equation, say $H = \Delta F$:[58]

The new Göttingen theory of the anomalous Zeeman effect runs roughly as follows:
(1) The model representations have in principle only a formal sense, they are the classical analogues of the 'discrete' quantum theory,
(2) Up to now it was usual to go over from model symbols to the real radiation frequencies by taking over the energy $H(J_1, \ldots, J_n)$ from the symbols to the ν_{qu} through the difference equation . . . .
(3) This is only a special case, which is right for Hydrogen. In other problems one must take from the symbols other functions than H. A general theory, as to *which* functions of the $J_1, \ldots, J_n$ is still outstanding.
(4) For the anomalous Zeeman effect, the function in question reads $F(k, r, j, m) = \int H(k, r, j, m)\, dj = \int H\, dj$. From F one gets to the H_{qu} through $\Delta F = H_{qu}$, from H to ν through $\Delta H = \nu$.

Heisenberg gave no real justification for this approach, but he stressed throughout the letter the fundamental role of the difference equations. He always referred either to "our" approach or to "the Göttingen" approach, rather than claiming it as his own, and he noted Born's summing up of their future programme as the "discretisation of atomic physics."[59] Combined with Heisenberg's later recollection of a seminar in Hilbert's department at this time on the very subject of difference equations,[60] this all goes to suggest that it was the concept of difference equations that lay behind the new formalism. If the seminar did play a crucial role, then Heisenberg and Born would indeed have shared in the idea to develop this concept. And it is possible, moreover, to reconstruct the argument in the letter to Pauli from the difference equation basis. Whereas Bohr's original quantum condition was indeed a difference equation, the more general Bohr-Sommerfeld conditions, Equation (1), were not. Having decided to express everything in difference equation form, Heisenberg would therefore have had to search for a new quantisation condition, and the requirement that it gave the correct g-factor was in fact sufficient to determine his choice. Earlier in 1923 Landé had given a general formula for the g-factor, and this gave for the required H in terms of the quantum numbers,[61]

$$H = \frac{\omega(j^2 - r^2 - k^2)}{2kr} + \frac{\nu m\,\{1 + (j + \tfrac{1}{2})(j - \tfrac{1}{2}) + r^2 - k^2\}}{(j + \tfrac{1}{2})(j - \tfrac{1}{2})}.$$

This could be obtained from the formula proposed by Heisenberg in the letter to Pauli with

$$F = \frac{\omega(\tfrac{1}{3}j^3 - r^2 j - k^2 j)}{2kr} + \nu m \left(j + \frac{j^2 - r^2 + k^2}{2j}\right),$$

i.e., as the difference between two terms each expressible in terms of a single clearly identifiable set of quantum numbers.

Although it caused quite a stir, the Heisenberg-Born formalism was not at first universally well received. From Bohr's point of view it merely compounded the felony of the core model and half-integer quantum numbers, and went in completely the wrong direction, even further away from any link with the classical conceptions.[62] Heisenberg's happy acceptance of, even advocacy of, the fact that his approach abandoned any attempt to make physical sense out of the basic formulae was not for Bohr.[63] Even Pauli, who unlike Bohr was favourably predisposed to the idea of a new discrete form of mechanics, could only despair at the lack of any conceptual foundation for the new ideas, writing to Landé that[64]

> I don't in any way share your opinion of Heisenberg's new theory. I even hold it for ugly. For despite radical assumptions it provides no clarification of the half quantum numbers or the failure of the Larmor theorem (especially the magnetic anomaly). I don't think much of the whole thing.

Faced with such opposition, Heisenberg withheld publication of his new formalism, and concentrated first on establishing the need for the half-integral quantum numbers, and on trying, unsuccessfully, to derive these from the formalism instead of having to put them in to it.[65] And by the time it was published, in the Summer of 1924, it was in many respects out of date. But by then it had already served its purpose, for it had provided Born and Heisenberg with a vital link between the problems of stationary states and the idea of discreteness on the one hand, and the theory of transition intensities and Bohr's emphasis upon continuity on the other.

THE BOHR-KRAMERS DISPERSION THEORY

Whereas the part of quantum atomic theory dealing with stationary states and frequencies was already well established by the end of the Great War, the part dealing with transition probabilities, or intensities, could not get under way until after the enunciation of Bohr's correspondence principle in 1919. Even then it developed only slowly as the problems with the basic model, some of which we have outlined above, dominated the researches of the physicists. One aspect of the intensities problem did however receive some attention, and that was the theory of dispersion.

In principle, any theory of dispersion had to be related in the Bohr theory to the orbital model of the atom and to the theory of transition frequencies derived therefrom. The Bohr atomic model did not itself incorporate any mechanism through which the transition probabilities could be predicted, and they had therefore to be derived from the classical intensities by way of the correspondence principle. But they could only be derived in this way as functions of frequencies. However, although the correspondence principle rested upon an application of the orbital theory in a limit in which the quantum transition frequencies were comparable with the classical absorption and emission frequencies, the rigorous application of the orbital theory had failed and the only cases of physical importance lay outside the limit of comparison. The only feasible approach to the quantum theory of dispersion was therefore to take the observed frequencies and intensities, to express the latter in terms of quantum transition probabilities, and to try thereafter to reconcile both with the quantum theory of the atom.

In 1921 Ladenburg took the first step in this direction by showing that, given the observed absorption and emission frequencies, both the observed absorption intensities and the observed dispersion coefficients corresponded to those predicted by the classical theory.[66] He could only draw very limited connections with the quantum theory of the atom, relating transition probabilities very roughly to statistical weights of atomic stationary states for a few simple cases. But his work did provide strong evidence that the classical theory of dispersion could be carried over to the quantum context, dispersion being of course a fundamentally wave-type of phenomenon. Working with Reiche in 1923, Ladenburg was able to extend this evidence and to draw the explicit conclusion that:[67]

> We believe on the grounds of the observed phenomena that we must consider the end result of a process in which a wave of frequency ν is incident upon the atom as not fundamentally different from the effect that such a wave exerts on classical oscillators: ... Even the force of scattered waves seems repeatedly to agree with that from an oscillator.

The dispersion theory of Ladenburg and Reiche was restricted to dispersion by atoms initially in their ground state, that is to dispersion as a pure absorption process, and apart from a vague reference to the correspondence principle it was without any foundations in the quantum theory of the atom. There was, however, no viable alternative to its basic approach, and in an unpublished manuscript of 1921 Bohr wrote encouragingly of Ladenburg's work:[68]

> Although it is at present an unsolved problem, how a detailed theory of dispersion can be developed on the basis of the quantum theory, a promising beginning on the indicated basis might nevertheless seem to be contained in the interesting considerations about this phenomenon, recently published by Ladenburg.

Writing to Darwin in December 1922, Bohr suggested that dispersion should be attributed to some "mechanism" called into play when the atom was illuminated, "with the effect that the reaction of the atom corresponds to that of a harmonic oscillator in the classical theory, with the frequency coinciding with that of a spectral line."[69] In a major survey of quantum theory, completed that November, he had expressed himself more carefully and at greater length:[70]

> On the one hand, as is well known, the phenomena of dispersion in gases show that the process of dispersion can be described on the basis of a comparison with a system of harmonic oscillators, according to the classical electron theory.... On the other hand, the frequencies of the absorption lines, according to the postulates of the quantum theory, are not connected in any simple way with the motions of the electrons in the normal state of the atom ... According to the form of the quantum theory presented in this work, the phenomena of dispersion must then be so conceived that the reaction of the atom on being subjected to radiation is closely connected with the unknown mechanism which is answerable for the emission of radiation on the transition between stationary states. In order to take account of these observations, it must be assumed that this mechanism, which is designated in the preceding paragraph the coupling mechanism, becomes active when the atom is illuminated in such a way that the total

reaction of a number of atoms is the same as that of a number of harmonic oscillators in the classical theory, the frequencies of which are equal to those of the radiation emitted by the atom in the possible processes of transition, and the relative number of which is determined by the probability of occurrence of such processes of transition under the influence of illumination.

Extracted from the familiar Bohr dressing, this long exposition amounts to the statement that one should use the classical theory for the calculation of transition probabilities, whether in dispersion phenomena or elsewhere, and that this is quite justifiable as the cause of the transitions is unknown anyway.

Considering X-ray absorption phenomena in a paper completed in the Autumn of 1923, Bohr's assistant Kramers noted that "the quantum theory in its present state tells us nothing about the mechanisms of absorption and does not therefore permit the direct calculation of the probability that an absorption process may occur."[71] In this context and in that of the X-ray emissions from electron-atom collisions, Kramers therefore adopted the procedure laid down by Bohr, and worked with the classical theory. In the latter context he explained that

the only procedure which offers itself at present seems to consist in estimating the statistical result of a great number of emission processes – in a way suggested by Bohr's correspondence principle – from the radiation which on the classical theory would be emitted by the free electrons in consequence of the change in motion produced by the forces owing to the electrical particles in the atom.[72]

The question remained as to how this classical treatment was to be tied in with the orbital quantum model of the atom. In November 1922 Bohr had been still prepared to hope that the orbital model might survive its current problems, and this hope was reflected in his reference to "a number of atoms" for comparison with the classical theory, it being impossible to incorporate all the possible transition frequencies within a single orbital atom. For atoms initially in a given stationary state, as in Ladenburg's theory of dispersion, it would have been desirable if all possible transition frequencies from that state could somehow have been contained in the appropriate electron orbit, and Kramers suggested in his X-ray paper that "one should expect that every possible transition corresponds to a certain frequency present in the motion of the electron."[73] But it was difficult to conceive how even this requirement might be satisfied. Kramers, who had not been directly involved with the problems of the core model and *unmechanischer Zwang*, appears to have held fast to a relatively literal interpretation of the Bohr model.[74] But to Bohr the pressures on this model were building up, and the need to incorporate the classical theory of dispersion on top of all the other problems prompted him to give further thought to the problem of conceptual foundations. Just as Heisenberg held back his new difference equation formalism from publication on account of criticisms from Bohr, so Bohr himself had refrained from publishing his thoughts on the *unmechanischer Zwang* while the implications of Heisenberg's ideas were worked out. And in the Winter of 1923–1924 the work of the previous Spring was finally superseded by a new manuscript draft which sought to accommodate both the problems of the core model and the results of dispersion theory, and which fell back on Bohr's ideas on energy conservation and causality.[75]

Both the general classical nature of the dispersion theory and the fact that it could not be reconciled with the orbital model of the atom other than on a statistical basis, for a large number of atoms, supported Bohr's earlier conclusions on statistical energy conservation. And in the manuscript of 1923–1924 he carried his argument slightly further than he had before, noting the need to "depart essentially from the causal description in space and time that is characteristic of the classical mechanical description of nature."[76]

PAULI, HEISENBERG, AND THE REJECTION OF ELECTRON ORBITS

While Heisenberg and Born pushed their discretisation of atomic physics and Bohr and Kramers pushed their classical theory of dispersion, with its emphasis upon the continuous wave formulation and associated changes in the laws of energy conservation and causality, Pauli concentrated on other things. But he did not refrain from all comment, and nor did he restrict his comments to criticisms. Thus in June 1923 he wrote to Sommerfeld with the idea that rather than adapting the dispersion theory to fit in with the hypothetical orbital model of the atom, it was the latter that should be adjusted to fit in with the former:[77]

> I often think that not only in dispersion, where they are under the influence of a simply harmonic periodic external force, but also in the mutual effects of the electrons in the atom, the individual electron orbits control themselves more as a system of oscillators in which the frequencies are associated not with the motion but with the transition.

Writing to Bohr in February 1924 he decided to press his view that the concept of electron orbits, upon which the whole atomic theory still depended, had to go:[78]

> The most important question appears to me to be this one, to what extent one may in general speak of fixed orbits of electrons in stationary states. I think that this can no way be assumed as self-evident, especially in view of your observations about the balance of statistical weights in coupling. Heisenberg has in my view hit the mark precisely when

he doubts the possibility of fixed orbits. *Doubts of this kind Kramers has never considered as reasonable.* I must nevertheless insist upon this, because the point appears to me to be very important.

Pauli's views on the inadmissibility of the concept of electron orbits within the atom were by this time well established, and they may be traced back to his criticisms of Weyl's attempt at a unified field theory. It was impossible to specify any operational means of defining the orbit, for once bound in the atom the electron could be 'observed' only by means of the transition intensities and frequencies. The suggestion of the letter to Sommerfeld, that the theory of the atom should be therefore based upon these observables rather than upon the hypothetical orbits, followed. But what of Heisenberg's opinions? From Pauli's strong and unbending criticisms of Heisenberg's work it would seem at first that the two physicists were in fundamental opposition, but this was not in fact so. Heisenberg recalled that the rejection of electron orbits in the atom had been a shared feature of their student days together in Munich, and although there is no direct evidence of Heisenberg discussing this matter before 1925, Pauli's letter to Bohr just cited confirms their agreement on it at an earlier date.[79] Heisenberg continued to work with his core model version of the orbital model and to use this, to Pauli's perpetual despair, without any regard for its foundations or physical consistency. But in correspondence with Pauli he recognised that "the model conceptions have principally only a symbolic sense."[80] And far from abiding by Bohr's insistence that the correspondence principle must be founded upon the orbital model, Heisenberg had claimed that "the correspondence principle renounces any model insight."[81] In 1921 he had actually compared the correspondence principle and the orbital model as rivals, and had concluded that the former was the more sound, since founded directly upon experiment.[82] In one sense, of course, Heisenberg's attitude to the correspondence principle in these examples represented the very "unphilosophical" approach that Pauli so lamented. But if Heisenberg did not provide sound foundations for his arguments he did not, at least, cling to unsound ones. And this ensured that while on one level he was subject to Pauli's criticisms on another, deeper, level he was open to them.[83] From the other side Pauli, though always critical, was far from dismissive:[84]

But if I talk to him, he strikes me as all right, and I see that he has all sorts of new arguments – at least in his heart. I therefore think of him – aside from the fact that he is also personally a very nice fellow – as very thoughtful, even a genius, and I think he will once again greatly advance science.

He did, but only after Bohr and Pauli had finally met head on, in the context of the virtual oscillator theory of radiation.

CHAPTER 5

FROM BOHR'S VIRTUAL OSCILLATORS TO THE NEW KINEMATICS OF HEISENBERG AND PAULI

INTRODUCTION

Between them, the various lines of thought current at the end of 1923 contained most of the ingredients that would be needed for the formation of a new quantum mechanics. But these ingredients had not yet been brought together. The two parts of the quantum theory of the atom were still largely independent. And the different approaches of Bohr, Pauli and Heisenberg still ran alongside each other rather than engaging in any fruitful union or dynamic conflict. For further developments to take place a new element was needed, and this was provided early in 1924 in the form of the Bohr-Kramers-Slater theory of virtual oscillators.[1] Paradoxically, although the theory has been widely treated as the central and most fundamental feature of the evolution of quantum mechanics, its basic principles were almost universally rejected. But this rejection was itself important, setting as it did the stage for the development of quantum mechanics upon the rival ideas of Pauli. Equally important, the theory also drew attention to ways in which the techniques of the quantum theory of dispersion could be applied to those parts of the theory previously in the domain of the orbital model.[2]

THE GENESIS OF THE VIRTUAL OSCILLATOR THEORY

In essence, the virtual oscillator theory entailed only a very slight, but very crucial, development of Bohr's position. Departing from the somewhat hesitant and reserved attitude to the problem of an intuitive picture that had characterised his previous publications, Bohr came out openly in support of the view that such a picture was essential, and that it would be possible if and only if causality and energy-momentum conservation were abandoned, and the classical oscillator representation of the atom, previously no more than a heuristic device tenuously justified on the basis of the correspondence principle, reinterpreted as a physically meaningful model. In describing this model in the Bohr-Kramers-Slater (BKS) paper, he was typically vague. He described the oscillators as having a "virtual" existence – a characterisation that was at best ambiguous and at worst quite meaningless – and he did not commit himself as to the relationship between the new oscillator model and the old orbital one, the validity of which he appeared to uphold. Even the rejection of causality, for which the paper is perhaps most famous, was left open to interpretation as a temporary measure.[3] But there can be little doubt, especially in view of his later reaction to its refutation, that Bohr took the new model interpretation very seriously, and that he saw his decision to concentrate on an intuitive description in terms of the classical wave conceptions as both fundamental and necessary.[4]

This decision was of course a natural one for Bohr to make, and the virtual oscillator paper continued the line of argument already manifest in his manuscript of 1923 – 1924, without introducing anything dramatically new. The new theory may indeed have been formulated independent of any further outside influences. But its public expression at least was stimulated by the arrival in Copenhagen of a young American physicist, Slater, and by his advocacy of a theory similar to, and possibly derived from, that which had been recently proposed by Louis de Broglie.

De Broglie had started work on the problems of quantum theory in 1921 and had since completed a systematic study of those phenomena that revealed most clearly the fundamental wave-particle duality of light. Influenced by his brother Maurice's work on the particulate properties of X-rays, and by his tremendous admiration of Einstein, he had started out convinced of the necessity of the light-quantum concept, but intrigued by the problem of how the frequency of such a particle could be defined. He first looked at the quantum phenomenon described by Stokes's law in which light, affected by matter, always passes from a higher to a lower frequency, and from this he drew an analogy between frequency and temperature, and between Stokes's law and the second law of thermodynamics.[5] But the concept of the temperature of a single material particle raised just as many problems as did that of the frequency of a light-quantum, and although this confirmed the strength of the analogy it did not actually get him anywhere. De Broglie therefore turned to the problem of Planck's law, of which there was still no satisfactory interpretation, and having derived Wien's law from the hypothesis of independent light-quanta he investigated the dependence necessary to modify this to Planck's law.[6] The research was still unproductive so far as any fundamental insight was concerned, but it did lead to the introduction of a rather unusual conception. For de Broglie treated his derivations of the radiation laws as exercises in straight-forward relativistic particle mechanics, and the light-quanta as traditional material particles of negligible (but not necessarily zero) rest mass and with velocities approaching (but not necessarily equal to) *c*. The aim of this approach appears to have been to sharpen the wave-particle paradox by treating the light-quanta as thoroughly traditional particles.[7] But for whatever reason they may have been introduced, the light-quanta that did not move at the speed of light remained a permanent feature of de Broglie's endeavours. And in the course of 1923 these resulted in his famous wave theory of matter.

Although de Broglie's theory was of crucial importance for the detailed development of wave mechanics it actually contributed little to the generally accepted conceptual foundations of quantum mechanics, and had no direct influence on the development of matrix mechanics.[8] To consider it in detail in this context would therefore be out of place. But before writing his full *Thèse*, de Broglie published a series of short papers describing his main results, and an English summary of these was sent to Fowler in Cambridge for publication early in 1924 in the *Philosophical Magazine.*[9] And in December 1923 Slater wrote to Kramers from Cambridge, where he was under the supervision of Fowler, proposing a treatment of light-quanta along the lines used by de Broglie.[10]

It would seem that before leaving Harvard in the Autumn of 1923, Slater had been committed to the classical wave theory of light. He recalled that he had been unable to accept the abandonment in the Bohr theory of the classical relationship between the width of the spectral lines (interpreted by Bohr as being due to a statistical spread of the energies for each stationary state over different atoms) and the length of a finite wave train (or the period of its emission).[11] He found the concept of instantaneous transitions between stationary states "quite silly", and was convinced that a finite emission period was needed.[12] In December, however, he wrote to Kramers that he had come to the "rather surprising conclusion that the only possible way of getting a consistent explanation was in the direction of light-quanta."[13] And he expounded an idea as to how this might be done:[14]

> Of course, the quanta can't travel in a straight line with the speed of light: but it seems possible to suppose that there is an electromagnetic field, produced not by the actual motion of the electrons, but with motions with the frequency of possible emission lines (or, in an impressed field, of possible absorption lines), and amplitudes determined by the correspondence principle, the function of this field being to determine the motion of the quanta. If this motion is determined by the condition that Poynting's theorem shall hold over an average taken over a long period of time, definite patterns are described, and the probability of moving along the paths is such, for example, as to account for interference, many quanta being led to the bright spots in the field.

Slater's idea involved a remarkable fusion of diverse elements. His new found preference for the light-quanta could reflect a number of influences, including that of American discussion of the Compton effect.[15] The concept of light-quanta governed by a guiding wave field had featured repeatedly in Einstein's speculations, recently expounded in California by Lorentz,[16] as well as in de Broglie's theory, in which the quanta moved, as for Slater, at less than the speed of light. Which if any of these was Slater's source remains a matter for speculation.[17] But he had proceeded to fuse the guided light-quantum concept with the oscillator theory of dispersion advocated by that concept's strongest opponent, Bohr, at which he had been looking prior to his visit to Copenhagen.[18] Throwing in Poynting's theorem, which he took from some of Cunningham's lectures in Cambridge,[19] he derived a theory in which his original problem of spectral widths was, somewhat spectacularly, solved.[20]

The first recipient of Slater's idea was presumably Fowler, his host, who seems to have been quite pleased with it.[21] Back in America, Kemble was also impressed.[22] But in Copenhagen the reception was mixed. Kramers either never mastered the full subtlety of Bohr's approach or never quite accepted this approach, but as his student and assistant he was strongly influenced by it, and in a popular introduction to quantum theory completed in 1922 he had followed Bohr's rejection of the light-quantum concept, which he likened to "medicine which will cause the disease to vanish and kill the patient".[23] On receiving Slater's idea he apparently reacted warmly to the extension of the oscillator approach to include some kind of wave field, but rejected outright the introduction of light-quanta.[24] Bohr then incorporated the idea into his own developing conceptualisation, accepting the guiding field as a physical realisation of the oscillator technique, and again rejecting the light-quantum component. When Slater arrived in Copenhagen, and showed them a paper describing his idea, Bohr and Kramers first edited this paper so as to present Slater's idea as leading on to Bohr's conception:[25]

> An atom may, in fact, be supposed to communicate with other atoms all the time it is in a stationary state, by means of a virtual field of radiation, originating from the oscillators having the frequencies of possible quantum transitions, and the function of which was to provide for stationary states conservation of energy and momentum by determining the probabilities of quantum transitions.

Slater recalled that this published version was a third draft, written under the strong influence of Bohr and Kramers. Moreover, writing to van Vleck that July he claimed, albeit somewhat emotionally, that this draft was actually written by them.[26] They then wrote up Bohr's conception more fully, adding Slater's name gratuitously to the paper:[27]

We will assume that a given atom in a certain stationary state will communicate continually with other atoms through a time-spatial mechanism which is virtually equivalent with the field of radiation which on the classical theory would originate from the virtual harmonic oscillators corresponding with the various possible transitions to other stationary states. Further, we will assume that the occurrence of transition processes . . . is connected with the mechanism by probability laws which are analogous to those . . . in Einstein's theory . . . . The occurrence of certain transitions in a given atom will depend on the initial stationary state of the atom itself and on the states of the atoms with which it is in communication through the virtual radiation field, but not on the occurrence of transition processes in the latter atoms . . . .

We abandon . . . any attempt at a causal connexion between the transitions in distant atoms, and especially a direct application of the principles of conservation of energy and momentum, so characteristic for the classical theories.

In citing Slater as a joint author of this paper, Bohr and Kramers probably meant nothing but kindness and respect. But Slater, who seems to have been quite opposed to the new theory, was naturally a little disturbed, and he left Copenhagen prematurely.[28]

THE REJECTION OF BOHR'S INTERPRETATION

Slater was not the only physicist to reject Bohr's interpretation. Sommerfeld and Compton, for example, insisted that the Compton effect provided definite evidence of the necessity of the light-quantum concept and of the energy-momentum conservation associated with it.[29] Bohr did attempt an explanation of the effect in the BKS paper, but this involved a velocity of the virtual oscillators different from that of the particles to which they were supposed to be attached, and was not remotely convincing.[30] Sommerfeld also referred contemptuously to the BKS "compromise",[31] while Stoner, hitting on the limited achievements of the theory in comparison with its assumptions, argued that "it seems unnatural to assume that [conservation] does not hold in individual processes when there is no definite evidence of its breakdown, unless the supposition leads to a much more complete and satisfying explanation of observed phenomena than has hitherto been put forward."[32] Einstein objected violently to the absence of conservation and causality, arguing among other things that "a box with reflecting walls containing radiation, in empty space that is free from radiation, would have to carry out an ever increasing Brownian motion,"[33] and Ehrenfest wrote to him that "this time, as an exception, I firmly believe you are right."[34] As reported by Pauli, Einstein also objected that there were now *two* explanations of spectral widths, from the decay time and from the state uncertainty, and that the theory thus needed a "pre-established harmony" that he did not like.[35] Unlike most of Bohr's critics, Pauli himself must have been at least partially aware of the thinking behind the interpretation. But from his own philosophical viewpoint this interpretation, restricted as it was to a classically intuitive description, evaded the fundamental issue of a new conceptual framework. His first reaction, in response to a preprint of the paper, was mocking:[36]

I have tried on the basis of the definition of the two words [kommunisieren, virtuell] to guess what your work is really about. But it is not easy. In any case, it is very interesting to me and if I can be of any help with the grammar I shall gladly oblige.

In October he wrote to Bohr that he could not reject the theory on scientific grounds, that Einstein's objections did not worry him, and that his own were not strictly logical. But he had to admit that he was completely opposed to the theory and that he shared this opposition with "many other physicists, perhaps even the majority" – in fact a considerable understatement.[37]

Bohr's interpretation did receive some support, from Schrödinger, who had also recently abandoned causality and conservation for the sake of a pure field theory,[38] and possibly from Kramers. But even Kramers's support is uncertain. In response to Breit's criticism that the emission component of his formula corresponded to classical oscillators in which e^2/m was negative (a feature later incorporated in the theory of holes), he described the virtual oscillators as "meant only as a terminology"; on the other hand, working with Heisenberg later, he ignored their virtual nature altogether and treated the oscillator model as naively as he had the orbital model, and in none of his own work did he mention the causality issue.[39] Of those who expounded or developed the new technique, Born's assistant Jordan was the most sympathetic to a pure continuum treatment, but he was also a strict positivist so far as physical interpretations were concerned.[40] Neither Fowler nor Becker, who both discussed the technique, made any reference to Bohr's interpretation.[41] Born adopted the technique "independent of the critically important and still disputed conceptual framework",[42] and Heisenberg became interested in it only after Born's work had tied it in with the discretisaton programme they had worked on together the previous Autumn;[43] his first reaction had been that "Bohr's work on radiation is indeed very interesting, but I do not really see it as an essential progress."[44] Ladenburg wrote to Kramers that he and Reiche were glad his work coincided so well with their own considerations,[45] but this was a response to a further development of the dispersion theory by Kramers, which was effectively independent of the interpretation, and not to the BKS theory itself.[46]

In summary, it seems clear that Bohr's interpretation did not have the most enthusiastic of receptions. And before it was even conceived, Ramsauer had published the results of his experiments on the penetration of atoms by slow electrons, results that posed similar problems to those encountered in radiation phenomena, but in a context where the rejection of conservation was impossible.[47] These results disturbed Bohr more and more. By the time Geiger and Bothe announced in April 1925 that coincidence counting of X-rays and recoil electrons confirmed Compton's light-quantum explanation of X-ray scattering,[48] implying that conservation of energy and momentum had to be upheld in radiation phenomena too, Bohr was already anticipating this refutation of his own interpretation.[49] It still came hard, especially when accompanied by yet another exposition of the advantages of de Broglie's theory, this time by Born.[50] But with Pauli's help he managed to recondition himself, and admit that his "revolution" was over.[51] As Pauli wrote to Kramers in July, in the wake of Heisenberg's new kinematics,[52]

> [The ideas of BKS] thus move in completely the wrong direction: it is not the energy concept that is to be modified but the concepts of motion and force. One can indeed define no fixed paths for the light-quanta where interference phenomena are present, but nor can one define any such paths for the electrons in atoms; and to doubt the existence of light-quanta on the grounds of interference phenomena is just as little justified, therefore, as to doubt the existence of the electron would be.

HEISENBERG, BORN AND THE DEVELOPMENT OF THE OSCILLATOR TECHNIQUE

Fortunately for the development of the oscillator technique, Bohr's closer colleagues knew of his prejudices and were not, apart from Pauli, put off by his interpretation. Toward the end of 1923, before the intervention of Slater's idea, Kramers had generalised the old quantum theory of dispersion to incorporate dispersion by emission from atoms not originally in the ground state,[53] and following the discussions with Slater and Bohr he dressed up the new formula in virtual oscillator language and presented it as the first derivation from the new theory.[54] Ladenburg's formula based on pure absorption had taken a form equivalent to

$$(1) \qquad \mathfrak{M}(t) \propto \left\{ \sum_a \frac{f_a}{(\nu_a^2 - \nu^2)} \right\} \cos 2\pi\nu t$$

for the scattering moment with frequency ν, and Kramers simply generalised this to

$$\text{(2)} \qquad \mathfrak{M}(t) \propto \left\{ \sum_a \frac{f_a}{(\nu_a^2 - \nu^2)} - \sum_e \frac{f_e}{(\nu_e^2 - \nu^2)} \right\} \cos 2\pi\nu t,$$

where the coefficients f_a and f_e, and the transition frequencies ν_a and ν_e, corresponded to absorption and emission processes respectively.

Kramers did not at first give any derivation of his new formula, which was almost certainly guessed from its classical equivalent,[55] but in a second paper, in July, he did sketch a derivation, apparently taken from the first important development of the oscillator technique, by Born.[56] Born had not shown any previous interest in dispersion phenomena, but he was struck by the similarity between Kramers's formula and the difference equations occurring in the discretisation programme he had been developing with Heisenberg.[57]

This difference equation formalism was, as we have seen, completely symbolic, all questions of physical interpretation having been explicitly set aside.[58] It was also completely general, with implications for radiation phenomena such as dispersion as well as for the structure of the atom.[59] And it was thus a matter of course that Born and Heisenberg should reinterpret Kramers's dispersion formula (2) as a difference equation,

$$\text{(3)} \qquad \mathfrak{M}(t) \propto \sum \left\{ \frac{f_a}{(\nu_a^2 - \nu^2)} - \frac{f_e}{(\nu_e^2 - \nu^2)} \right\},$$

and seek to incorporate it in their general programme. That this is indeed what happened is confirmed by Heisenberg's letter to Pauli reporting his first serious interest in the oscillator approach, where he wrote that the difference equations were the key to the whole thing,[60] and by both Born's and Heisenberg's recollections.[61] The problem of how and where to apply the difference equation approach so as to derive Kramers's formula became a recurring subject of seminar discussions, through which Born's new theory gradually emerged. On 13 June two papers, one by Born on the new theory and one by Heisenberg on the (previously unpublished) difference equation quantum conditions, were submitted for publication simultaneously.[62]

Born's paper acknowledged the help not only of Heisenberg but also of Bohr, who visited Göttingen during its preparation.[63] Having been composed effectively in seminar, the theory in its final form shows few traces of its genesis, and any attempt to isolate either the order of ideas or their origin in individual minds would here be so speculative as to be worthless.[64] But the outcome was the use of perturbation theory, at which Born was the expert among quantum physicists,[65] to extend the oscillator treatment of dispersion, in terms of difference equations, to the general behaviour of the atom. Born argued that,[66]

> Since one knows that . . . atoms react to light waves completely 'non-mechanically', it is not to be expected either that the interactions between the electrons of one and the same atom should comply with the laws of classical mechanics; this disposes of any attempt to calculate the stationary orbits by using a classical perturbation theory complemented by quantum rules. For as long as one does not know the laws for the interaction of light with atoms, i.e. the connection of dispersion with atomic structure and quantum jumps, one is left all the more in the dark about the laws of interaction between several electrons of the same atom.

He therefore considered whether it might be possible to extend Kramers's treatment of dispersion, closer study of which "leads one to investigate whether the method of quantisation used by him is not based on some general property of perturbed mechanical systems."[67] This introduction followed the basic Göttingen approach: given the failure of classical mechanics (noted explicitly by both Born and Heisenberg the previous year), and the failure of the orbital model within this mechanics (as stressed by Heisenberg), the success of the oscillator approach to dispersion suggested a search for new quantum conditions based upon this approach, interpreted in terms of difference equations. The reference to electron-electron interactions was presumably a hangover from the seminar discussions, where Pauli's idea of extending the oscillator approach to these interactions would have been discussed, perhaps in connection with some of Born's work on the orbital model published in 1923, in which he ran into considerable difficulties with high frequency electron-electron coupling in the atom.[68] In fact, the generalisation proposed by Born did not yet extend to such interactions, the oscillators continuing to be applied in effect to the atom (any interactions being represented as perturbations), rather than to the individual electrons.

To derive his new "quantum mechanics" Born first reviewed classical perturbation theory; using a Fourier expansion of the Hamiltonian, suggested presumably by the oscillator approach, he brought the formulae for systems with and without external forces into the same form, and then went over classical dispersion theory as an application of these formulae. Next he introduced the oscillator representation, associating a given stationary state with "emission resonators", $\nu(n, n')$, and "absorption resonators", $\nu(n', n)$, each corresponding to a higher harmonic of that state in the classical theory

$((\nu\tau); \tau = |n - n'|)$.[69] The division of the oscillators into two classes, one of which could only emit and the other only absorb, was conceptually unsatisfactory, but it was a natural consequence of the reinterpretation of Kramers's formula (2) as a difference equation, (3), and since Born did not afford the oscillator treatment too much physical significance he was no more disturbed than Kramers had been by. Breit's objections. Physically, an atom in a given state could undertake infinitely many absorptions, but only finitely many emissions (in the ground state, for example, none), and given the mathematical symmetry between emissions and absorptions in (3) the condition had to be imposed that the relevant transitions were possible, or that the relevant absorption or emission oscillators existed. Next, guided by the quantisation process, $J_k = n_k h$, of the old theory, Born attempted to connect the quantum and classical frequencies of an unperturbed system, H_0:[70]

The following quantitative connection exists between the classical frequency $(\nu\tau)$ and the quantum-theoretical absorption frequency $\nu(n', n)$. Let us imagine that the transition $n_k \to n_k' = n_k + \tau_k$ is performed in a 'linear' way; i.e. let us set for the action integrals $J_k = h(n_k + \mu\tau_k); 0 \leqslant \mu \leqslant 1$. Then we obtain on the one hand,

$$(\nu\tau) = \sum_k \nu_k \tau_k = \sum_k \frac{\partial H_0}{\partial J_k} \tau_k = \frac{1}{h} \sum \frac{\partial H_0}{\partial J_k} \frac{\partial J_k}{\partial \mu} = \frac{1}{h} \frac{dH_0}{d\mu},$$

and on the other,

$$\nu(n', n) = \frac{1}{h} [\mathrm{H}_0(n + \tau) - H_0(n)];$$

therefore $\nu(n + \tau, n) = \int_0^1 (\nu\tau)\, d\mu$ .... One can say that the ways in which $\nu(n + \tau, n)$ and $(\nu\tau)$ are obtained from H_0 stand in the same relationship as differential coefficients stand to difference quotients.

Born next considered the interaction process described by the perturbation function λH_1. For the classical perturbation energy he had obtained a Fourier series expansion that was characterised, like the classical frequency, by the operator $\sum_k \tau_k \frac{\partial}{\partial J_k} = \frac{1}{h} \frac{d}{d\mu}$, and he concluded that "we are as good as forced to adopt the rule that we have to replace a classically calculated quantity, wherever it is of the form $\sum_k \tau_k \frac{\partial \phi}{\partial J_k} = \frac{1}{h} \frac{d\phi}{d\mu}$, by the liner average or difference quotient $\int_0^1 \sum_k \tau_k \frac{\partial \phi}{\partial J_k} d\mu = \frac{1}{h} [\phi(n + \tau) - \phi(n)]$." Applying this transformation to the perturbation results he obtained Kramers's dispersion formula and an analogy in simple cases with Heisenberg's difference equation quantum conditions.

Heisenberg himself began to work with the oscillator technique in the Autumn, when he went to Copenhagen for a semester. His first paper there was on a simple extension of the dispersion theory to allow for additional spontaneous atomic transitions during the scattering process, and was written jointly with Kramers.[71] He recalled that both the idea, in response to a treatment by Smekal based on the light-quantum concept, and the redaction were due to Kramers,[72] and this was reflected in the terminology adopted: opposed as he was to the light-quanta, Kramers relapsed into the terminology of purely classical oscillators and waves. No mention was made of the virtual oscillator concept, and BKS was referred to only as an example of how the correspondence principle, on which the treatment claimed dependence, might be applied.

Heisenberg's role in the dispersion paper seems to have been merely to contribute some rigour to Kramers's physically inspired guesswork,[73] but at the same time he was also working by himself, extending the application of the oscillator technique and removing it still further from Bohr's interpretative context. This work, which was to be critical for the development of the new kinematics, was on fluorescent polarisation.[74]

Wood and Ellett had found that if a polarised light source was used to stimulate fluorescent resonance radiation from mercury, the stimulated radiation showed about 100% polarisation in a weak magnetic field, a result that seemed compatible with the classical theory but not with that based on the orbital quantum model of the atom.[75] In the absence of a magnetic field, the polarisation of the incident light was maintained in the resonance radiation on both theories, but the empirical and classically acceptable extension of this result to the presence of a magnetic field clashed with the natural assumption of equal statistical weighting of the magnetically induced multiplet states in the orbitals model. In a short paper on the subject, Bohr had suggested that the results might be obtained from the virtual oscillator theory, and Heisenberg now took up this possibility.[76] But whereas Bohr, though preferring the virtual oscillator model explicitly here to the orbital model, had continued to view the two models as compatible, and had insisted that the results did not contradict his theory of atomic structure,[77] Heisenberg took a slightly different approach. He had always treated the virtual oscillator approach and its parent the correspondence principle as constituting an empirical approach, neutral as to any physical models. He had earlier compared the correspondence principle, interpreted in this way, with the orbital model, and he had since come to the personal conclusion that the model was untenable. What he did now was to demonstrate this conclusion, comparing the orbital model with the oscillator representation, as physical model approach against symbolic phenomenological approach.

Heisenberg based his oscillator treatment on some work by Dorgelo, Ornstein and Burger, who had been studying related problems of multiplet structure by analogy with the classical theory.[78] Consideration of this work led him to the empirically-based assumption that the multiplet degeneracy revealed in the presence of a magnetic field was a permanent feature of the atom, the multiplet states being present (though indistinguishable) even in the absence of a field. His use of this idea (and with it the suggestion of solving a problem in the absence of a field by first introducing one, then applying the usual quantisation, then setting its strength to zero) seems to have led to an argument with Bohr, who criticised him as usual for ignoring the fundamental principles of quantum theory,[79] but it was undoubtedly a major breakthrough in the treatment of complex spectra. He introduced it together with the orbital model approach, in the context of the polarisation problem with unpolarised incident light and no field; then, considering the introduction of a weak field he used the new situation to decide between his approaches. Classically, the resonance radiation remained unpolarised, and this conclusion was carried through in the oscillator approach, being interpreted statistically as equal intensities of parallel and perpendicular polarised components. On the orbital theory, however, the field acting on the orbit produced a polarisation effect. And as Heisenberg wrote, "we have every reason to believe that polarisation is not present."[80] Having justified the use of the oscillator approach and rejected that of the model, he reproduced the Wood and Ellett results, obtaining predictions in full agreement with experiment.

Heisenberg interpreted both his own work and that of Dorgelo, Ornstein and Burger as extensions of the empirically based correspondence principle approach to the problem of statistical weights, previously the domain of the orbital model. And, successful as it was, he clearly found this extension encouraging. He could not reject the orbital model on principle, for it was still theoretically essential for both the calculation of frequencies and the justification of the correspondence principle itself. But he could conclude that any new theory should be developed from the "symbolic", or essentially phenomenological, nature of the oscillator treatment, this being the feature that ensured its success.[81] The models were no longer useful, but restrictive.

THE RETURN OF PAULI AND THE GENESIS OF HEISENBERG'S NEW KINEMATICS

Meanwhile, Pauli had refused to have anything to do with the virtual oscillators. But in the course of 1924 he had come back to quantum atomic theory from his self-imposed exile, and by developing his earlier "phenomenological" approach he had begun to unravel some of its problems. Before turning to the virtual oscillator theory Heisenberg and Born, together with Landé, had published a series of papers in support of the core model of the atom and the associated use of half-integral quantum numbers.[82] Pauli had never been happy with this model, and when he returned it was to show that it was innately inconsistent. Looking once again at the anomalous Zeeman effect he was able to show that according to the core theory the Zeeman splitting had to depend on the atomic number, which it empirically did not.[83] To get round this he suggested that the anomalous angular momentum and magnetic moment, previously attributed to the hypothetical structure of the core, should instead be transferred to the outer electron. Instead of counting the core twice in its contributions to atomic properties, as Heisenberg had done, Pauli counted the electron twice:[84]

> According to the interpretation suggested here, Bohr's 'Zwang' does not manifest itself in a violation of the permanence of the quantum numbers in the coupling of a series electron to the atomic core, but only in a characteristic Zweideutigkeit ["two-valuedness"] in the quantum-theoretical characteristics of the individual electrons in the stationary states of the atom.

Soon after having reached this conclusion, Pauli read a paper by Stoner containing a new scheme for the shell structure by which the atom was related to the periodic table of elements, a scheme that was different from Bohr's but completely natural and in excellent agreement with experiment.[85] One of Stoner's innovations was to assign values of the "inner" quantum number, $j + \frac{1}{2}$, to each electron rather than treating it as part of the electron-core interaction, a procedure that was clearly similar to Pauli's own treatment of the anomalous Zeeman effect. This gave a set of three quantum numbers for each electron, (n, l, j), and Stoner found that the number of electrons in each shell was equal to twice the inner quantum number of that shell, $2(j + \frac{1}{2})$. From this Pauli saw that the whole shell structure could be obtained very naturally, independent of any atomic model, and without any recourse to arguments from the correspondence principle, by giving each electron a fourth quantum number, m_j, $-j \leqslant m_j \leqslant j$, and insisting that each state defined by a set of the four quantum numbers (n, l, j, m_j) could represent, or be occupied by, just one electron. The *Zweideutigkeit* was absorbed into the new quantum number, and the whole idea was expressed as his famous exclusion principle, that no two electrons should occupy the same state in the atom.[86]

Given this tremendous success of his approach, it was with greatly increased fervour and confidence that Pauli once again expounded his views on the future of quantum theory to Bohr in December 1924.[87] He suggested that for "weak" people, who needed the support of well-defined orbits and mechanical models, one could justify the exclusion principle on the basis that electrons in the same orbit would crash.[88] But he explained that he had consciously avoided the use of such terminology in his paper, and that he thought that the future would involve not only an abandonment of the orbital concept but also some fundamental changes in the kinematic concepts themselves:[89]

> The relativistic doublet formula [the *Zweideutigkeit*] appears to me to show unquestionably that not only the dynamic concept of force but also the kinematic concept of motion of the classical theory shall have to undergo fundamental changes (it is for this reason that I have avoided entirely in my work the designation 'orbit') . . . I think that the energy and momentum values of stationary states are something much more real than 'orbits'. The (still unattained) goal must be to deduce these and all other physically real, observable characteristics of the stationary states from the (fixed) quantum numbers and quantum theoretical laws. However, we should not want to clap the atoms into the chains of our preconceptions (to which in my opinion belongs the assumption of the existence of electron orbits in the sense of the usual kinematics), but must on the contrary adjust our ideas to experience.

In a footnote to the first sentence of this quotation Pauli, referring to the "children" who lapped up Kramers's "picture book" (and meaning children in terms of wisdom rather than years), noted that

> Even though the demand of these children for visualisation [*Anschaulichkeit*] is partly legitimate and healthy, this demand should still never count in physics as an argument for the retention of a certain set of concepts. When the system of concepts is once clarified, then will there be also a new visualisation.

The fundamental ideas of Pauli's proclamation were of course those that he had advocated consistently over the years: the rejection of orbits, the use of a phenomenological approach in the absence of anything better, and the operational basis of concepts. But the last idea, which was the most fundamental, was here expressed more clearly and positively than it ever had been in the context of the quantum thoery. Whereas Bohr stood by the permanence of those fundamental physical concepts corresponding to the classical visualisation, Pauli insisted that this visualisation must be put aside and new concepts derived from experience. He even suggested an equation between "physically real" and "observable" properties. Moreover, Pauli was also able for the first time to make positive recommendations as to how these ideas should be brought to bear on the quantum problem, adding to the above suggestions the thought that the solution to the whole problem would come through the hydrogen atom. If this relatively simple problem could be solved through a new mechanics, he argued, then the principles of that mechanics would allow the solution to be extended to the more general case.[90]

This new and specific expression of Pauli's ideas was well timed, for it coincided with Heisenberg's move in the very same direction. And in March 1925 Pauli visited Copenhagen for a few weeks, and argued out his views on the future of quantum theory with both Heisenberg and Bohr.[91] Following his conclusion on the fluorescent polarisation problem, disturbed by Sommerfeld's recent discovery that the use of half-integer quantum numbers failed after all for helium,[92] and not yet happy with Pauli's treatment of the anomalous Zeeman effect, Heisenberg had turned back to this problem.[93] And considering the application of the correspondence principle to the theoretical derivation of Zeeman intensities he noted in his work that:[94]

The application of the correspondence principle to the derivation of the selection rules and intensities is legitimately only possible through the possession of unequivocal mechanical models.

But as Heisenberg had himself demonstrated the existing mechanical models were themselves invalid. And he argued that whether Pauli's *Zweideutigkeit* was attributed to the core or the outer electron was essentially immaterial, as both representations were equally artificial.[95] To obtain a satisfactory treatment, he reasoned, the approach had to be an essentially empirical one, but at the same time a proper theoretical derivation required some kind of model. The problem was therefore to derive a new model, and a radically new one at that, from the empirical results. And it was to this task that he turned next.

Heisenberg's programme, based on a rejection of the existing model of the atom and a reliance primarily upon the observed results, was now close to that advocated by Pauli. And it was presumably at Pauli's suggestion that he decided to concentrate on hydrogen and so arrived at Göttingen in April armed with a book of Bessel functions with which he hoped to improve the mathematics of the correspondence principle. He then intended to "guess" a symbolic scheme for the reaction of hydrogen to an external field, and so to deduce a new model for hydrogen.[96]

In fact, Heisenberg had to give up on hydrogen as being too difficult. But by June, as is well known, he had found his scheme and completed his paper on 'A theoretical reinterpretation of kinematic and mechanical relations'.[97] In preparing this paper, he adopted both Pauli's phenomenological approach and, after talking out the interpretation of the scheme with him during a short visit to Hamburg, his operational ideas as well.[98] In its final form, Heisenberg's new presentation was based on a restriction to quantities that were in principle observable, and on a complete revision of kinematics. The electron orbits were finally abandoned, and the electrons themselves were replaced, as Pauli had earlier suggested, by systems of complex oscillators. Pauli himself could write, following the completion of this paper, that he and Heisenberg were as much in agreement as any two individuals could be.[99]

Bohr, meanwhile, turned away from the world of publication to that of contemplation.

CHAPTER 6

THE NEW KINEMATICS AND ITS EXPLORATION

HEISENBERG'S NEW KINEMATICS

The avowed aim of Heisenberg's paper was "to try and establish a theoretical quantum mechanics, analogous to classical mechanics, but in which only relations between observable quantities occur."[1] The attempt was a confused one, if only on account of the variety of conflicting notations,[2] but the fundamental idea was clear: to take over the classical equation of motion,

$$(1) \qquad \ddot{q} + f(q) = 0,$$

but to replace the classical acceleration $\ddot{q}$ and potential $f(q)$ by quantum-theoretical representations derived from their series or integral Fourier expressions. In the case of the position function $q(t)$ of the periodic motion of an electron in an atom, the position itself was not observable: as Heisenberg and Pauli had long recognised, the kinematic conception of an orbit was in this case operationally meaningless.[3] But the terms of the Fourier expansion of the position could be directly related to observables. Classically the position vector of an oscillating electron could be expanded as a Fourier series,

$$(2) \qquad q(n, t) = \sum_{\alpha=-\infty}^{\infty} q_\alpha(n)\, e^{i\omega(n)\alpha t},$$

and the radiation corresponding to each harmonic was proportional to the real part of the Fourier component, $\mathbb{R}\{q_\alpha(n)\, e^{i\omega(n)\alpha t}\}$. In quantum theory the Bohr frequency condition,

$$(3) \qquad \nu(n, n-\alpha) = \frac{1}{h}\{W(n) - W(n-\alpha)\},$$

with $W(n)$ the energy of the nth state, led to the requirement that the harmonic components take the form of expressions $q(n, n-\alpha)\, e^{i\nu(n, n-\alpha)t}$, corresponding to pairs of states or transitions. Heisenberg assumed that the observable radiation was again given by their real parts. In order to construct a mechanics of observable quantities he extended this form of representation as an ensemble of functions of possible transitions to a general quantity, deriving for the product of two quantities (now assumed to be scalars)

$$
\begin{aligned}
& x(t) = \{x(n, n-\alpha)\, e^{i\nu(n,\, n-\alpha)t}\}, \\
& y(t) = \{y(n, n-\alpha)\, e^{i\nu(n,\, n-\alpha)t}\}, \\
(4) \qquad & z(t) = x(t)y(t) = \{z(n, n-\beta)\, e^{i\nu(n,\, n-\beta)t}\} \\
& = \{\sum_{\alpha=-\infty}^{\infty} x(n, n-\alpha)\, y(n-\alpha, n-\beta)\, e^{i\nu(n,\, n-\beta)t}\}.
\end{aligned}
$$

Since the occurrence of quantum transitions was necessarily ordered, the multiplication was non-commutative, and Heisenberg noted that one would therefore have to replace ambiguous classical products by symmetric alternatives, $x(t)y(t) \rightarrow \frac{1}{2}(xy + yx)$. But apart from this the classical expressions, both scalar and by implication also vector, could simply be replaced in the classical equations of motion by their quantum equivalents,

$$q_\alpha(n)\, e^{i\omega(n)\alpha t} \longrightarrow q(n, n-\alpha)\, e^{i\nu(n,\, n-\alpha)t}, \text{ etc.}$$

The solution to the equation of motion (1) had been given in the old quantum theory by a quantisation of the action,

$$(5) \qquad J = \oint p\mathrm{d}q = \oint m\dot{q}\mathrm{d}q = nh,$$

which could not be translated into the new quantum terminology. But by differentiating with respect to η Heisenberg was able to obtain a form that could be so translated, giving a new quantisation condition,[4]

$$
\begin{aligned}
(6) \qquad h = 4\pi m \sum_{\alpha=0}^{\infty} \{ & |q(n, n+\alpha)|^2\, \nu(n, n+\alpha) \\
& - |q(n, n-\alpha)|^2\, \nu(n, n-\alpha)\}.
\end{aligned}
$$

This condition had already been derived from the virtual oscillator theory from which, on the technical level, Heisenberg's own theory had developed. But in the former case one had had to represent the electron by a set of classical oscillators with the observed transition frequencies and then, working with these classical oscillators, replace the resulting differential equations of a certain type by difference equations. This procedure had had no real theoretical foundation, and the use of the oscillator model, on which the oscillator radiation took the classical continuous form but the transitions remained discrete, had entailed a departure from energy-momentum conservation that had now been shown, by the experiments of Geiger and Bothe, to be impossible.[5] In Heisenberg's energy-conserving theory, the quantum-theoretical solution, given by the equation of motion (1) and the quantisation condition (6), followed straight from the replacement of the kinematic expressions.

The new kinematics constituted a major breakthrough in the treatment of quantum phenomena, introducing explicitly the break from classical kinematics and restriction to observables that had long been advocated by Pauli. Despite these conceptual innovations, however, it was still in many ways unclear and lacking in direct applicability. Heisenberg was able to derive the dispersion formula of the old virtual oscillator theory as well as the approximate energy levels of the one-dimensional anharmonic oscillator and rotator, but even these derivations lacked rigour. He was unable to establish energy conservation as a general feature of the theory (though he had established it for his simple examples and convinced himself of its general existence),[6] and was unable to apply this theory even to the relatively simple problem of the hydrogen atom, let alone to systems of more than one electron.[7] The physical significance of the restriction to observables was unclear and, as he wrote to Pauli, the fundamental problem remained as to "what the equations of motion really mean, when one treats them as relations between transition probabilities."[8] If the traditional kinematics was invalid, how could the equations of motion, derived conceptually from this kinematics, be justified?

THE RECEPTION OF HEISENBERG'S THEORY, AND BORN'S MATRIX MECHANICS

Not surprisingly, since he was himself responsible for most of its conceptual innovations, Pauli greeted Heisenberg's theory with delight, reporting to Kramers that "on the whole I believe that I am now close to Heisenberg in my scientific opinions, and that our opinions agree in everything as much as is in general possible for two independently thinking men."[9]

Also enthusiastic was Born, under whom Heisenberg was nominally working at the time. Born later recalled that he had discussed with Heisenberg and Jordan, before Heisenberg's innovations, the possibility that "[transition amplitudes] might be the central quantities and be handled by some kind of symbolic multiplication",[10] and on seeing Heisenberg's paper, that[11]

I began to ponder about his symbol multiplication, and was soon so involved in it that I thought the whole day and could hardly sleep at night .... And one morning ... I suddenly saw the light: Heisenberg's symbolic multiplication was nothing but the matrix calculus, well known to me since my student days.

In fact Born, who was trained as a mathematician, must have been rather more familiar with matrices than these well-known recollections suggest,[12] and this suggests that his initial concern may have been with the physical implications of Heisenberg's work, and with its advantages over the virtual oscillator theory and de Broglie's wave theory of matter, rather than with its "symbolic multiplication".[13] But however we may reconstruct the details of his reaction it is clear that it was favourable and that it did lead him within a few days to the matrix formulation in terms of which Heisenberg's theory came to be known. Born had long ago expressed a willingness to abandon the space-time description of the inside of the atom;[14] and like Heisenberg he had appreciated and contributed to the heuristic value of the virtual oscillator theory, though without attributing to this any physical significance.[15] He had anticipated the Geiger-Bothe results on the preservation of energy conservation, and had responded to Einstein's extension of the Planck-Bose statistics to a material ideal gas, the electron scattering results of Davisson and Kunsman, and the barrier penetration by slow electrons demonstrated by Ramsauer by adopting a version of de Broglie's theory of matter waves as a physical alternative to the virtual oscillators.[16] Just before assimilating Heisenberg's paper he had expressed the view that "the wave theory of matter could be of very great importance."[17] The results of Einstein, Ramsauer, Davisson and Kunsman all displayed a wave-particle duality of matter that found natural expression in de Broglie's theory, and it was with these results that Born and Franck had been largely concerned during the Spring,[18] and with the Planck-Bose statistics of light that Born and Jordan had been concerned in their most recent applications of the virtual oscillator theory.[19] But Bohr had rejected the matter-wave hypothesis outright and Born, as he recalled, had also been following closely Heisenberg's attempt to derive a new and consistent physics from the starting point of the observable transition amplitudes.[20] His immediate sympathy with Heisenberg's theory, despite the confusion and complication of its presentation, was therefore natural; and given his mathematical ability the re-expression of this theory as a matrix mechanics was more or less inevitable.

Adopting the matrix notation and looking at the physical implications of non-commutativity, Born saw at once that the quantisation rule (6) gave the value $h/2\pi i$ for the diagonal elements of the position momentum commutator,

$pq - qp$. Convincing himself that the off-diagonal elements of the expression would probably be zero, he suggested the resulting commutation relationship,[21]

(7) $$pq - qp = \frac{h}{2\pi i}\, 1,$$

and retired to Switzerland for a much needed holiday,[22] leaving his assistant Jordan to prepare a joint paper that was completed on his return and received on 27 September.[23] In this paper Heisenberg's sets of Fourier coefficients were expressed as Hermitian matrices, $x(n, n - q) \rightarrow x(nm)$ and the commutation relationship, the frequency law and energy conservation were derived as general results for non-degenerate systems.[24]

By defining an artificial 'symbolic differentiation' of one matrix with respect to another, Born and Jordan were able to express the classical equations of motion, taken as matrix equations, for a general Hamiltonian energy function, H, in the canonical form

(8) $$\dot{q} = \frac{\partial H}{\partial p}, \qquad \dot{p} = \frac{-\partial H}{\partial q}.$$

For non-degenerate systems, where any change of state was associated with a non-zero change of energy, they were able to show that the time derivative of a general quantity $g(nm)\, e^{2\pi i\nu(nm)t}$, given by $\dot{g} = 2\pi i\nu(nm)\, g(nm)\, e^{2\pi i\nu(nm)t}$, was zero if and only if its matrix $g(nm)$ were diagonal. From this and from Heisenberg's quantisation rule (6) they deduced the commutation relationship (7) and, from substitution in the general result,

(9) $$\dot{g} = \frac{2\pi i}{h}(Hg - gH),$$

the frequency law (3) and energy conservation,

(10) $$\dot{H} = 0.$$

In the last section of the paper, Jordan noted Heisenberg's implicit assumption that $|q(nm)|^2$ determined the transition probabilities (i.e., that Heisenberg's $\mathbb{R}\{q(n, n - \alpha)\, e^{i\nu(n,\, n-\alpha)t}\}$ did in fact correspond to the observable radiation), and proceeded "to see in what way this assumption can be based upon general considerations".[25] Applying the matrix mechanics to the electromagnetic field, he found that the mean radiation, identified as the diagonal sum of the radiation matrix, was indeed determined by the $|q(nm)|^2$.

THE MATRIX TRANSFORMATION THEORY OF BORN, HEISENBERG AND JORDAN

Born and Jordan noted in their paper that the canonical equations of motion (8) and Heisenberg's quantum condition (6) could be replaced as foundations for the theory by the equivalent assumptions of energy conservation (10) and the commutation relationship (7). But such a fundamental role for the commutation relationship is first apparent in a letter from Heisenberg to Pauli of 18 September, and this letter also contains the first attempt to express the new theory in terms of a theory of transformations, an attempt which was to have far-reaching consequences.[26] Heisenberg had received details of the new matrix formulation by 13 September and in his reply to Jordan he had set out immediately to extend its applicability, drawing for this purpose on classical perturbation theory.[27] By the time he wrote to Pauli a few days later he had based this treatment on the supposition that any p, q satisfying the commutation relationship and for which the Hamiltonian was diagonal would represent a solution to the problem, deriving such a p, q for the perturbed problem by a transformation of known unperturbed solutions.[28] Heisenberg's transformation theory, derived from that used in the classical case, was somewhat unwealdy, but on receiving the idea Born quickly suggested a simpler form of transformation,[29]

$$(11) \qquad p = Sp_0S^{-1}, \qquad q = Sq_0S^{-1}.$$

This was the standard form for a matrix transformation leaving the matrix equations invariant, and had indeed appeared as such in the introduction to the Born-Jordan paper.[30] It preserved the commutation relationship, and together with a natural extension of this relationship to several degrees of freedom it formed the core of the definitive formulation of matrix mechanics in the 'three-man-paper' of Born, Heisenberg and Jordan.[31] The problem, first stated by Heisenberg and reformulated by Born, could now be expressed as follows:[32]

Given the canonical equations of motion for a known Hamiltonian,

$$(12) \qquad \dot{q} = \frac{\partial H}{\partial p}, \qquad \dot{p} = \frac{-\partial H}{\partial q},$$

and any Hermitian matrix quantities p_0, q_0 satisfying the commutation relations,

$$p_{0j}q_{0k} - q_{0k}p_{0j} = \frac{h}{2\pi i}\,\delta_{jk},$$

to find a transformation matrix S such that $p = Sp_0S^{-1}$, $q = Sq_0S^{-1}$ gave

$$H(pq) = SH(p_0, q_0)S^{-1} = W,$$

a diagonal matrix.

For the case of small perturbations, $H = H_0 + \lambda H_1 + \lambda^2 H_2 + \ldots$, $p = p_0 + \lambda p_1 + \ldots$, $q = q_0 + \lambda q_1 + \ldots$, the solution $S = 1 + \lambda S_1 + \ldots$ could be found by calculating the terms successively.

PHYSICISTS AND MATHEMATICIANS: DISAGREEMENT AND DIVISION

The three-man-paper, first drafted by Born and Jordan and finally completed on 26 November, after Born had left on 28 October for America, by Jordan and Heisenberg, was presented as a joint effort. But there were in fact strong differences of opinion between the authors, and these were manifest both in their paper and in their ensuing research. Born recalled that he had originally asked Pauli to help with the matrix mechanics, but that he had been given a "cold and sarcastic refusal" on the lines of: "Yes, I know you are fond of tedious and complicated formalisms. You are only going to spoil Heisenberg's physical ideas by your futile mathematics."[33] Whether accurate or not, this recollection conveys a true impression. Pauli had past experience of working with Born,[34] and about this time he noted of that earlier work, which he had not found very satisfying, that "the effort expended did not correspond to the results achieved, especially as these were chiefly negative."[35] He wrote to Kronig in October that "one must next attempt to free Heisenberg's mechanics from the Göttingen *Gelehrsamkeitsschwall* [literally, "torrent of erudition"] and expose still further its physical crux,"[36] and in November Heisenberg wrote to him that[37]

> I'm still pretty unhappy about the whole theory, and was thus glad that you were so completely on my side in your views on mathematics and physics. Here I'm in an environment that thinks and feels the exact opposite, and I do not know if I'm not just too stupid to understand mathematics.

A major factor in the success of Göttingen as a centre for theoretical physics lay in its unique composition: the three departments of experimental physics, theoretical physics and mathematics were run by three men, Franck, Born and Hilbert, all of whom were outstanding in their respective fields, and who worked in extremely close collaboration with each other. Much of Born's work on the quantum theory seems to have been prompted by the influence of Franck (who himself worked closely with Bohr),[38] while the idea of his earlier difference equation formulation of this theory had most probably stemmed from a seminar in Hilbert's department.[39] Later, the extremely close links between Hilbert and Born were to be of crucial importance for the formulation of quantum mechanics. But while bringing great advantages, such a situation also led naturally to tensions. Now, as Heisenberg wrote to Pauli,[40]

> Göttingen splits into two camps, one which with Hilbert ... speaks of great success, achieved through the introduction of matrix rules to physics, the other which with Franck says that we still cannot understand the matrices.

In the Hilbert camp, Heisenberg placed Weyl and, by implication, Born, who had been carried away by the mathematics of the new theory. Jordan, as far as may be gathered from his subsequent research, took something of a middle path. But Heisenberg was firmly in the physical camp. The problems resulting from his fundamental innovations, problems as to observability and the physical significance of the new kinematics and commutation relations, seemed to play no part in Born's attempts to elaborate on the mathematical theory. To Heisenberg, indeed, the whole concept of a "matrix mechanics" represented a rejection of the physical problem, and he expressed to Pauli a serious intention of replacing the terminology by something physically meaningful, such as "mechanics of quantum-theoretical quantities".[41] He also expressed serious doubts about the fundamental role that Born and Jordan attributed to the transformation theory, and wrote to Pauli on this point that "you have shown in Hydrogen how one actually integrates" (in a mathematical *tour de force* Pauli had obtained a direct solution from the theory for the hydrogen atom) "and so that rest is just formal rubbish."[42] Pauli, mathematically more experienced,[43] was less upset by the new developments; but he seems nevertheless to have agreed wholeheartedly with Heisenberg about the need for physical comprehension, and once again to have found himself on the opposite side of the fence from Weyl.

Given the above division, it is not surprising that within the context of the three-man-paper, the discussion surrounding it and the work consequent upon it, there were several quite distinct and divergent lines of development. In the three-man-paper itself, Heisenberg and Jordan derived the laws of conservation of linear and angular momentum and thence, for a non-degenerate system, the standard selection rules and normal Zeeman intensities.[44] In a paper completed in January, Pauli applied the theory successfully to the hydrogen atom,[45] and those applications between them consolidated the prospects for the theory's success. In two papers completed the following Spring, Jordan set the transformation theory upon a rigorous footing, providing that every canonical transformation could be put in the form $p = Sp_0S^{-1}$, $q = Sq_0S^{-1}$, and that every point transformation was canonical.[46] Although the transformation theory remained practically restricted to the case of small perturbations where the unperturbed solution was known, it was thus established as a theoretically generally valid approach to the solution of a general physical problem, and this was to be important for the further development of the theory. Of more immediate importance in this respect, however, were those developments reflecting the particular concerns of the authors: Jordan's application of matrix mechanics to radiation fluctuations and quantum statistics, Born's development of the mathematics of the theory so as to generalise its applicability, and the discussion by Heisenberg and Pauli of some of the theory's physical implications.

PAULI, HEISENBERG, AND THE PHYSICAL SIGNIFICANCE OF THE NEW KINEMATICS

Heisenberg had written to Pauli on 16 November that "I have taken great pains to make the work more physical than it was",[47] and his efforts had born some fruit. He had persuaded Born to include a physically based form of matrix differentiation alongside the physically incomprehensible "symbolic differentiation" of the Born-Jordan paper,[48] and he had written a wholly physical introduction to the three-man-paper in which the word "matrix" was completely avoided.[49] He here layed emphasis upon both the observability criterion and the new kinematics, and posed the fundamental problem of establishing a general relationship between the symbolic quantum theory and the classical conceptions of the observed world. But the physical insight to the new theory that he sought continued to elude him. On 17 November, however, Pauli wrote a letter to Bohr, for general consumption, in which he included something "on the principal questions that are still left open in the new theory":[50]

This theory is so far cut out only for those cases in which all points remain in finite spaces. In its present formulation, it is still not fitted for example, either to include impact phenomena or to include the problem of understanding simultaneously coupling and interference. Generally we still have no logically uniform theory that includes *all* the applications of classical theory in the borderline case of high quantum numbers.

So far, Pauli was simply voicing generally recognised facts. But he then proceeded to offer some ideas of his own:

Perhaps the following hits the right direction for possible further progress. In the new theory, all physically observable quantities still don't really occur. Absent, namely, are the time instants of transition processes, which are certainly in principle observable (as, for example, are the instants of the emission of photoelectrons). It is now my firm conviction that a really satisfying physical theory must not only involve no unobservable quantities, but must also connect *all* observable quantities with each other. Also, I remain convinced that the concept of 'probability' should not occur in the fundamental laws of a satisfying physical theory. I am prepared to pay as high a price as you like for the fulfilment of this desire, but unfortunately I still do not know the price for which it is to be had.

Expanding on his statement that the time instants of transition processes were absent from the new theory, he added a short footnote:

Instead of $e^{2\pi i\nu t}$ one could write just as well $e^{2\pi i\nu}$ or something like that, and *define* time derivatives of a quantity q as $2\pi i(Wq - qW)/h$. The general problem is then to set up wider concepts that embrace the actual applications one has made of the classical space-time picture. The position is now that the concept 'duration' [*zeitlicher Ablauf*] of a process, and in particular the concept of 'time period' or 'frequency of oscillation' has become wholly formal. *The formal character of the frequency condition . . . is a consequence of the formal character of time, and not of the formal character of energy.*

On where this led, Pauli was still unclear, but he thought that one might try defining time in terms of energy and proceed from there:[51] since all time measurements depended upon periodic processes, this was also possible within the existing quantum mechanics, and consistent with his general operationalist ideas.[52]

In due course Heisenberg read Pauli's letter,[53] and writing to him on 24 November he offered sincere thanks for such a clear and helpful exposition of Pauli's views, together with some thoughts of his own prompted by these views:[54]

Your problem of the 'duration' naturally plays a fundamental role, and I've thought over several matters for domestic use. First, I believe that one can distinguish between a 'coarse' and a 'fine' duration. When, as in the new theory, a point in space has no longer a fixed place, or when this place is still only defined formally and symbolically, then the same is true also of the time-point of an event. But there is always given a rough duration, as also a rough place in space: with our geometric picture we shall still be able to achieve a *rough* description of the phenomena. I think it is possible that this *rough* description is perhaps the only one we may ask for from a formalism. Now the beautiful thing is that for purely periodic motion evidently not even such a coarse lapse can be defined; it seems to me that the formulae do not admit of such an interpretation (i.e. one knows of the electron only that it is somewhere close to the core). But if one has an aperiodic orbit, i.e. a Fourier integral, then the – let us say – infrared part of the spectrum agrees with the classical theory, the usual calculating rules being valid eo ipso in good approximation (the better the longer the wave) – and just this infrared part indeed gives the *coarse* duration! A motion sufficiently like uniform rectilinear motion would thus be as classical as ever possible. But as soon as purely periodic motions are superposed our space-time presentation again fails completely (Compton effect).

In the earlier derivation of the new kinematics it had been Pauli who had provided the (operationally based) conceptual innovations, and Heisenberg who had put these into a practically viable form. Now, in their struggle to understand the physical implications of this kinematics, a similar procedure was repeated. Pauli raised the fundamental problem of the definition of time in the new theory, and Heisenberg linked this problem with that of space, and with Bohr's correspondence principle. The latter was no longer an integral part of the new theory, which had its own foundations built upon an altogether different type of correspondence. But it had played a crucial part in the development of this theory, and in Heisenberg's search for an understanding of the new type of correspondence, and of the new kinematics, it was to be a valuable guide. Looking at the new theory in the light of Pauli's remarks, Heisenberg saw that it included no such things as space and time in the classical sense. But in order to correspond with the classical theory and classical visualisations it had to provide for approximate space-time specifications, and for time-dependent quantities these took the form of time averages, given by the diagonal sum of the matrix. These quantities were thus specified the better, classically, the smaller the contribution of off-diagonal terms to the whole matrix. In particular the quantity position, whose off-diagonal coefficients corresponded to transition probabilities, was most accurately specified when only small transitions were possible, i.e., for long wavelengths (low frequencies) and especially for largely aperiodic motions. The closer one got to uniform rectilinear motion, the closer one got to zero radiation and a purely diagonal position matrix, allowing in principle (though the theory could not yet cope with it) a precise position specification. As to the specification of time, however, Heisenberg still seems to have been confused; for while defending the overall classical nature of uniform rectilinear motion, he followed Pauli's suggestion of identifying long wavelengths (large oscillating periods) with coarse time specifications.

The full relationships between space, time, momentum and energy were not yet apparent in matrix mechanics, and it would be over a year before Heisenberg could see things clearly enough to formulate his uncertainty principle. But some of the ideas behind that principle were already present in his letter to Pauli. In particular, the idea that a rough classical description was all that could be expected from the formalism, combined with the basic hypothesis that this formalism was concerned with a theory of observables, was to be of crucial importance.

JORDAN AND HEISENBERG ON QUANTUM STATISTICS

Before he was interrupted by the arrival of Heisenberg's new theory, Jordan had been largely concerned with the problem at the centre of the wave-particle paradox of light in quantum theory, that of the statistics of black-body radiation. In a joint paper with Born he had derived Planck's law (using Bose's form of the statistics) from the virtual oscillator theory,[55] and he had followed this up with a paper drawing on Einstein's theory of the ideal gas.[56] Now, in the three-man-paper, he re-examined Planck's law and Einstein's energy fluctuation formula from the point of view of matrix mechanics.[57]

One of the most striking features of black-body radiation theory was that it had remained, over the years, almost entirely unconnected with the atomic theory on which it should, ideally, have depended. Whether formulated as a theory of waves or of light-quanta it was dependent upon a non-classical, physically incomprehensible, and apparently arbitrary assignment of *a priori* probabilities, and this was as true for the recent derivation proposed by Bose as for all the earlier attempts.[58] Only in de Broglie's theory was the assignment of probabilities given any physical justification, and even this was vague, and dependent upon the highly speculative hypothesis of matter waves.[59] In the light-quantum terminology of Bose's theory, one had to assume for the distribution of light-quanta over unit cells, of volume h, in phase space a formula that was based upon the non-classical assumption that the light-quanta be treated as indistinguishable (and thus somehow non-independent) for the calculation of probabilities. Working independently, de Broglie had associated each light-quantum with a phase-wave, and had calculated the distribution of quanta from the requirement that the resulting wave pattern should be stationary. Since the number of waves thus associated with a region in phase space was equal to the number of unit cells into which that region could be divided, de Broglie's derivation closely paralleled Bose's, but with the advantage that the non-independence of light-quanta required could be given some sort of physical explanation:[60]

If two or more atoms or light-quanta have exactly superposing phase waves so that one can say in consequence that they are transported by the same wave, their movements

cannot be considered as entirely independent and the atoms can no longer be treated as distinct unities in the calculation of probabilities.

Following Einstein's extension, inspired by de Broglie's work, of the Bose statistics to the distribution of an ideal gas, de Broglie's theory generally and his derivation of Planck's law in particular had received close attention from Born and Jordan in Göttingen.[61] Now, wishing to investigate the physical significance of the new kinematics and its implications for the fundamental wave-particle problem of light, it was natural that Jordan and Heisenberg should turn to the black-body context and to the question as to how de Broglie's conceptualisation of the Bose statistics carried over into matrix mechanics: might these statistics even be derivable from the new kinematics? The derivation of Planck's law itself, being fundamentally an exercise in statistical mechanics, did not prove amenable to this path of attack, but the derivation of Einsteins's fluctuation formula,

$$(13) \qquad \overline{(E - \overline{E})^2} = \frac{\overline{E}^2}{z_\nu V} + h\nu\overline{E},$$

with the bars signifying mean values, V a volume of phase space and $z_\nu = 8\pi\nu^2 \mathrm{d}\nu/c^2$, did. The quantum-theoretical formula for the mean square energy fluctuation of a radiation field had been derived by Einstein through consideration of Planck's law, but it was fundamentally, in the terminology of classical theory, a formula for the fluctuation due to interferences in a wave-field. Generalising the identification between de Broglie's phase waves and Bose's cells in phase space, Jordan noted that "if waves are propagated with a phase velocity v in an s-dimensional isotropic part of space, $V = L^s$, the number of eigenvibrations for the frequency range $\mathrm{d}\nu$ is equal to the number of [unit] cells . . . , and this in fact holds for arbitrary s, hence, e.g., also for vibrating . . . strings."[62] The number of quanta in an appropriate cell corresponded to the quantum number of an oscillator. If one could derive Einstein's fluctuation formula from the interferences in a vibrating string, Jordan reasoned, one could in principle extend the calculation to more general cases of wave fields, and thence to black-body radiation, eventually deriving from the quantum theory of the wave field all the phenomena associated with light-quanta.[63] And in the three-man-paper he was indeed able to derive the required fluctuation for a vibrating string.

In the light of its possible implications, Jordan's result was one of the most important achievements of matrix mechanics to date. It brought black-body radiation phenomena in principle within the scope of the new theory, and opened the way for the future quantum field theory. But it did not as yet lead to the Bose statistics and it did not as yet answer the conceptual questions that were bothering Heisenberg. It offered no insight to the extension of the wave-particle problem to matter and the ideal gas, and did not even approach the physical significance of this problem in respect of the new kinematics. Heisenberg wrote to Pauli in November 1925 expressing the hope that from an analysis of the grounds on which the new kinematics led to Einstein's fluctuation formula all the essential features of the light-quantum theory might be rediscovered.[64] But the conceptual problems entailed were too much for him, and besides, as he had written earlier when first reporting Jordan's work, he wished he knew more about statistics.[65]

Given this dual obstacle to the development of his understanding, Heisenberg could apparently make little of Jordan's achievements, but he was determined to make more progress toward comprehending the quantum statistics. The following Spring, at the end of a paper in which he extended the matrix mechanical transformation theory to the realm of many-body problems, he came back to the statistical problem in the material context of the distribution of electrons in an atom – the 'statistical weights' riddle of Bohr's atomic theory.[66] This riddle had previously been solved by Pauli's exclusion principle, but Heisenberg noted that neither this principle nor the Bose-Einstein statistics had yet been derived from matrix mechanics. And from the consideration of a characteristic resonance phenomenon he deduced that both were in fact compatible with it and indeed – applied to the same particles – with each other. Considering two identical particles, he argued that for any solution to a given problem there must be a second solution, obtained by switching the bodies round, and that[67]

> If only one of the two systems occurs in nature, then on the one hand this admits a reduction of the statistical weights [as in the Bose-Einstein statistics]; . . . [but] on the other hand Pauli's exclusion of equivalent orbits is of itself fulfiled.

Generalising to a system of n identical particles, he noted that there was again a reduction of statistical weights, from $n!$ to 1, and he again found the Bose-Einstein statistics to be "in harmony with Pauli's exclusion".[68]

Though Heisenberg was apparently unaware of the fact, Fermi had already demonstrated that the exclusion principle led to the completely different set of statistics associated with his name,[69] and Heisenberg's effort emphasises that, as he later admitted,[70] he was still very confused indeed about statistics.[71] But it also emphasises his continuing concern with their interpretation, which was later to feature in his struggles toward the uncertainty principle.

BORN'S MATHEMATICAL APPROACH AND THE OPERATOR FORMALISM

While Heisenberg concerned himself with the meaning of the new theory and Jordan with its applications, Born concentrated upon extending its applicability and rigour through a generalisation of the mathematics. In the three-man-paper he replaced the matrices $a(nm)$ with bilinear forms, $A(x, y) = \sum_n \sum_m a(nm)\, x_n x_m^*$. This presentation had the advantage of bringing the theory into the realm of a more familiar algebra, and it enabled Born to prove the uniqueness and – for finite variables or for bounded forms of infinitely many variables – the existence of an energy solution to the matrix mechanical transformation problem.[72] Through the introduction of a traditional eigenvalue terminology it also pointed the way toward a treatment of degenerate cases, associating these with the familiar problem of the occurrence of multiple eigenvalues W_n in the solution, $W = (\delta_{nm} W_n)$. Continuous spectra, corresponding to aperiodic phenomena, still caused problems, for the theory was still based upon the periodic behaviour of electrons in atoms; uniform rectilinear motion, for example, was completely outside its framework. But cases in which a Fourier integral representation was possible could be handled by the simple replacement of the summation sign of the bilinear form with an integral,

$$\sum_n W_n x_n x_n^* \longrightarrow \int W(\phi)\, y(\phi)\, y^*(\phi)\, \mathrm{d}\phi.$$

Apart from the problem of relating transition probabilities to amplitude densities instead of to discrete amplitudes, all went through as in the discrete case. In both cases the mathematical theory needed for the existence proof had been developed, by Hellinger, only for bounded forms. But Born felt justified in carrying the theory over to unbounded forms also, arguing that "Hellinger's methods obviously conform exactly to the physical content of the problem posed."[73]

At the end of October, Born went as a guest lecturer to MIT, where Wiener, who had visited Göttingen the previous year, was professor of mathematics. Wiener recalled that Born had arrived very excited and searching for a further generalisation of matrix mechanics.[74] By divine plan, coincidence or whatever, Wiener had only six months previously completed a paper on "the operator calculus",[75] and this paper, which was very much in the Göttingen tradition of the fusion of rigorous mathematics with the established techniques of physics, provided just the generalisation required. Operators shared the non-commutative property of matrices but not their restrictions, and Born and Wiener were able to build up a general formulation of the new quantum mechanics in which even those aperiodic phenomena that had not been covered by Born's integral formulation could now be treated.[76] They defined an operator q in a completely general way as "A rule in accordance with which we may obtain from a function $x(t)$ another function $y(t)$, which we symbolise by $y(t) = qx(t)$."[77] Using the function

$$(14) \qquad x(t) = \sum_n e^{2\pi i W_n t/h} x_n,$$

they showed that an operator q could be derived from any matrix or Hermitian form (q_{nm}) by putting

$$(15) \qquad q = \lim_{T\to\infty} \frac{1}{2T} \int_{-T}^{T} ds \cdot q(t, s),$$

where

$$(16) \qquad q(t, s) = \sum_n \sum_m q_{nm}\, x_m(t) x_n^*(s) = \sum_n \sum_m q_{nm}\, e^{2\pi i (W_n t - W_m s)/h}.$$

If $y_m = \sum_n q_{mn} x_n$, then $y(t) = qx(t)$. To derive a matrix from the operator q, they applied the operator $e^{-2\pi i W t/h} q$ to the function $e^{2\pi i W t/h}$, generating

$$(17) \qquad q(t, W) = e^{-2\pi i W t/h} q\, e^{2\pi i W t/h}.$$

They then defined

$$(18) \qquad q_{VW} = \lim_{T\to\infty} \frac{1}{2T} \int_{-T}^{T} q(t, W)\, e^{-2\pi i (V - W)t/h}\, dt$$

$$= \lim_{T\to\infty} \frac{1}{2T} \int_{-T}^{T} e^{-2\pi i V t/h}\, q\, e^{2\pi i W t/h}\, dt,$$

which was the reverse process of the above. Not all such integrals converged, and they showed later that uniform rectilinear motion in fact gave divergent (oscillating) integrals and so had no matrix representation. But since the operators were themselves quite generally defined, this motion was as answerable to the operator mechanics as was any other.[78]

Born and Wiener showed that their operators obeyed exactly the same rules as did matrices for multiplication, etc., and that since the operators $D = \partial/\partial t$ and $2\pi i W/h$ acted equally on the functions used to connect matrix and operator representations one could replace the time derivative of a matrix, $\dot{q} = \frac{2\pi i}{h}(Wq - qW)$, by that of an operator, $\dot{q} = Dq - qD$. The canonical equations of motion and commutation relationship were also reinterpreted as operator equations.[79]

CHAPTER 7

WAVE MECHANICS AND THE PROBLEM OF INTERPRETATION

SCHRÖDINGER'S WAVE MECHANICS AND ITS RECEPTION

Born's operator formulation opened up a host of new avenues, which he clearly intended to explore in further publications.[1] But before he could do this, events took a dramatic turn with the advent of Schrödinger's wave mechanics.[2] In his first communication of this new theory, submitted at the end of January, Schrödinger transformed the time-independent Hamiltonian partial differential equation of motion,

$$(1) \qquad H\left(q, p = \frac{\partial S}{\partial q}\right) = E,$$

into a form that could (at least for the one-electron problem with constant mass) be expressed as a quadratic form of a new unknown ψ, $S = k \log \psi$, and of its derivative $\partial\psi/\partial q$, set equal to zero:

$$(2) \qquad F = H\left(q, \frac{k}{\psi}, \frac{\partial\psi}{\partial q}\right) - E = 0.$$

Seeking solutions in which ψ was a product (S a sum) of functions of individual coordinates, he then treated the problem as a variational one, asking for real finite continuous unique-valued and twice differentiable eigenfunction solutions ψ, and corresponding eigenvalues E, for which the integral of the quadratic form was an extremum,

$$(3) \qquad \delta \int F \, \mathrm{d}q = 0.$$

Applied to the hydrogen atom, this gave a continuous spectrum of possible solutions for all $E > 0$ (hyperbolic orbits), but only a discrete finite set of solutions, corresponding to the Bohr energy levels, for $E < 0$.

In his second communication, a month later, Schrödinger expanded on his treatment, placing it in the context of Hamilton's and de Broglie's work, and demonstrating for the general (time-dependent) case the previously assumed connection between the Hamiltonian problem (1) of particle mechanics and the variational problem (3). He noted that the variational problem had to be conducted in multi-dimensional configuration space rather than in ordinary

Euclidean space, and the solutions ψ, uninterpreted in his first communication, corresponded to a wave form in configuration space. Showing that, as for the de Broglie phase waves, the group velocity of his configuration-space waves corresponded to the velocity in the particle picture, he adopted de Broglie's conclusions from the analogy with geometric and wave optics and suggested that the 'true' mechanical phenomenon should be seen as represented by wave processes in configuration space, the particle picture being merely an approximation in cases of well-defined wave groups. In his first communication he had not mentioned that, since his solutions ψ were physically undefined, he had in fact abandoned the orbital model of the atom and replaced it as in matrix mechanics by a purely symbolic scheme. But he now stated this fact explicitly and explained that in his view the abandonment of the orbital model was necessitated by the fundamental wave nature of phenomena, to which a particle approximation could not in this case be given. And while the matrix-mechanical oscillators remained symbolic and physically incomprehensible, his own wave oscillations constituted a perfectly viable physical picture. He then built up the new wave mechanics from scratch, deducing the form of wave equation he had previously derived for hydrogen from the periodic solutions of the simple second order wave equation,

$$(4) \qquad \nabla^2 \psi - \frac{1}{u^2} \ddot{\psi} = 0.$$

Finally, he explained that in this approach the discreteness of quantum theory arose naturally from the imposition, as in classical vibration theory, of boundary conditions, and without any need for a discrete quantisation condition.

In a paper completed in March, Schrödinger established a formal connection between his theory and matrix mechanics,[3] obtained by identifying the matrix P with the operator $\frac{h}{2\pi i}\ \partial/\partial q$; a general function $F(p, q)$ led to an operator $[F]$, and he associated this operator with a matrix F^{kl},

$$(5) \qquad F^{kl} = \int \rho(x)\, u_k(x)\, [F]\, u_l(x)\, dx,$$

where the functions $u_i(x)\sqrt{\rho(x)}$ constituted a complete normalised orthogonal system in configuration space. In this paper Schrödinger also pursued the physical interpretation of his theory, and suggested that $\mathrm{IR}\,(\psi \partial \psi^* / \partial t)$ might correspond to the charge density, the idea being to derive the intensities of emitted radiation (Heisenberg's transition amplitudes) from the wave equation, and this from a modified electromagnetic theory.

In two further papers completed in May and June Schrödinger pursued the application of the wave mechanics, including the time-dependent case, and in the latter paper he returned to the question of the interpretation of ψ. He now identified the charge density as $\rho = e\psi\psi^*$ and explained that[4]

> $\psi\psi^*$ is a kind of *weight function* in the configuration space of the system. The *wave-mechanical* configuration of the system is a superposition of several – strictly *all* – kinematically possible point-mechanical configurations. If you like paradoxes, you can say that the system is, as it were, simultaneously in all kinematically conceivable positions, but not 'equally strongly' in all of them. In the case of macroscopic motions, the weight function contracts in practice to a small region of practically indistinguishable positions the centre of gravity of which in the configuration space covers macroscopically detectable distances.

In this "reinterpretation",[5] the fundamental conceptual role of the ψ themselves as the ultimate physical reality was explicitly played down, and the wave motions were reinterpreted in terms of possible particle motions. But, reflecting an ambiguity already present in de Broglie's work,[6] Schrödinger nevertheless held fast to the priority of the wave over the particle picture.[7]

The reception of Schrödinger's wave mechanics was mixed. Many physicists who, like Einstein or Wien, had shared Schrödinger's repugnance of the physically incomprehensible "transcendental algebra" of matrix mechanics, were thoroughly delighted.[8] But even Einstein and Lorentz soon became critical of Schrödinger's interpretation as practically untenable,[9] and among the physicists who had been responsible for matrix mechanics the reaction was more immediately and more strongly ambivalent. Even before Schrödinger had drawn the explicit connection between the two theories it had been clear to all concerned that they must be closely related. Schrödinger's replacement of the classical electron particle motion in the atom by the motion of a complex wave form in configuration space constituted a clear and explicit rejection of classical kinematics parallel to that in matrix mechanics. The necessary complexity of his wave form corresponded to that of the matrix-mechanical oscillators as noted by Born and Wiener. And the preliminary results of the theories were the same. Moreover the formal connection between the theories was implicit in the Born-Wiener operator formulation of matrix mechanics, for once the identification $p \sim \partial/\partial q$ had been made, then the ordinary Hamiltonian equation of motion,

$$H = \frac{p^2}{2m} + V(q) = E$$

if viewed as an operator equation, led directly to Schrödinger's wave equation.

The identification itself could be inferred by analogy from the results $H \sim \partial/\partial t$, $pq - qp \sim 1$, and $Ht - tH \sim 1$, the latter not explicit in the Born-Wiener paper but easily derivable from the formula for time differentiation first established by Born and Jordan,

$$\dot{g} = Hg - gH.$$

Born later cursed himself for having missed the $p \sim \partial/\partial q$ identification, claiming that it would have led him to the wave mechanics before Schrödinger.[10] But this seems unlikely in view of the conceptual gulf separating the two approaches, and there is nothing whatsoever to suggest that Born even connected the two theories in this way before Schrödinger's own demonstration of the connection. Early in April, before Schrödinger's demonstration had been published, Pauli independently produced a fuller and more careful analysis of this connection.[11] But although he had adopted the operator formulation in January, and had then derived explicitly the crucial energy-time commutation rule, he appears, like Schrödinger, to have begun his analysis from the wave mechanics end.[12]

There could be little doubt in the minds of the matrix-mechanical physicists that the wave mechanics was both relevant and, since it attained the generality of the Born-Wiener formulation without recourse to abstract mathematics and through the portrayal of the physical problem in a classically familiar and apparently visualisable form, important. Pauli rated Schrödinger's work as "among the most important . . . of recent times"[13] and Heisenberg wrote that "Schrödinger's mathematics clearly signifies a great advance."[14] From a conceptual viewpoint, however, they were less impressed. Pauli was troubled by the conviction that had dominated his physical thinking since he was a student, that a pure continuum field theory of physics was impossible on operational grounds.[15] Writing to Schrödinger in May, he first praised the latter's theory, but then added that "I have generally the strongest doubt in the feasibility of a consistent wholly continuous field theory of the de Broglie rays. One must probably still introduce into the description of quantum phenomena essentially discontinuous elements as well."[16] Writing again the following November he expressed a "real conviction that the quantum phenomena in nature show facets that cannot be covered by the concepts of continuum (field) physics alone",[17] and in December he insisted yet again that "quantum phenomena can never be explained" – whatever success Schrödinger might achieve with his formulation – "in terms of continuum physics."[18]

Concerned with more immediate problems, Heisenberg had meanwhile written to Dirac in May that "I quite agree with your criticism of Schrödinger's theory with regard to a wave theory of matter. This theory must be inconsistent, just like the wave theory of light."[19] And while Dirac does not seem to have been too incensed by Schrödinger's interpretative claims – which he later dismissed as "metaphysics" – Heisenberg was. As a mathematical generalisation of matrix mechanics Schrödinger's theory was fine, but otherwise it was a misleading delusion. It did not lead to a consistent wave mechanics in de Broglie's sense,[20] and its treatment of the wave function ψ, complex and defined in multi-dimensional configuration space, as somehow constituting a visualisation [*Anschaulichkeit*] or classical physical representation was simply "rubbish".[21] Mocking the idea of a "rotating electron, its charge distributed over the whole of space with axes in a fourth and fifth dimension", he wrote to Pauli in June that he found this part of the theory the more detestable the more he thought about it.[22] Upon meeting Schrödinger personally in July, Heisenberg came to the conclusion that any element of deception was quite unconscious, and that Schrödinger was simply a nice chap twenty-six years out of date,[23] and this opinion seems to have been widely shared. Sommerfeld, at whose institute the meeting had taken place, drew the "general conclusion that the wave mechanics is truly an admirably worthy micro-mechanics, but that the fundamental quantum riddles are not remotely solved by it."[24] Some months later, after Schrödinger had lectured also in Copenhagen, Bohr wrote to Kronig that Schrödinger seemed to think that he had ridded physics of the quantum hypothesis altogether, but that "this appears, however, to be a misunderstanding, as it would seem that Schrödinger's results so far can only be given a physical application when interpreted in the sense of the usual postulates."[25] Jordan, whose "slightly tactless" behaviour had to be followed by apologies from Born,[26] wrote to Schrödinger that "all of the quantum-mechanical people known to me are convinced that the basic concepts of Bohr are still to be retained."[27]

BORN'S STATISTICAL WAVE MECHANICS

Born himself wrote to Schrödinger in 1927 that, although he had "meanwhile returned again to Heisenberg's standpoint", he had originally disagreed with Heisenberg and had thought that "your wave mechanics signified more physically than our quantum mechanics."[28] He could not follow Schrödinger's interpretation of the theory but he was, he wrote, convinced of its superiority to matrix mechanics by its simple handling of aperiodic processes, and he had therefore adopted it for his own researches.[29]

After visiting MIT, Born had spent January through March 1926 on a lecture tour of the United States, and on his return to Germany he had gone straight to Frankfurt where his wife was convalescing after an illness. By the time he got down to work again in Göttingen the first installments of Schrödinger's wave mechanics had already been published and the connection with matrix mechanics derived. Born was naturally attracted to the new theory. It had been developed from the wave theory of matter to which he had himself been attracted the previous year, and it was coextensive with the theory he had just formulated with Wiener. But it was physically more suggestive than that theory (albeit misleadingly so), and was expressed in terms of familiar classical mathematical physics. As he wrote later in the year, wave mechanics also had the advantages of permitting "the retention of the conventional ideas of space and time in which events take place in a completely normal manner",[30] a statement with which Heisenberg would not have agreed, and of offering a natural origin for the quantum behaviour: the commutation relations arose naturally as operator equations in this theory, but had to be imposed as axiomatic upon matrix mechanics.[31] But it seems to have been the familiar classical form and wide applicability that appealed most immediately. In his first paper based upon wave mechanics, completed in June 1926, Born considered its application to collision processes, a subject with which he had been concerned before the arrival of Heisenberg's new kinematics, which combined the main problems – aperiodic effects and transition processes – with which quantum theory was faced, and which he had presumably intended already to attack on the basis of his own operator formulation.[32] He concluded that[33]

> Of all the different forms of this theory only Schrödinger's has proved suitable here, and I may directly on these grounds take it as the most profound comprehension of the quantum problem.

Applying wave mechanics through a first order perturbation theory of the collision between an atom and a free electron, Born obtained an asymptotic solution at infinity ('Born approximation'),

$$(6) \qquad \underset{\text{electron atom}}{\psi'_{n\tau}(x, y, z; q_k)} = \sum_m \iint_{\alpha x+\beta y+\gamma z>0} d\omega\, \Phi_{\underset{\tau}{nm}}(\alpha, \beta, \gamma) \cdot \sin\{k_{nm}(\alpha x + \beta y + \gamma z + \delta)\}\, \psi_m^0(q_k),$$

where $\psi_m^0(q_k)$ was the mth eigenstate of the unperturbed atom, $\Phi_{\underset{\tau}{nm}}(\alpha,\beta,\gamma)$

the wave function for the electron (originally incident from the z-direction with energy τ), and

$$W_{\underset{\tau}{nm}} = h\nu^0_{nm} + \tau,$$

with ν^0_{nm} the transition frequencies of the unperturbed atom, the solution for the final energy of the electron.

The problem was how to interpret this physically. Born claimed that[34]

If one is to reinterpret this result in terms of particles, only one interpretation is possible: $\Phi_{\underset{\tau}{nm}}$ [corrected in proof to $|\Phi_{\underset{\tau}{nm}}|^2$] indicates the probability that an electron incident from the z-direction will have been sent in the α, β, γ direction (and with phase δ), its energy τ having been increased in the process by a quantum $h\nu^0_{nm}$, at the expense of the atomic energy. . . .

Schrödinger's quantum mechanics thus gives a complete answer to the question as to the effect of a collision, but there is no question of a causal relationship. One cannot answer the question "what is the state after the collision" but only the question "what is the probability of a given effect of the collision" (in which quantum-mechanical energy levels must naturally be preserved).

Here the whole problem of determinism presents itself. From the standpoint of our quantum mechanics there is no quantity that remains causal in the case of an individual collision effect; but also in practice we have no grounds to believe that there are inner eigenstates of the atom which stipulate a determined collision path. Should we hope to discover such eigenstates later (such as phases of internal atomic motions), and to determine them for the individual case? Or should we believe that the agreement of theory and experiment on the impossibility of giving a stipulation of the causal lapse is a preestablished harmony, resting on the non-existence of such stipulations. My own inclination is that determinism is abandoned in the atomic world. But that is a philosophical question, for the physical arguments are not conclusive.

Coming from someone who had shown no previous inclination toward such a philosophical stance (but had on the contrary been viewed as a staunch determinist), and in a context where his generally more philosophically-minded colleagues were asking questions as to the meaning of space and time that threatened to render the concept of causality meaningless, Born's declaration was remarkable. More remarkable still was his subsequent insistence upon acausality. In a second paper, completed in July, he returned to the possibility of hidden phases providing a causal description of individual events and argued that[35]

It appears to me a priori improbable that quantities corresponding to these phases can easily be introduced into the new theory, but Mr. Frenkel has told me that this may perhaps be the case. However this may be, this possibility would not alter anything

relating to the practical indeterminacy of collision processes, since it is in fact impossible to give the values of the phases; it must in fact lead to the same formulae as the 'phaseless' theory proposed here.

In a subsequent short note to *Nature*, Born again stressed that for practical purposes microscopic coordinates did not exist.[36] Classical theory, he claimed, introduced such coordinates (for example those relating to the motions of individual molecules) only to ignore them and take their statistical aggregate. Quantum theory, on the other hand, did not bother with this charade: one could not dismiss the possibility of microscopic coordinates existing, but they were of no significance unless one could measure them, which one could not.

We shall return to these arguments. But first we must look more closely at the statistical interpretation itself. In his second paper Born noted that whereas matrix mechanics "started from the idea that an exact representation of the processes in space and time is quite impossible and is therefore statisfied with the establishment of relations between observable quantities", Schrödinger attempted to assign to his wave process "a reality of the same type as light waves possess".[37] Neither conception appeared to him satisfactory, however, and he preferred "to adhere to an observation of Einstein, . . . that the [light] waves are present only to show the corpuscular light quanta the way".[38] He interpreted the Schrödinger waves in this way as a "ghost field" (which term he attributed to Einstein) "or, better, guiding field", the amplitude corresponding to a (determined) probability as in the first paper.[39] As might be expected, Einstein in fact rejected Born's acausal interpretation,[40] but Born was clearly upset by this rejection and seems genuinely to have assumed that his and Einstein's interpretations were the same.[41] How, then, did this confusion arise, and how did Born reach his own interpretation?

The first stated requirement of Born's interpretation was that it should retain the corpuscular nature of the electron, and in his recollections he linked this with Franck's experiments.[42] The most recent of these experiments had in fact demonstrated the wave nature of the electron, and Born himself had been strongly attracted to de Broglie's wave theory of matter.[43] But there could be no doubt in his mind that beyond collision phenomena electrons (as also light) regained a particulate nature, and it must have already been clear from his analysis of collision effects that the Schrödinger wave packets would not remain sufficiently localised for this to be possible. The particles could therefore not be seen simply as constructed from wave packets, the interpretation given them by Schrödinger. Nor, in view of their complex and multi-dimensional nature, as well as of experiments conducted with single quanta or electrons in the apparatus at any time, could the wave be treated simply as the statistical result of many individual particles.[44] Given his rejection of these possibilities, Born was virtually forced by his acceptance of the physical significance of the Schrödinger waves into the interpretation he adopted. The waves had to be seen as somehow guiding the motion of particles, and the wave amplitudes had therefore to be identified with probabilities in the way that he said. But what was not forced, and what Einstein disagreed with, was the abandonment of any hope that the guidance of the particles by the waves might ultimately be causally determined. The idea of a guiding field had indeed been mooted by Einstein in the course of his efforts to understand the wave-particle duality in the early part of the decade,[45] and his failure to construct a causal model on these lines may well have been a factor in the subsequent rejection of causality by Schottky and others.[46] But Einstein had been concerned only with real waves in 3-dimensional space, and he himself had clung to the hope of a causal theory – in fact a causal field theory with the corpuscular singularities arising from boundary restrictions much as Schrödinger had suggested.[47] At the time, Born appears to have shared this hope, but he now rejected it with such conviction that he expected Einstein to follow him. Why?

There are possibilities for many influences here, including those of Weyl – who, having taken an active part in both matrix mechanics and wave mechanics,[48] may reasonably be supposed to have been in communication with Born – and, in view of Born's emphasis on the fundamental "fusion of mechanics and statistics" in his theory,[49] of Spengler or Reichenbach. These influences, which have been considered elsewhere, may well have played a part in Born's thinking.[50] But there were also internal factors operating. Firstly there was Heisenberg's earlier suggestion that the quantum-mechanical formalism might necessarily offer only limited information.[51] Secondly, there was the fact that the theory appeared in some way to generate statistics. Quantum theory had long involved an element of uncertainty, in the location of an orbit, the moment of transition, etc., but this had always been a case of uncertain conclusions following from uncertain data. In Born's analysis of the electron-atom collision, however, uncertain conclusions (the atom in a superposition of states, the electron spread throughout space) appeared to follow from definite data on the initial motion of the electron and state of the atom. Finally, Born's denial of the existence of further observable microscopic coordinates raises a particularly interesting consideration. For the most striking features of his presentation were the identification of utility and profundity, the confident assertion that the theory was final and immune to further experimental advance, and the related assertion that it was complete:[52]

Many take it that the problem of the transitions of quantum mechanics . . . cannot be comprehended, but that new concepts will be needed. I myself came, through the impression of the completeness of the logical foundations of quantum mechanics, to the opinion that the theory is complete and that the transition problem must be contained in it.

Born's insistence upon acausality, resting as it did on the non-existence and non-observability of further microscopic coordinates, was equivalent to his insistence that his theory was final, and the above remarks tie in this insistence with that of the superiority of a mathematical to a physical approach, and thus with the divisions that had arisen on this score the previous Autumn. In concentrating on the practical applicability and mathematical formulation of the theory (his second paper incorporated some of the mathematics worked out with Wiener), Born had arrived at a theory the predictive range of which coincided perfectly with that of existing experimental observations. If this coincidence was taken as final – but only in this case – then his theory could be taken *independent* of any underlying epistemological considerations, and would constitute a full justification of his formal approach.

REACTIONS TO BORN'S THEORY

Born was, as Heisenberg noted,[53] a "mathematical methods man". His basic approach to physics was to take a physically clear problem and to seek the mathematically rigorous solution with the widest applicability, not to concern himself with the philosophical subtleties of problem definition. In his view a useful formulation of a problem was indeed a profound one, and he must have considered acausality a small price to pay for the combination of physical clarity and wide applicability in Schrödinger's theory. But to Heisenberg and Pauli, who felt that through a reexamination of the kinematical concepts they were at last getting to the core of the fundamental problems of quantum theory, Born's attitude seemed retrogressive. His theory contained distinct advances. His probabilistic interpretation of the wave function was clearly an important extension of the concept of transition probabilities, and combined with the wide-reaching mathematical theory it allowed the prediction of non-classical electron collision phenomena (barrier penetration, etc.) previously outside the scope of the new quantum mechanics. But of the observability criterion and new kinematics there was no longer a trace. They were indeed explicitly rejected and in Pauli's eyes at least Born's theory must have shared the same faults as had Bohr's virtual oscillator theory. The need for a redefinition of the fundamental concepts was again shelved for the sake of utility, and the failure of the existing kinematics once again expressed through a concept of acausality that could only be defined in terms of those kinematics.[54] Pauli himself was apparently too interested in the advances of the new theory, or too busy with his monumental encyclopedia article on quantum theory, to waste words in criticism,[55] but Heisenberg was less reticent. He recalled that he was very angry with the relapse into pseudo-classical terminology,[56] and writing to Pauli in July he mocked Born's use of the wave terminology, likening his statement of the wave theory to that of the apostolic creed.[57] But he too had to recognise the advances of the theory. The problem now to be faced was not that of a choice between matrix mechanics and wave mechanics, but rather that of bringing together the conceptual innovations of the one with the practical achievements of the other and with the variety of additional ideas to which matrix mechanics had already led.

QUANTUM MECHANICS IN THE SUMMER OF 1926

By the Summer of 1926, the innovations of Heisenberg, Pauli and Schrödinger had already led to a wide range of theories, results and ideas, and many of these had to be brought together before a single fully-fledged 'quantum mechanics' could be evolved. Among developments of particular importance we have already noted the reflections of Heisenberg and Pauli on kinematics and the thoughts of Pauli on observability, the development of matrix mechanics as a theory of transformations, and Born's probabilistic interpretation of Schrödinger's wave function. There had also been other developments due to Lanczos, who had given a continuous integral equation formulation of matrix mechanics,[58] to Fermi, and especially to Dirac.

Working independently, Dirac had derived from Heisenberg's new kinematics many of the results of matrix mechanics. But he had done so through a particularly elegant, and conceptually clear, formulation. His approach had been to take the notion of "quantum-mechanical quantities", as defined by Heisenberg, and to develop a consistent calculus of them. This had led him in late 1925, through a comparison with the classical theory of action and angle variables, to the conclusion that the quantum-mechanical equations could be derived from the classical by substituting the quantum quantities

$[x, y] = (xy - yx)\, 2\pi/ih$ for the classical commutators, $(x, y) = \left\{ \frac{\partial x}{\partial \omega} \frac{\partial y}{\partial J} - \frac{\partial y}{\partial \omega} \frac{\partial x}{\partial J} \right\}$. For classical canonical variables, p, q, this gave him the commutation relations, frequency condition, and indeed all the formulae of matrix mechanics.[59] In a second paper, completed in January 1926, he had clarified the relationship between the classical and quantum theories by introducing the concepts of "q-numbers" (quantum-mechanical quantities obeying quantum-mechanical rules) and "c-numbers" (ordinary classical numbers).[60] He admitted that "at present we can form no picture of what a q-number is like",[61] but the terminology helped him toward a clear expression of the physical problem within the new theory: having solved the equations of motion in their quantum-mechanical form, one could only get results comparable with experiment if one could represent the q-numbers by c-numbers corresponding to experimentally observable values.

Dirac's formulation had the great virtue of preserving the significance of the observability criterion and new kinematics. And in August 1926 he pursued the former concept into the field of quantum statistics that had so intrigued Jordan and so confused Heisenberg. He argued that[62]

> Heisenberg's theory . . . enables one to calculate just those quantities that are of physical importance, and gives no information about quantities such as orbital frequencies that one can never hope to measure experimentally. We should expect this very satisfying characteristic to persist in all future developments of the theory.

The criterion of observability became relevant when, for an atom of several electrons, "the positions of two of the electrons are interchanged", and "the new state of the atom is physically indistinguishable from the original one."[63] Labelling the two states (m, n) and (n, m), Dirac argued that one could not observe the individual transition intensities from a third state, $(m', n') \rightarrow (m, n)$ and $(m', n') \rightarrow (n, m)$, but only the sum of the intensities of the two transitions. The two states (m, n) and (n, m) therefore had to be treated in quantum mechanics as interchangeable, and this requirement led to two possible solutions for the whole system,

$$(7) \qquad \psi_{mn} \sim \psi_m(1)\, \psi_n(2) \pm \psi_n(1)\, \psi_m(2).$$

One solution corresponded to $\psi_{mm} \sim 0$, or Pauli's exclusion principle. The other gave $\psi_{mn} = \psi_{nm}$, corresponding to (m, n) and (n, m) referring to one and the same state: extended to many particles this gave the Bose-Einstein statistics. Fermi had demonstrated earlier in the year that Pauli's exclusion principle led directly to a different set of statistics,[64] and Dirac now derived the same conclusion independently, showing that if the Bose-Einstein statistics were expressed in the form

$$(8) \qquad N_s = A_s/(e^{\alpha}\, e^{\beta E_s} - 1),$$

then those following from the exclusion principle gave

$$(9) \qquad N_s = A_s/(e^{\alpha}\, e^{\beta E_s} + 1).$$

He noted that the theory did not as yet allow one to determine which set of statistics applied to the ideal gas, but suggested that this was more likely to behave like electrons (as Fermi had in fact assumed) than like light-quanta.

In the same paper Dirac also drew on the matrix formulation of quantum mechanics to generalise Schrödinger's theory. Schrödinger had restricted himself to the classical problem of finding eigenfunctions ψ_n corresponding to energy eigenvalues E_n. Dirac now pointed out that in matrix mechanics there was nothing special about energy, and that one could equally well ask for eigenfunctions Ψ_n corresponding to eigenvalues of any function of the space-time and momentum-energy coordinates.[65]

PAULI'S IDEAS ON QUANTUM MECHANICS

Dirac's generalisation of wave mechanics was a crucially important first step in the fusion of the various aspects of quantum mechanics into a coherent whole. This fusion was largely completed in the Autumn and Winter of 1926–27, during which time Dirac was in Copenhagen with Heisenberg, whose ideas on quantum statistics he quickly put straight, and Bohr. It was catalysed by a letter from Pauli to Heisenberg written on 19 October, in which Pauli, stimulated by Dirac's considerations, tried to bring together some of the work of recent months under the two headings of quantum statistics and collision theory.[66] Having thanked Heisenberg for a letter, no longer extant, Pauli remarked that he had been thinking about the Fermi-Dirac statistics, and he raised the question of zero-point energy, the existence of which Jordan had taken the previous year to be one of the most fundamental features of quantum theory,[67] and of which of the two sets of quantum statistics should be applied to the ideal gas:[68]

So far there exists a distinction between a crystal lattice and radiation, that, namely, in respect of the zero-point energy $\frac{1}{2}h\nu$. Now there are *a priori* reasons for supposing that a zero-point energy is also *present* in the ideal gas. [Pauli noted that he had discussed this with Stern.] But the Einstein-Bose theory only allows it to be plugged on artificially (c.f. the relevant statements of Schrödinger in *Berliner Berichte* and *Physikalische Zeitschrift*), and that speaks from the first against this theory and for Fermi-Dirac. To make matters clearer for myself I have carried through the fluctuation considerations of Einstein's work, Berlin Academy S8, §8, from the standpoint of the Fermi-Dirac theory. . . .

Pauli continued to give a lengthy discussion of this problem and also to consider the question, again raised by Jordan in the wake of an analysis by Ehrenfest,[69] of additive entropy. Neither discussion led to any really significant new insight, but in a letter of 15 November Heisenberg, following his instruction in quantum statistics by Dirac, extended the latter's analysis to establish the usual characterisation of particle statistics.[70] Following up Dirac's characterisation of the statistics as relating to symmetrical or anti-symmetrical solutions he showed that an overall symmetry requirement led to an association of the statistics with spin: half-integral spins gave anti-symmetric space functions and Fermi-Dirac statistics, while integral spins gave symmetric space functions and Einstein-Bose statistics.

The second part of Pauli's letter, concerning collision theory, was more provocative and deserves extensive quotation.[71] What he had to say was, he admitted, something of an "undigested dumpling", but it was certainly a very rich dumpling. He began by considering a one-dimensional collision:

A mass point runs over an obstacle, characterised by a potential eigenfunction $V(x)$, falling off sharply to zero on both sides from a fixed point x_0. The maximum of V is finite. Further, the incident energy E of the mass point is large compared with the maximum value of V, so the unperturbed uniform rectilinear motion must be taken as a zeroth approximation, and successive perturbation theories can then be applied. Naturally according to classical mechanics, if $E > |V_{max}|$, the mass point always runs over the obstacle. But according to Born's quantum mechanics . . . it sometimes happens even in the first approximation that the mass point is reflected back, i.e. reverses its velocity direction at the obstacle. And indeed there arises according to Born as a standard for the probability of reflection the square of the amplitude of the wave for $x < x_0$, represented by

$$\text{(A)} \qquad \psi_1 = \frac{1}{2i}\frac{1}{k}\,e^{ikx}\int_{-\nu}^{\nu} e^{2ik\xi}\,V(\xi)\,d\xi.$$

Pauli's next step was to treat a one-dimensional rotator perturbed by a force field characterised by a potential $V(\vartheta)$:

The case is periodic and can be treated by means of ordinary matrix mechanics. . . . The essential thing now is that the system is degenerate, in that the given (quantised) energy of the mass point, E_n, can rotate to the left (quantum number $+n$) as much as to the right (quantum number $-n$). A time-constant quantity f would have in its matrix representation not only diagonal elements f_{+n+n}, f_{-n-n} but also elements of the form f_{+n-n}, f_{-n+n}. According to Schrödinger's recipe, the matrix elements are now given in the unperturbed rotation by some function $F(\vartheta)$, real and periodic in the angle ϑ:

$$F(\vartheta)\, e^{in\vartheta} = \sum_{m=-\infty}^{\infty} F_{nm}\, e^{im\vartheta}$$

or

$$F_{nm} = \frac{1}{2\pi} \int_0^{2\pi} F(\vartheta)\, e^{i(n-m)\vartheta}\, d\vartheta.$$

That the numerical values F_{nm} here depend only on the difference $n - m$ arises from the fact that the system is *exactly* force-free (everything goes the same in the three-dimensional case).

The time average of the perturbation energy $V(\vartheta)$ in the state $\pm n$ is a Hermitian matrix of the form

$$\begin{matrix} V_{+n+n} & V_{+n-n} \\ V_{-n+n} & V_{-n-n} \end{matrix}$$

wherein, according to (B),

$$V_{+n+n} = V_{-n-n} = \frac{1}{2\pi} \int_0^{2\pi} V(\vartheta)\, d\vartheta,$$

$$V_{+n-n} = V^*_{-n+n} = \frac{1}{2\pi} \int_0^{2\pi} V(\vartheta)\, e^{2in\vartheta}\, d\vartheta.$$

Putting $V_{+n+n} = C_0$, $|V_{+n-n}| = C_1$, $V_{+n-n} = C_1\, e^{i\delta}$, $V_{-n+n} = C_1\, e^{-i\delta}$, one gets the 'secular' equation for energy E,

$$\begin{vmatrix} C_0 - E & C_1\, e^{i\delta} \\ C_1\, e^{-i\delta} & C_0 - E \end{vmatrix} = 0.$$

$E = C_0 \pm C_1$, so the original energy values split into two. This results further in the fact that in both states the perturbed rotator corresponds to standing waves:

$$\psi = \cos(n\vartheta - \delta/2) \quad \text{or} \quad \sin(n\vartheta - \delta/2).$$

That is naturally your favourite unadulterated resonance phenomenon (released moreover from the mess of equivalent electrons; here there is only *one* particle). The energy swings to and fro between the left- and the right-turning oscillators. And in each of the 'secular' quantised states of the perturbed rotator, the particle must rotate as often in the positive as in the negative sense. But how is that possible? It is only possible, as Born has rightly shown in his treatment (in defiance of all classical mechanical presentations) if the particle is reflected back from the obstacle despite its greater kinetic energy E.

Once again, the same typically quantum behaviour was manifest. The next step, "and now comes the beauty of it", was to connect the wave-mechanical formulation of the one case with the matrix-mechanical formulation of the other:

The integral, which according to Born's formula (A) gives the probability of reflection, is none other than the matrix element V_{+n-n} , . . . , when one goes to the limit of an infinitely great radius of the rotator. $|V_{+n-n}|^2$ is (to within the factor $1/k$) proportional to the probability, where $2|V_{+n-n}|$ is the energy splitting of the secular disturbance.

Naturally this connection can also be extended to higher approximations. *The essential thing is this: one carries out the matrix perturbation calculation, but not that for the secular perturbation, so that the energy is derived from a non-diagonal time-constant matrix of the type*

$$\begin{pmatrix} E_{+n+n} & E_{+n-n} \\ E_{-n+n} & E_{-n-n} \end{pmatrix}$$

Then $|E_{+n-n}|^2$ *gives the collision probability.*

As Pauli explained, the same recipe could also be generalised to the three-dimensional case:

The mass point now runs through a central field and the potential energy $V(r)$ falls off quickly from the centre. Now each such space-function falling off quickly with x, y, z can be represented classically in the unperturbed rectilinear motion as a time-dependent Fourier integral – corresponding to a continuous matrix. One must only decide which canonical variables to take for the *unperturbed* (rectilinear) motion. The ps (analogous to the action variables J) must be constant in time and the energy must always be a function of the ps; the qs (analogous to the angle variables ω) must be constant or a linear function of time. When the system is degenerate they are *not* unequivocally determined.

Example 1: the ps are the ordinary cartesian momentum components, the qs the cartesian coordinates.

Example 2: the ps are E, angular momentum P, angular momentum Q parallel to z. The qs are t, perihelion angle β, node angle γ.

These are also defineable for rectilinear motion.

Now comes the dim point. The ps must be taken as *controlled*, the qs as *uncontrolled*. That is to say that one can always calculate only the probabilities of fixed variations of the ps from given initial values and *averaged over all possibe values of* q . . . .

We remain now in classical coordinates. Each space-function $F(x, y, z)$, which falls of sharply from the centre, would then correspond (according to Schrödinger's procedure for the calculation of matrices) to the continuous matrix

$$F^{p_x p_y p_z}_{p'_x p'_y p'_z} = \int F(X, Y, Z)\, e^{2\pi i \{(p'_x - p_x) X + (p'_y - p_y) Y + (p'_z - p_z) Z\}/h}\, dX\, dY\, dZ.$$

Then $\left| V^{p_x p_y p_z}_{p'_x p'_y p'_z} \right|^2$ corresponds again to the probability of the deflection from p_x ... to p'_x ..., ..., where the boundary condition of energy conservation, $\Sigma\, p_x^2 = \Sigma\, p'^2_x$ holds.

In respect of the higher approximations, these too presumably give no essential difficulties. One can also proceed from the matrix corresponding to the acceleration dp/dt and then observe that *classically* the Fourier coefficient corresponding to the zero value of the frequency determines the deviation. But quantum-mechanically the square of the Fourier coefficient of dp/dt that corresponds to zero energy *change* determines (to within a harmless factor) the probability of the deflection. In inelastic collisions it is in principle the same.

So much, wrote Pauli, for the mathematics. But what of the physics?

The physics of it still continues to be unclear to me. The first question is, why may only the *p*s, and in any case not both the *p*s and also the *q*s be described with any accuracy. This is the old question that occurred if the velocity direction and asymptotic distance of the orbit from the core were given (at least with a certain accuracy). On this I know nothing that I have not already long known. It is always the same: there is not on account of the bending [of the orbit] any weak radiation in the wave optics of ψ fields, and one may not at the same time relate both the '*p*-numbers' and the '*q*-numbers' to ordinary '*c*-numbers'. One can see the world with *p*-eyes and one can see it with *q*-eyes, but if one opens both eyes together then one goes astray.

The second question is how the above matrix elements come to determine the collision probabilities. The direction in which one must here steer, I believe to be the following. The historical development has involved the connection of matrix elements with the observation of accessible data being undertaken in a roundabout way, through the emitted radiation. But I am now convinced with the whole fervour of my heart that *the matrix elements must be connected with in principle observable kinematic (perhaps statistical) data of the particles concerned in the stationary states*. This is quite apart from whether anything in general and what (electromagnetically) is radiated (the velocity of light may be set to ∞). Also I do not doubt that behind it is hidden the key to the treatment of unperiodic motions. Now we have that all *diagonal* elements of the matrices (at least of functions of *p* alone or of *q* alone) can already generally be interpreted kinematically. So one can indeed ask first for the probability that in a fixed stationary state of the system the coordinates q_k of a particle lie between q_k and $q_k + dq_k$. The answer is then $|\psi(q_1 \ldots q_f)|^2\, dq_1 \ldots dq_f$ with ψ the Schrödinger eigenfunction. (From the *corpuscular* standpoint it thus already makes sense for it to lie in multi-dimensional space). We must look at this probability as in principle observable, just as the light intensity as a space-function in standing light waves. It is thus clear that the diagonal elements of the matrix of each *q*-function must be

$$F_{nn} = \int F(q_k)\, |\psi_n(q_k)|^2\, dq_1 \ldots,$$

physically interpreted as the 'mean value of *F* in the *n*th state'. Here one can make a mathematical quip: there is also a corresponding probability density in *p-space*: for this one puts (formulated in one dimension for simplicity)

$$p_{ik} = \int p\, \phi_i(p)\, \phi_k^*(p)\, dp$$

$$\frac{2\pi i}{h} q_{ik} = -\int \phi_i \frac{\partial \phi_k}{\partial p} dp = +\int \frac{\partial \phi_i}{\partial p} \phi_k^* dp, \ldots$$

You see that I have switched to the opposite of the usual prescription for the construction of the matrix elements p_{ik} and q_{ik} from the eigenfunction. From the matrix equation of energy conservation,

$$\frac{p^2}{2m} + V(q) = E,$$

one gets

$$\left[\frac{p^2}{2m} + V\left(-\frac{h}{2\pi i}\frac{\partial}{\partial p}\right)\right]\phi = E\phi;$$

V is thought of as an operator, say in powers of $\partial/\partial p$. In the harmonic oscillator, where the Hamiltonian is symmetric in p and q, ϕ is also Hermite's polynomial. One can also operate a perturbation theory with the ϕ.

Similarly in the Hydrogen atom, ϕ must be a simple function, but I have not yet worked it out. In any case, there is thus also a probability that in the nth state p_k lies between p_k and $p_k + dp_k$, and it is given by $|\phi_n(p_1 \ldots p_f)|^2\, dp_1 \ldots dp_f$, so that

$$F_{nn}(p) = \int F(p)\, |\phi_n(p_k)|^2\, dp_1 \ldots dp_f.$$

When, accordingly, the diagonal elements of the matrices follow physically from the kinematic statements contained in the $\phi_k(p_1, \ldots, p_f)$ and $\psi_k(q_1, \ldots, q_f)$, I do not believe that your fluctuation treatment can say anything new and important for the intepretation of unperiodic motions beyond this. For there it is always a question of time averages. In the collision phenomena it is likewise a question of time-constant elements, but of non-diagonal elements of the matrix in degenerate systems (the physical significance of which is still not clear to me).

But my chief question is what the other matrix elements mean purely point-kinematically, and wholly unconnected with electromagnetic radiation. I have proposed a statement, in which is asked: I know that at time t the particle has position coordinates q_0. What is then the *probability* that it has the coordinate q at time $t + \tau$? But I am no longer certain whether that is a reasonable question. I think in any case in kinematics of statistical data, which convey the time development of the behaviour of particles. These data could throughout be of such a kind that one cannot speak of a definite 'path' of the particles. Also one again hits here on the fact that, as elsewhere, one may not ask after p and q together.

This second part of Pauli's letter does not have the coherence of a published paper, is not easily split up, and needs reading several times. It seems to have been stimulated, as was the first part, by the work of Dirac. Pauli also drew on the thoughts of Kramers,[72] who had been developing what later became known as the Wentzel-Kramers-Brillouin approximation in the context of Schrödinger's theory, and he referred to some work of Heisenberg's on fluctuations. But when Heisenberg published this work he himself drew heavily on Pauli's thoughts,[73] and these would appear to have been largely original. They were also still somewhat confused, but Pauli had nevertheless been able to make distinct progress toward a fusion of Born's theory with the results – and problems – of matrix mechanics. Following Dirac's comments on the extension of Born's theory to include predictions of quantities other than energy, Pauli had shown, among other things, that this theory could be expressed in momentum space as well as in configuration space. He had suggested how the theory might lead to probability predictions of p, q, and functions of these, and he had stressed the identification, previously assumed by Jordan, of the diagonal elements of a matrix with the time mean of the quantity represented. He had stressed repeatedly the impossibility of simultaneous predictions of p and q and had linked this with the necessarily statistical nature of classical kinematic pictures. All this constituted substantial progress.

CHAPTER 8

TRANSFORMATION THEORY AND THE DEVELOPMENT OF THE PROBABILISTIC INTERPRETATION

TRANSFORMATION THEORY

On 28 October, Heisenberg replied to Pauli with profuse thanks.[1] The letter had been handed round to Bohr, Dirac and Hund, and had been generally discussed. Above all, Heisenberg wrote, he had been inspired by the discussion of collision processes, and he now understood much better the significance of Born's formulation. In particular, Pauli's discussion of the rotator indicated generally that "wherever in classical mechanics one type of motion changes discontinuously into another, quantum mechanics supplies a continuous transition which, so far as it may be thought of graphically, signifies a probability dictum".[2] In the wake of Schrödinger's controversial lecture,[3] and of the arrival of Dirac, attention at Copenhagen was focussed very much on the problem of relating Born's formulation to matrix mechanics, and on that of demonstrating that Schrödinger's theory could not be continuously interpreted but must share the essential discreteness of matrix mechanics. Pauli's letter clearly contributed much to both problems, and on 4 November Heisenberg wrote to him again, declaring himself "more and more inspired by the content of your last letter every time I reflect on it."[4] Two days later, he submitted a paper on the energy fluctuations of a gas, in which both problems were brought together.[5] Drawing on his earlier application of matrix mechanics to many-body problems, Heisenberg now looked at the energy interchange between two particles and showed that for observable quantities (such as the time means of energy and of the energy fluctuation squared) the new quantum mechanics agreed with the discrete conclusions of the light-quantum theory. The solutions for a system of two particles were derived, on the new theory, through a matrix transformation S of the solutions to the 'unperturbed' problem of independent particles. In place of the continuous oscillations of classical theory this led to the states of interacting atoms being superpositions of their unperturbed states, which could only be interpreted probabilistically. The discrete states were thus maintained, the atoms oscillating between them as in Pauli's discussion of the rotator and spending a determinable proportion of time in each. Following Pauli's recipe, the probability that a system initially in state α

should be found in a new state β was given by the appropriate element of the transformation matrix. The time means of observable quantities (and thus the proportion of time for which they took each allowed value) followed as the diagonal sums of their respective matrices.

Heisenberg's main conclusion, as stressed in his paper, was that "a continuous interpretation of the quantum-mechanical formalism, including that of the de Broglie-Schrödinger waves, would not correspond to the essence of the known formal interaction."[6] But of greater importance for the progress of the theory were the identification of the transformation matrix with a set of probabilities, which brought together Born's interpretation and the transformation theory; the explicit identification of the diagonal sum of a matrix with its time mean; and the application of this to the prediction of the proportion of time for which a system occupied a given state. In addition to these achievements, Heisenberg had written to Pauli in his letter of 4 November that he had realised that "in general, every scheme that satisfies $pq - qp = h/2\pi i$ is correct and physically useful, so one has a completely free choice as to how to fulfil this equation, with matrices, operators or anything else."[7] He had moreover recognised the wave function ϕ of Pauli's p-space representation as the Laplace transform of Schrödinger's wave ψ in q-space, and had concluded that "the problem of canonical transformations in the wave representation is thereby as good as solved."[8] The problem was not yet actually solved, as Heisenberg, like Pauli, was still too confused about the various possible formulations to make the necessary generalisation. But this generalisation was provided a few weeks later by Dirac.

Dirac had already been looking, at the end of October 1926 at the matrix interpretation of Schrödinger's charge density $\psi\psi^*$.[9] In early November he appears to have been somewhat diverted by the appearance of Klein's extension of Kaluza's five-dimensional unification of general relativity theory and Maxwellian electromagnetic theory.[10] Klein was working in Copenhagen under Bohr, and Dirac, who was interested in relativity theory as well as in quantum theory, naturally explored the possibilities of his work. But by 23 November Heisenberg could write again to Pauli informing him that "Dirac has managed an extremely broad generalisation of my fluctuation paper" – which is to say of the ideas Heisenberg had developed through consideration of Pauli's earlier letter:[11]

The fundamental idea of Dirac is perhaps this: let p, q, be any canonical conjugate quantities, $f(p, q)$ a function of the same.
Question: what can one say about f physically, in quantum mechanics?
Answer: one chooses, for example, q as a c-number (e.g. $q = 10$). Then it is *classically*

possible to calculate the function $f(p, 10)$. Quantum-mechanically this *cannot* be done. But one can specify how great the range of p is for which f lies between the c-numbers f and $f + df$. If one puts for p and q the variables E and t, with $f = W^a$, you see already that the Dirac problem amounts to my fluctuation note. But one can also, for example, choose E and t and put $f = q$. So one can deduce the fraction of time for which q lies between q and $q + dq$. The solutions of this problem are the Schrödinger $\psi(q)\ \psi^*(q)$. Dirac has generally succeeded in solving the above problem mathematically as well. As the probability function there appears a matrix S, of which the indices in the general problem (f, p, q) are $S_{f,q}$ (in the special case E, t, W^a, then as for me, $S_{W^a, E}$). The $S_{f,q}$ are, as always, solutions of a principal axes transformation of a Hermitian matrix. The principal axes transformation can be reduced to a differential equation à la Schrödinger. One therefore has the means to actually determine S quite generally (this last point corresponds very closely with your idea of p-waves). In the special case E, t, q, the matrix has indices q and E, so $S_{q,E}$ is the Schrödinger function, $S_{q,E} = \psi_E(q)$. The Schrödinger function is thus identical to the matrix S in Born's principal axes transformation for the determination of the eigenvalue W. But the above-named physical interpretation of S contains at the same time all the physical statements that may actually be made at present: e.g. Born's collision processes, Jordan's canonical transformations, etc. Dirac's work certainly has many points of contact with yours and mine. But it is nevertheless very general, as it functions, for example, just as easily and surely in continuous variations as in periodic ones. I hold Dirac's work to be an extraordinary advance.

Dirac's theory was indeed a great advance, and it did indeed have much in common with the ideas of Heisenberg and Pauli, from whose combined considerations it seems to have arisen. Heisenberg and Pauli had long been asking just what measurements the new quantum mechanics could predict, and in their recent correspondence it had become clear that, as Dirac noted explicitly, a joint specification of p and q, or of E and t, was impossible. They had concluded that what could be predicted were probabilistic results, and they had specified these results for a variety of individual cases: Pauli had specified the probability distributions of functions in a given energy state, Heisenberg had specified the probability distribution of energies resulting from a given perturbation, and both had linked the probability functions with transition matrices. Dirac had always shared their interest in the observability criterion and the associated applicability of the theory, and all he had to do now was to apply his much more abstract mathematical view of the theory and generalise. The question he asked was how, "when one has performed all the calculations with the q-numbers and obtained all the matrices one wants, ... one is to get physical results from the theory, i.e. how can one obtain c-numbers from the theory that one can compare with experimental values."[12] Generalising the viewpoint of Pauli and Heisenberg by treating p, q and E, t simply as pairs of canonical coordinates, he deduced that the only questions that could be posed in quantum mechanics were of the form:

Given ξ_r, what do we know about any variable g as a function of η_r, the canonical conjugate of ξ_r?

To answer these questions one had to transform from one scheme of matrices (e.g. that in which the ξ_r were determined, i.e. diagonal) to another scheme (in which the required function was diagonal). From consideration of Lanczos's theory Dirac had developed for himself a continuous matrix formulation in which to work, and using this he developed the theory of transformations required, deducing the existing formulations as special cases. Following the recipe of Born, Pauli and Heisenberg, he identified the probability distribution of g as a function of η for given ξ with the diagonal sum of the matrix transforming from a scheme in which ξ were diagonal to one in which g were diagonal.

A fortnight after Dirac's paper was completed, a similar treatment was reached independently by Jordan, who had also been in communication with Pauli, and who had also, like his colleagues, been searching for a fusion of wave mechanics and matrix mechanics.[13] To base a generalised quantum mechanics upon transformation theory was natural to Jordan, who had been primarily responsible for the development of that theory in matrix mechanics. Working under Born and generally favourable to a continuous formulation of quantum theory – he had devoted the larger part of his research to date on deriving the light-quantum phenomena from oscillator theories[14] – he naturally approached the problem, however, from the wave mechanical formulation rather than from the matrix mechanics. He thus based his treatment on Pauli's form of the probabilistic interpretation rather than Heisenberg's and on the transformation from one set of variables to another (as discussed in Pauli's letter) rather than from one scheme to another representing the same variables. He set himself the problem:[15]

[If] in place of p, q, new variables, P, Q, may be introduced by a canonical transformation such that $H(p, q) = \bar{H}(P, Q)$ . . . we wish to construct the new wave equation with $\bar{H}$.

If the old equation were $\left\{H\left(\frac{\epsilon\partial}{\partial y}, y\right) - W\right\}\phi(y) = 0$, and the new $\left\{\bar{H}\left(\frac{\epsilon\partial}{\partial x}, x\right) - W\right\}\psi(x) = 0$, Jordan asked how the new function ψ was related to the original function ϕ, and he based his answer of Pauli's suggestion that

If $\phi_n(q)$ is normalised, then $|\phi_n(q)|^2\,dq$ is the probability that when the system finds itself in the state n, the coordinate q takes a value in the interval $(q, q + dq)$. If q, β, are two Hermitian quantum-mechanical quantities, which we shall here take for convenience as both constantly varying, then there will always exist a function $\phi(q_0, \beta_0)$, such that $|\phi(q_0, \beta_0)|^2\,dq$ gives the (relative) probability that for a given value β_0 the quantity q will take a value in the interval $(q_0, q_0 + dq)$. The function $\phi(q, \beta)$ of Pauli denotes the probability amplitude.

From this suggestion, Jordan deduced that two postulates should be expected: the function $\phi(q, \beta)$ should be independent of the mechanical nature of the system and dependent only on the kinematic relations between q and β; and the functions should combine as $\Phi(\psi_0, \beta_0) = \int \psi(Q_0, q)\,\phi(q, \beta_0)\,dq$. He noted that it was the probability amplitude ϕ (and not the probability, $|\phi|^2$) that followed the usual combination law for probabilities, and he related this to the interference of the probability waves.

Guided by the above considerations, Jordan postulated that for any two quantum-mechanical quantities, q, β, standing in a determined kinematic relationship with each other, there should exist a probability amplitude $\phi(x, y)$ as defined by Pauli, together with a complementary amplitude $\psi(x, y)$, such that $\overline{\phi}(x, y) = \psi^*(y, x)$, *et cetera*. He postulated further that the amplitudes should combine interferingly, in a way that led to the relationship above. He then defined two variables as canonically conjugate if their probability amplitudes were given by $\rho(x, y) = e^{-xy/\epsilon}$, from which he deduced that for a given value of one variable, all values of the conjugate variable were equally probable. He postulated that for any given variable a canonical conjugate variable should always exist, and proceeded to develop the theory of operators transforming from one set of canonical coordinates to another. The Schrödinger and momentum space wave equations followed as special cases, as did the other formulations so far developed.

THE FORMULATION OF QUANTUM MECHANICS

The theories of Dirac and Jordan were developed independently, but they shared the same conceptual roots and were mathematically equivalent. On 24 December, Dirac wrote to Jordan, giving a detailed account of his own theory, and apologising for once again having duplicated Jordan's results. On the relationship between the two theories he wrote that [16]

Dr Heisenberg has shown me the work you sent him, and as far as I can see it is equivalent to my own work in all essential points. The way of obtaining the results may be rather different though .... In your work I believe you consider transformations from one set of dynamical variables to another, instead of a transformation from one scheme of matrices representing the dynamical variables to another scheme representing the same dynamical variables, which is the point of view adopted throughout my paper. The mathematics would appear to be the same in the two cases however.

Despite their intimate connection, the two theories did in fact differ in some respects, and in particular in respect of the relationship proposed between formalism and interpretation, an aspect of quantum mechanics of which there was as yet no clear understanding. Born had claimed that his interpretation was a necessary consequence and integral part of the wave-mechanical formalism, but this attitude does not appear to have been generally shared, and was clearly rejected by both Heisenberg and Pauli as glossing over the fundamental problems involved. But neither of these physicists was clear as to what to substitute for it. Pauli was committed to a philosophy according to which all terms in the theory should be operationally defined, but the radical conceptual revision that this would entail was clearly beyond him, and in his letter of 28 October 1926 he concentrated on the lower order requirement that all the terms be observationally (but if necessary statistically) interpreted. Heisenberg too had concentrated on the limits of applicability of existing concepts, which he had suggested might be fundamentally statistical, and he too had thus committed himself, though he would clearly have liked to see the theory physically founded, to working *from* the formalism *toward* the physical interpretation. Dirac seems to have viewed the foundations of the theory as essentially mathematical; but he took great care to kept formalism and interpretation distinct, and emphasised that the probabilistic interpretation did not follow from the formalism, as Born had suggested, but must rather follow from a separate association of theoretical and physical terms that included probabilistic assumptions. In his letter to Jordan he stressed that Heisenberg's form of the probability interpretation followed "if all points in η-space are equally probable (*and only when* this is so)";[17] and in his published paper he again emphasised that, as Pauli had in fact required,[18]

The notion of probabilities does not enter into the ultimate description of mechanical processes; only when one is given some information that involves a probability (e.g. that all points in η-space are equally probable for representing a system) can one deduce results that involve probabilities.

Jordan, on the other hand, incorporated his statistical postulates into the foundations of his theory, which was built up explicitly of functions satisfying these postulates. At first sight this had the great advantage that as in classical theory a physical interpretation of the symbols preceded their mathematical theory, but in fact the physics and mathematics were hopelessly confused. The statistical postulates were first required in respect of general mathematical functions, the physical interpretation of which was only deduced later.

The problem of the relationship between mathematics and physics is of course a philosophical one, and the issue could never be settled to everyone's liking. Through the earlier debate on the relative importance of mathematics and physics in matrix mechanics, and through a continuing struggle between the operationally conditioned corpuscular approach of Pauli and Heisenberg and the field-theoretical preferences of Jordan and Bohr,[19] the old disagreement of 1918–1919 between Weyl, Pauli and Einstein on the construction of physical theories had remained very much alive and unresolved.[20] But a generally acceptable and workable formulation of quantum mechanics was badly needed, and since a full understanding of the observability criterion and new kinematics was still wanting this had necessarily to take some sort of axiomatic form. In the Spring of 1927 Hilbert, the doyen of the axiomatic method in physics, recovered sufficiently from a severe illness of the previous year to take up once again his early interest in the new quantum mechanics. And with his assistants, Nordheim and von Neumann, he provided a clear axiomatic formulation of transformation theory.[21] In the interests of clarity and in order to allow the maximum freedom for the development of the formalism, they followed Dirac in keeping formalism and interpretation quite distinct. But developing their theory primarily from Jordan's (Jordan was of course in Göttingen) they based the interpretation upon a set of physical axioms akin to Jordan's statistical postulates. For any two mechanical quantities, $F_1(p, q)$, $F_2(p, q)$, it was required that there should exist a function $\phi(x, y; F_1, F_2)$ such that $\phi\phi^* = \omega(x, y; F_1, F_2)$ was the relative probability that, given $F_2 = y$, F_1 was in the range $(x, x + dx)$. The probabilities were required to be independent of the mechanical system and coordinates, to be reflexive, $\phi(x, y; x, y) = \phi(y, x; y, x)$, and to combine as $\phi(x, z; F_1, F_3) = \int \phi(x, y; F_1, F_2)\, \phi\, (y, z; F_2, F_3)\, dy$. To obtain a quantum mechanics, it was postulated that the physically defined $\phi(x, y; q, F)$ should be associated with the kernel of the canonical transformation: $q \to F(q)$.

A second difference between the theories of Dirac and Jordan, and one that was not alleviated by the Hilbert-von Neumann-Nordheim exposition, was in respect of the relationship between discrete and continuous formulations. Jordan's theory, and in particular his statistical postulates, were expressed in the continuous wave-mechanical formulation. Dirac's theory, on the other hand, was based on the discrete formulation of matrix mechanics. At first sight this difference appeared to be relatively unimportant, for Jordan was able to derive the matrix mechanics, and Dirac the wave mechanics. But Dirac's derivation rested upon the use of the delta function

$$\delta(x - y) = 0, \qquad x \neq y; \qquad \int_{y-\alpha}^{y+\alpha} \delta(x - y)\, \mathrm{d}x = 1,$$

which, as he well knew, was not a true mathematical function at all.[22] Jordan, and Hilbert, von Neumann and Nordheim, did not use the delta function as such. But their derivations depended upon the existence of transformations which could only in fact be completed using the delta function.[23] Jordan did not explore this problem. Dirac recognised it but appears to have thought it insignificant given what he saw as a natural equivalence between continuous and discrete treatments.[24] Von Neumann, however, combined an insistence on mathematical rigour with a liking for mathematical generalisation, and in 1927–28 he quickly achieved both.[25]

Rather than relating the continuous and discrete functions themselves von Neumann considered the spaces on which these functions were defined. He was then able to draw on a famous theorem by Fischer and Riesz, to the effect that the space of all Lebesgue integrable complex valued *functions* $f(q)$, with finite norm $N = \int |f|^2 \, \mathrm{d}q$, was isomorphic to the space of all complex valued *sequences* $\{q_i\}$, with finite $\sum_i |q_i|^2$.[26] Defining a more general Hilbert space as any space isomorphic to these he then reformulated the transformation theory in this more general space, and so overcame the delta function problem.

At the same time, von Neumann also generalised Dirac's probability interpretation to its definitive form.[27] Given that the system was in a state $\psi(q)$ he defined the probability that the measurement of a quantity corresponding to the operator A should give a value between a and b as the norm of the projection corresponding to the spectral resolution of A. If

$$A = \int_{-\infty}^{\infty} \lambda \, \mathrm{d}P(\lambda), \qquad 1 = \int_{-\infty}^{\infty} \mathrm{d}P(\lambda),$$

then the projection was defined as

$$P_{[a,\, b)} = P(\mathrm{b}) - P(a),$$

and the required probability was

$$\int |P_{[a,\, b)}\, \psi(q)|^2 \, \mathrm{d}q.$$

From this von Neumann could derive the earlier expressions of the probability interpretation by appropriate choices of the space over which the functions were defined. Making further use of the spectral resolutions of operators he was also able to generalise the interpretation to give the probabilities that, given that the values of quantities corresponding to operators B_i lay in the ranges J_i, those corresponding to the operators A_i should lie in the ranges I_i.[28]

CHAPTER 9

THE UNCERTAINTY PRINCIPLE AND THE COPENHAGEN INTERPRETATION

THE UNCERTAINTY PRINCIPLE

The problem of founding quantum mechanics upon a new conceptual framework, replacing that of classical kinematics, was never fully solved. But concurrent with the development of transformation theory, and working from the same set of conceptual insights, Heisenberg was able to make some progress toward understanding the limits of applicability of the existing kinematical concepts in the new theory. Much of Pauli's October letter had been devoted to this latter problem, and in his reply of 28 October Heisenberg took up some of Pauli's leads:[1]

But the most interesting of your observations is naturally the so-called 'dim point'. I should like to believe that your p-waves have just as great a physical reality as the q-waves; only naturally not so great a practical significance. But I am very sympathetic to the equivalence on principle of p and q. The equation $pq - qp = hi$ thus corresponds always in the wave presentation to the fact that it is impossible to speak of a monochromatic wave at a fixed point in time (or in a very short time interval). But if one makes the line less sharp, the time interval less short, then that very truly does have a meaning. [Pauli noted here that in the contrary case, given a short time interval, it was meaningless to speak of a precise energy value.] Analogously, it is meaningless to talk of the position of a particle of fixed velocity. But if one accepts a less accurate position and velocity, that does indeed have a meaning. So one understands very well that it is macroscopically meaningful to speak more approximately of the position and velocity of a body.

Confirming that his analysis so far drew on Pauli's ideas as well as his own, Heisenberg added in parentheses at this point that "all this is throughout naturally nothing new to you." Both Pauli and Heisenberg had recognised the impossibility of a joint specification of a pair of canonical coordinates, and in particular of p and q or E and t, and they had linked this with the commutation relations for such coordinates. From this letter of Heisenberg's it is clear that they had also recognised that the degree to which one could specify one coordinate of a pair was inverse to the degree of specification of the other coordinate, and that by specifying neither of them too precisely one could get an approximate, macroscopic, joint determination. But what did this mean? The previous year Heisenberg had speculated on the possibility of "coarse" determinations of space and time, and he now suggested[2]

that space and time are actually only statistical concepts, as, perhaps, are temperature, pressure, etc. in a gas. I mean, that space-like and time-like concepts are meaningless for *one* particle, and that they become more and more meaningful the more particles are treated.

But this could only be the first step to a new kinematic understanding, and Heisenberg admitted that although Pauli's comments had raised his hopes somewhat he had not, despite frequent attempts, got any further in this direction. He had also got no further than Pauli on the kinematical significance of the off-diagonal matrix elements.[3]

The next significant step in Heisenberg's thought appears to have arisen in response to some considerations of Pauli's, now lost, on ferromagnetism.[4] Writing to Pauli on 15 November, Heisenberg approved the general tendency of his work, but suggested that the application of the ideas was "incautious":[5]

The general division of phase space into cells the volume of any of which is the quantity h is certainly a correct principle. But, and now comes my objection, if you specify the cell walls sharply, and can then determine how many particles are in each cell, can you not then, through the choice of neighbouring cell walls, find the number of atoms in as small a cell as you like? . . . I mean, is the choice of *fixed* cell walls physically meaningful?

He continued, illustrating his argument with a diagram (see Figure 1), to suggest that

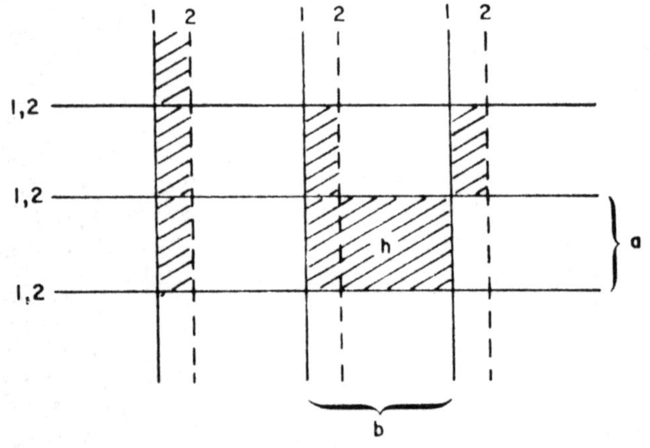

Fig. 1.

Perhaps it is so that one can only specify, for example, the ratio of the two cell walls, a/b, but not the position of a fixed cell wall.

One could in other words specify the shape of the cells, but not their precise location, for the latter would allow the consideration of overlapping cells of arbitrarily small volume, and would thus be equivalent to a precise joint specification of p and q. Heisenberg continued:

The same objection also holds now for the E, t distribution. But for the special case of fixed t, everything is again in order. I hold it to be very reasonable to study even this special case more closely, out of which will perhaps come something of the kinematic meaning of the matrices.

In conclusion, I am also of your opinion that at the end of the dim point will be a very clear point. I mean: if once space-time is already somehow discontinuous, it is then very satisfactory that it makes no sense to speak, for example, of the velocity $\dot{x}$ at a fixed point x. For to define velocity one always needs at least two points, which can lie in a discontinuous relationship but *not* infinitely close.

Heisenberg's "special case" is not in fact special in the way suggested, but Pauli, who had noted as much upon Heisenberg's previous letter, would no doubt have put him right on that.[6] What is important is that by taking the problem of the limitations of the classical kinematical concepts back into the foundations of the statistical theory of gases, on which Pauli was now working, Heisenberg had reached a new picture of this problem in which the degree of possible joint specification of p and q was precisely defined in terms of the cells in phase space. For an atom in a given cell, the joint specification of p and q was subject to the restriction $p_1 q_1 \sim h$, where p_1, q_1, corresponded to the "gauge" of the measurements, or to the walls of the cell.

A second feature of Heisenberg's letter, and one that was to be of great importance for the further development of his ideas was his association of the problem of specifying p given q with the problem of measuring or defining p in a discrete quantum theory. In his last paper, and in his current discussions with Bohr, he had emphasised repeatedly the essential discreteness of quantum theory, and he now saw that if this discreteness operated on the space-time metric itself, then this would lead directly to the impossibility of a joint position-momentum specification.[7] To define a momentum or velocity required two position measurements, which could classically be taken infinitely close to each other; if in quantum theory they could not, then a precise velocity measurement would leave an uncertainty as to where between the two positions it applied. In the letter to Pauli reporting Dirac's work, a week later, Heisenberg followed up this insight with a related but more general reflection:[8]

I often reflect on the true meaning of this whole formal connection, but it is horribly difficult to be clear about it. That the world might be continuous I am more than ever convinced is completely out of the question. But so long as it is discontinuous then all the words that we apply to the description of an event have too many *c*-numbers. One no longer knows what the words 'wave' or 'particle' mean.

As we shall see, this viewpoint was also to be important, both for Heisenberg's ideas and also for Bohr's.[9]

At about the time he wrote this last letter, Heisenberg was also writing a non-technical paper on quantum mechanics for the scientific journal *Die Naturwissenschaften*, and in this he again stressed the discreteness of the quantum-mechanical world and the need to free quantum mechanics from the constraint of visualisability (*Anschaulichkeit*), or visual pictures (*anschauliche Bilder*), if contradictions were to be avoided. In late November, however, Heisenberg's thoughts were diverted by the development of the Dirac-Jordan transformation theory, and he does not seem to have returned to the kinematic problem until February 1927 when, he recalled, Bohr's absence on a holiday gave him the opportunity to sort out his ideas in peace and quiet.[10] He may also have been stimulated in response to a new paper on the interpretation of quantum mechanics by Jordan.[11] On 5 February he wrote to Pauli that he was "again occupied all the time with the logical foundations of the whole $pq - qp$ swindle",[12] and criticised Jordan for talking of such things as "the probability of an electron being at a determined point" (as in the classically-oriented interpretation of Born), when "the concept 'path of an electron' is not properly defined." Heisenberg had already established the previous Autumn a rough theoretical uncertainty in terms of the phase space cells, and with Pauli and Dirac he had always viewed the impossibility of a joint theoretical specification of p and q as corresponding to an operational problem of their joint observability. He had also begun to relate the latter problem to a possible discreteness of space-time. But as yet, despite all the indications of a perfect correspondence between theory and observation, he had not investigated closely the problem of the measurement of the kinematic properties of an electron, and had not made the connection between theory and measurement a precise one. These he now did, and writing to Pauli on 27 February he asked:[13]

What does one understand by the words 'position of the electron'? This question can be replaced according to the well-known model by the other: 'How does one *determine* the position of the electron'? One takes perhaps a microscope with sufficiently good resolving power and looks at the electron. The accuracy depends on the wavelength of the light. For a sufficiently short wave-length of light the position of the electron at a fixed time (and if necessary its size) can be ascertained to any accuracy; that could equally be obtained through the impacts of very fast particles with the electron. It thus has *this* meaning also, if we designate the electron as a particle. According to experience we completely disturb the electron's mechanical behaviour through such an observation of its position, through the Compton effect or collision effect respectively. At the moment when the position is 'q', the momentum is wholly undertermined:

$$pq - qp = h/2\pi i.$$

. . . Analogous considerations may be applied to the velocity of the electron. The following experiment might perhaps be applied to the definition of the words 'velocity of the electron': at a known time one suddenly makes all forces on the electron zero, then the electron runs on linearly, and one deduces the velocity perhaps from the Doppler effect of the reddest possible light. The accuracy would be the greater the redder the light concerned, but then the electron must run correspondingly *longer* without external forces. Then one switches on the forces again. The [position] accuracy depends on the distance for which the electron remains without forces:

$$pq - qp = h/2\pi i.$$

. . . Such considerations may be repeated in the same way for all canonical coordinates. One will always find that all thought experiments have this property: when a quantity p is pinned down to within an accuracy characterised by the average error p_1, then the canonical coordinate q can only be given at the same time to within an accuracy characterised by the average error

$$q_1 \approx h/2\pi p_1.$$

Heisenberg's derivation of observational uncertainty could not be called rigorous. Even when, writing it up in his paper on the subject,[14] he filled out the discussion of the position measurement to include explicitly the momentum uncertainty resulting from the Compton effect, Bohr could still object that the finite aperture of the microscope had been ignored.[15] And even when this and other points had been taken into account the derivation remained philosophically unsatisfactory: on the one hand it was operationally based, but on the other hand it was phrased entirely in terms of concepts that were operationally undefined. As a demonstration of the limited applicability of the classical concepts in practice, and as an order of magnitude estimation of the resulting uncertainty, it was however convincing and effectively valid. Moreover, it was accompanied by a derivation of the quantum-theoretical uncertainty with which it was in complete agreement. Working from the Dirac-Jordan theory, Heisenberg interpreted the assumption that q should be determined to an accuracy q_1 as q' through the requirement that the probability amplitude of a function η, $|S(\eta, q)|^2$, fell at $q = q' \pm q_1$ to e^{-1} of its value at $q = q'$. Working out the probability $|S(\eta, p)|^2$ he found that the corresponding points $p' \pm p_1$ were then given by $p_1 q_1 = h/2\pi$, in agreement with the thought experiments.

In his derivation of the uncertainty relations, which he extended in his paper to the observation of energy and time, Heisenberg showed that in terms of the usual kinematics the essential quantum discontinuity imposed a restriction upon our observations, which restriction was accurately reflected by quantum mechanics. But while this specified the limits within which the classical concepts could be consistently applied, it did nothing to define the relationship between theory, observation and concepts. In another part of his letter, however, and again in his paper, Heisenberg did approach this more fundamental problem.[16] Arguing from his thought experiments on the measurement of position, velocity, etc., he asserted that these concepts *were* operationally well defined, and therefore perfectly valid.[17] The problem lay in combined conceptions, such as those he had earlier characterised as having "too many *c*-numbers": "particle", equivalent to particle path, a series of joint position-momentum specifications, or "wave". In the letter to Pauli he wrote that[18]

> It is therefore meaningless to talk, for example, of the $1-S$ 'orbit' of the electron in the Hydrogen atom.For if we wish to actually determine the position of the electron essentially *more accurately* than to within 10^{-8} cm, the atom will already be destroyed by a single observation. The word $1-S$ 'orbit' is thus, as it were, already purely experimentally, i.e. without knowledge of the theory, meaningless. On the other hand the imagined position determination can be repeated in many $1-S$ Hydrogen atoms. So there must be an exactly determined probability function (the well-known $\psi_{1S}(q)\ \psi^*_{1S}(q)$) for the given energy $1-S$. The probability function corresponds to the classical 'orbit' over all *phases*. One can say, with Jordan, that the laws of Nature are statistical. But one can say with Dirac, and this seems to me essentially more profound, that all statistics are first introduced through our experiments. The fact that we do not know in which position the electron will be at the moment of our experiment arises only, so to speak, from the fact that we do not know the phase in advance if we know the energy: ($J\omega - \omega J = h/2\pi i$) and in the classical theory it was in *this* respect no different. That we *cannot* know the phase, without once again destroying the atom, is characteristic of quantum mechanics.

One could not define or measure precisely the phase of an electron orbit of given energy, and similarly, Heisenberg noted in his paper, one could not specify the continuous path (joint position and velocity) of an effectively free electron.[19] Born's interpretation of the spreading wave packet in terms of possible continuous paths of particles was operationally unsound. But one could use a series of measurements to achieve an approximate specification of the electron orbit in an atom, and one could similarly measure and define an approximate, discontinuous, path of an electron as a time sequence of distinct position observations. Whether Dirac had anything like this in mind when asserting that statistics were introduced through the experiments is extremely doubtful – he tended as we shall see to avoid all questions of interpretation, and was probably only concerned, as in his transformation theory, to keep these questions and their statistical overtones out of the mathematical formalism. But turning this assertion to his own use, Heisenberg now claimed, on the basis of his analysis, that "the electron path comes into existence only when we observe it."[20]

Heisenberg also disagreed with the approach of Born and Jordan on the related problem of causality. Born had declared his belief in the ultimate rejection of causality.[21] Jordan had argued that the existing quantum laws could only be interpreted statistically (a fact that he associated with the inclusion of the imaginary constant in the fundamental equations), refusing to commit himself as to whether an underlying causal description might or might not exist.[22] But both had argued, and were arguing at this time,[23] entirely within the assumption of the classical kinematical conceptions. To Heisenberg, Dirac and Pauli these conceptions themselves were invalid, or at least restricted in their validity, and this meant that the classical concept of causality was effectively undefined. In a discussion strongly reminiscent of the earlier analysis of Senftleben,[24] Heisenberg now insisted that "we *cannot* know, as a matter of principle, the present in all its details" and argued that "since physics has to confine itself to the formal description of the relations between perceptions", the causality concept was simply irrelevant.[25]

In his letter to Pauli, Heisenberg explained that he was still very unclear on many points, and that he was writing to clarify his own thoughts, and he asked for Pauli's "relentless criticism".[26] A fortnight later, having written up the first draft of his paper, he repeated the request.[27] But Pauli apparently had little criticism to offer, and the final version of the paper, submitted on 23 March, differed little in essential content from the letter of 27 February. But bringing all his earlier ideas together, Heisenberg did introduce as a foundation for his argument the essential discreteness of quantum mechanics, and the implications for the measurement and definition of velocity of a discrete space-time.[28] He also compared, in the introduction of this paper, the uncertainty relations and their place in quantum mechanics with Einstein's principle of the constant velocity of light and its place in special relativity theory.[29] And in later recollections he associated the origins of the uncertainty principle with Einstein's observation that, to quote Heisenberg, "it is the theory that decides what we can observe."[30] But the close relationship between theory and observation was hardly new to Heisenberg, and the model of special relativity seems to have been introduced after Heisenberg had completed his paper rather than before. There is nothing of it in the February letters to Pauli, and having been introduced to the first draft of the paper it was subsequently omitted. On 9 March Heisenberg wrote to Pauli on the subject:[31]

I do not believe at all that one can somehow make the quantitative laws plausible through the equations $p_1 q_1 \sim h/2\pi i$. But that is no different in quantum mechanics from anywhere else. For example in relativity theory, the principle of constant light velocity is also unfounded. Why should not the light velocity depend ultimately on the masses at infinity? The statement of *constant* light velocity is only the simplest if one accepts Einstein's definition of simultaneity. So I believe also: if one once knows that p and q are not simple numbers, but that $p_1 q_1 \sim h/2\pi i$, then the statement that p and q are matrices is the simplest conceivable. I am, naturally, well aware that this formulation might appear unsatisfactory, but is it not the same arbitrariness as is met in all physical theories? I have written something on this in the conclusion. But this conclusion is generally still very dubious, and I can imagine that I might still change it completely or leave it out completely.

COMPLEMENTARITY

In the past, Heisenberg had often agreed with Pauli on the need for firm physical foundations of a theory, but had often been carried away by the success of his own unfounded speculations. Since the introduction of his new kinematics he had held firmly by Pauli's belief that quantum mechanics must ultimately receive a clear physical foundation based on the operational definition of all the terms involved. But the magnitude of this task had forced both physicists to concentrate on the limits of the old concepts rather than on new foundations, and successful in his investigation of these limits Heisenberg was once again carried away. The title of his paper may be translated as "On the perceptual (*anschaulichen*) content of quantum-theoretical kinematics and mechanics", and in it he claimed that with the addition of the uncertainty principle quantum mechanics became in some sense visually, perceptually or intuitively – his use of the word *anschaulich* was now shifting slightly – consistent. He claimed in other words that the fundamental problem of establishing a new conceptual foundation appropriate to the theory was now more or less solved. In fact, of course, Heisenberg's interpretation did not constitute a proper foundation and he, and the presumably critical Pauli, appear from the last letter cited to have realised this. Although the fundamental quantum discontinuity was used in the paper as a foundation of the uncertainty principle there was no logical connection between the two, and the latter was more of an excuse than an explanation. But Heisenberg was content with what he had got. The defensive tone of Heisenberg's letter suggests that Pauli was not so happy, but he had no foundation to offer himself, and was besides deeply involved with technical aspects of the theory. The task of putting Heisenberg's ideas onto some sort of a physical basis thus fell on Bohr, his severest critic of the moment, who had yet to publish anything on the new quantum mechanics.

Of all those actively involved in the search for a new quantum mechanics in the 1920s, Bohr was at once the most radical and the most conservative. He had been initially responsible for the idea that classical mechanical and kinematical concepts were incapable of describing quantum phenomena, and he had continued to believe this throughout. But he had also held fast to the belief that these concepts, and especially those of the classical wave theory of light, could not be replaced.[32] While most of his colleagues kept an open mind on the issue of wave-particle duality, this belief had led Bohr to take a firm stance in support of the classical wave theory and against the light-quantum concept and, in late 1923, against Heisenberg and Born's programme of a discrete physics.[33] It had also led him, in early 1924, to the virtual oscillator model of the interaction between matter and radiation, and while Born and Heisenberg accepted this merely as a heuristic device, and Pauli rejected it outright, Bohr appears to have held to it as to a genuine physical theory.[34] Early in 1925, however, an accumulation of new experimental evidence had forced him to reconsider. First the results of Einstein on gas statistics, Ramsauer on barrier penetration by slow electrons, and Davisson and Kunsman on electron scattering appeared to extend the wave-particle duality to material collisions, for which energy-momentum conservation could not be rejected as it had been in the virtual oscillator theory.[35] Einstein's study of the statistical mechanics of a monatomic ideal gas showed that such a gas had to be treated as having elements of both wave and particle behaviour, just as had black-body radiation. In Ramsauer's experiments a slow electron incident upon an atom was shown to have a definite probability of passing right through the atom without any change in its motion. This, the first example of a quantum barrier penetration phenomenon, suggested that the electron might partake of wave as well as particle properties. Finally the results of Davisson and Kunsman on the scattering of slow electrons through crystals showed a periodic variation of scattering intensity with scattering angle, interpreted in Born's Göttingen department in the Spring of 1925 as electron diffraction, again a wave phenomenon of matter. Then, on top of all this, the results of Geiger and Bothe showed that, as most physicists by then expected, the rejection of energy-momentum conservation was non-viable even for radiative processes.[36]

Bohr's conviction was still sufficient for him to speak out against energy conservation in a talk delivered on 20 February 1925,[37] but on 18 April he wrote to Heisenberg that he was prepared for all eventualities, "even for an acceptance of a coupling process in distant atoms. The costs of this acceptance are of course so great", he wrote, "that they cannot be measured in the usual space-time description."[38] The results of Geiger and Bothe were in fact already in the post, and three days later Bohr replied to Geiger that[39]

I was completely prepared that our proposed point of view on the independence of the quantum processes in separated atoms should turn out to be incorrect . . . . Not only were the objections of Einstein [to the implications of the virtual oscillator theory] very unsettling; but recently I have also felt that an explanation of collision phenomena, especially of Ramsauer's results on the penetration of slow electrons through atoms, presents difficulties for our usual space-time description of a kind similar to those presented by a simultaneous understanding of interference phenomena and a coupling of the change of state in separated atoms through radiation. In general I believe that these difficulties so far exclude the maintenance of the ordinary space-time description of phenomena that in spite of the existence of coupling conclusions concerning an eventual corpuscular nature of radiation lack a satisfactory basis.

In this letter Bohr appeared calm, but he seems to have been struggling hard to adapt to the new situation and to comprehend and perhaps accept something of the more radical ideas of Pauli. He had written to Heisenberg that "particularly stimulated by discussions with Pauli, I am these days working slavishly to the best of my power to accustom myself to the mysteries of nature and to attempt to prepare myself for all eventualities."[40] And on the same day that he wrote to Geiger, he also wrote to Franck:[41]

I have long been intending to write to you again, for the uncertainty about the correctness of my reflections on collision processes, to which I gave expression in the postscript to my last letter, has since been more and more strengthened. In particular there are the Ramsauer results on the penetration of slow electrons through atoms, which do not appear to tie in with the accepted point of view. Indeed these results must offer difficulties for our usual space-time description of nature of a similar kind to the coupling of transitions in distant atoms through radiation. But then there are no longer any grounds for doubting such a coupling, or for doubting conservation laws in general . . . . Also the thermodynamic considerations of Einstein were very disturbing indeed. . . . Moreover, I've just now heard from Geiger that his experiments have decided in favour of coupling and there is really nothing to do but take our attempted revolution as painlessly as possible into oblivion. Nevertheless, we cannot forget our goals that easily, and in the last few days I've been tormented by all sorts of wild speculations, in order to find an adequate foundation for the description of radiation phenomena. On this I have talked a lot with Pauli, who is still here. . . .

On 1 May, in reply to Born's advocacy of the theory of matter waves, which he wholly rejected, Bohr again indicated the direction of his thoughts:[42]

Quite apart from the question of the correctness of such objections against your theory, I should like to stress that I am of the opinion that the assumption of coupling between stationary states in different atoms through radiation excludes any possibility of describing the physical situation by visual pictures. By my utterances about coupling in my letter to Franck I meant only that I had already come to suspect that for collision processes such models had shown a more inferior application than usual. In essence, this is indeed purely negative, but I feel, especially if the coupling should really be a fact, that one must then take refuge to an even higher degree than before in symbolic analogies.

Finally, in a postscript written in July to a paper based on the virtual oscillator theory, Bohr noted that the coupling issue did not involve a decision between the wave and particle pictures but[43]

rather the problem as to how far the space-time pictures in terms of which physical phenomena have so far been described can be applied to atomic processes. . . . In view of the recent results one should not be surprised if the required extension of classical electrodynamics should lead to a far-reaching revolution of the concepts on which the description of nature has so far been based.

Bohr could have made the last point ten years earlier. The rejection of visual pictures and restriction to symbolic analogies suggested in his letter to Born were nothing new, but just brought him back into line with Pauli and (by this time) Heisenberg, and indeed with his own earlier views. The difference was that whereas before his pessimistic statements as to the impossibility of a space-time description of quantum phenomena had always gone hand in hand with a more optimistic approach, according to which a wave-theoretical approach might be saved at the expense of the laws of conservation and causality, now the pessimism dominated. He could not forget his hopes, and must still have retained something of a wave-theoretical bias, but he could think of no new escape and as the new quantum mechanics developed in 1925–26 he remained silent and apart. According to Heisenberg's recollection it was only after Schrödinger's Copenhagen lecture that, in the provocative company of Heisenberg and Dirac, Bohr was once again stirred into action, and then it was apparently in opposition to the approach being adopted by Heisenberg and Pauli.

We have little evidence of Bohr's arguments during the Autumn of 1926, only Heisenberg's recollection that while he himself took part optimistically in the development and elucidation of quantum mechanics Bohr remained pessimistic, and that the relationship between them grew very tense.[44] But Bohr seems to have had two main objections to Heisenberg's work. First, he seems to have criticised it, as Heisenberg had earlier criticised Born's, for being too concerned with the formalism. Unable to develop a new conceptual foundation for the theory both Heisenberg and Pauli were forced to concentrate on the limitations of the old conceptions, and to approach this entirely – at least until Heisenberg sorted out his ideas in February – from the study of the formalism. Bohr, though unable to do any better himself, insisted that the limitations of the classical concepts should be founded directly on physical and philosophical principles, and the formalism then derived from the physics. Heisenberg would not have disagreed with this on principle, but as often before he preferred to go for results first. Bohr's second objection seems to have been to Heisenberg's emphasis upon the particle picture at the expense of the wave picture, and this was again related to earlier disagreements. Heisenberg had always emphasised the discreteness of quantum theory,[45] and now that he was won over to Pauli's views he presumably shared Pauli's operational preference for a particle theory and objection to a pure field theory.[46] Moreover, by concentrating on the problem of the observationally possible precision, or degree of localisation, in terms of the classical concepts he naturally adopted particle terminology. Since an observation always corresponds to an apparently localised event (observations of wave properties being second order deductions), any discussion of the measurement problem must emphasise the particle as reality, and even when he discussed the theoretical problem of the orbit concept in terms of spreading wave packets, in his uncertainty principle paper, Heisenberg interpreted all observations, as localisations or reductions of the wave packet, in terms of the particle conceptions. Operationally this view-point was perfectly reasonable, but it clearly did not do sufficient justice to the wave concept for Bohr's liking. Bohr had always preferred a field theoretical approach, and even now he appears to have supported his student Klein in his advocacy of an interpretation of quantum mechanics based upon Kaluza's analogue to Weyl's unified field theory – the theory in opposition to which Pauli had first developed his operationalist preference for the particle theories.[47] At the very least he seems to have insisted upon the symmetry of the wave-particle duality being maintained.[48]

In February, Bohr left Copenhagen for a holiday, and when he returned Heisenberg had formulated the considerations of his uncertainty principle paper.[49] Here the priority of the particle picture was maintained, and although his analysis of observations was lacking in rigour and divorced from any general physical principles (the vaguely supposed derivation from the quantum discreteness was a perfect example of what Pauli elsewhere called a "specifically Göttingen bad habit", into which Heisenberg relapsed when released from Bohr's supervision),[50] Heisenberg seemed perfectly happy with it. Bohr was not, and the arguments recommenced with renewed vigour,[51] the first manifestation being in an appendix to Heisenberg's paper in which he noted some objections raised by Bohr, including that[52]

Above all the uncertainty in the observation does not depend exclusively upon the occurrence of discontinuities, but is directly connected with the requirement that justice be done at the same time to the different experiences, expressed on the one hand by the corpuscular theory and on the other hand by the wave theory. For example, the necessary divergence of the wave-packet due to finite aperture has to be taken into account in the use of an imaginary γ-ray microscope. This first leads to the observation of the electron position in the direction of the Compton recoil being known only within an uncertainty, which leads to the relation $[p_1 q_1 \sim h]$. Next, it is not sufficiently stressed that the simple theory of the Compton effect is strictly applicable only to the free electron. The resulting caution in the application of the uncertainty relation is essential, as Professor Bohr has stated clearly, among other things for an overall discussion of the transition from micro- to macro-mechanics and for the discussion of the generation of the orbital path by observation. Finally the considerations on resonance fluorescence are not quite correct, as the connection between the phase of the light and that of the electron is not so simple as assumed.

Bohr's technical objections, though valid, did not affect Heisenberg's conclusions, and Heisenberg wrote to Pauli on 14 March that, in respect of the position experiment, "I believe that everything I wrote is correct."[53] But far more important were Bohr's criticisms on the lack of foundation of the analysis – it did not follow from the occurrence of discontinuities – and on the insufficient justice done to the wave picture. On 4 April Heisenberg wrote again to Pauli, reporting among other things on friendly discussions with Bohr:[54]

Otherwise there continues to be general harmony here, and constant discussion of thought experiments. I am arguing with Bohr as to how far the relation has its foundations in the wave or discontinuity sides of quantum mechanics. Bohr stresses that in the γ-ray microscope, for example, the inflexion of the wave is essential, I stress that the light-quantum theory and even the Geiger-Bothe experiment are essential. By overemphasising one the one side and one the other we can discuss much without anything new.

In another letter of 16 May, Heisenberg reported on another error in his γ-ray microscope experiment, which was not however crucial, and on his continued differences with Bohr:[55]

Since my return from the Easter holiday . . . we have talked a lot here about the quantum theory. Bohr will write a general work on the 'conceptual structure', from the viewpoint that 'there are waves and particles' – if one begins at once with that one can naturally also make everything contradiction free. Prompted by this work Bohr has called my attention to the fact that there was still something essential overlooked in my work (Dirac also asked me about it subsequently): in the γ-ray microscope one could first of all imagine: one determines the direction of a falling light-quantum and of the reflected light-quantum, then one knows according to the Compton effect both the position and velocity very precisely (more so than $p_1 q_1 \sim h$). But one can not in fact do this on account of the bending of the light (wave theory!) To give the accuracy λ the microscope must have an aperture of order 1. So sure enough the relationship $p_1 q_1 \sim h$ comes out naturally, only not quite as naturally as I had thought. Besides this several points could also be better put and better discussed in all particulars if one began a quantitative discussion immediately with the waves. Nevertheless, I am naturally now as before of the opinion that the discontinuities are the only interesting things in quantum theory and that one can never stress them enough. For this reason I am also, now as before, very happy about my last work – despite the known defects all the results of the work are indeed correct, and on them I am also in agreement with Bohr. Otherwise there are at present between Bohr and I essential differences of taste over the word *anschaulich*.

As we have seen, Heisenberg had maintained that with the uncertainty relations quantum mechanics was *anschaulich.*[56] But to Bohr the requirements of *Anschaulichkeit* were much stricter, and could be satisfied only by a fully consistent classical visualisation. Quantum mechanics was not *anschaulich* in this sense, and Bohr's complementarity principle was to be based on the insistence that it was not, and could not be so.[57] Unfortunately Bohr, emphasising the importance of the classical conceptions, seems to have emphasised especially, as he had so often before, the importance of the wave concepts. For writing again on 31 May Heisenberg explained to Pauli that there had in effect been a serious division between the adherents of the wave mechanics and those of the matrix-mechanical Dirac-Jordan theory.[58]

In earlier correspondence, Heisenberg and Pauli had joined up in opposition to a movement in support of the wave mechanics, led by Ehrenfest, Darwin and Klein.[59] More recently a misunderstanding between Heisenberg and Klein had lead to personal animosity, with Bohr weighing in on Klein's side, and calm discussion had given way to trenchant argument.[60] Pauli, however, had resolved the personal problem, everything was now calm again, and Heisenberg could look at things more dispassionately:[61]

So I came into a battle for the matrices and against the waves: in the heat of this battle I often criticised Bohr's objections against my work too sharply and thus, without knowing it or wishing it, offended him personally. If I now reflect on those discussions I can understand very well that Bohr was annoyed by them.

What led Heisenberg to take such a humble position as he now did is unclear, but it could be that in the wake of the personal reconciliation brought about with Pauli's help Bohr told him something of the content of his proposed paper on the "conceptual structure" of quantum mechanics. For in this paper, a version of which was first delivered at Como the following September, it became apparent that Bohr's insistence on the wave-particle duality did not directly threaten Heisenberg's stress on the particle picture for observations, and did not prevent Bohr from treating the quantum discontinuity as being of prime importance.[62] In the heat of the argument, Heisenberg seems to have confused Bohr's attack on the particle picture with an attack, such as Bohr would never in fact have made, on the fundamental quantum postulate, and realising this he may have felt a little ashamed.

What Bohr did was to fill in the logical gap in Heisenberg's analysis between the quantum postulate of discontinuity and the derivation of the uncertainty relations. He did this in a way that preserved the symmetry of the wave-particle duality, replacing Heisenberg's identification of observation and definition by a contrast between these two concepts. Heisenberg's thought experiments showed the impossibility of a joint specification and observation of position and momentum, and whereas Heisenberg had taken the fundamental contradiction as that between position and momentum Bohr took it as that between specification and observation. Looked at this way the problem was not a specific one relating to conjugate coordinates, but the general philosophical one of the observer and the observed, with which Bohr was familiar, to which he was sympathetic, and through which he could at last understand and derive from a physical basis the non-applicability of classical concepts and lack of a visualisable theory that had long bothered him. The argument as presented in his paper was that[63]

Our usual description of physical phenomena is based entirely on the idea that the phenomena concerned may be observed without disturbing them appreciably. . . . [But] the quantum postulate implies that any observation of atomic phenomena will involve an interaction with the agency of observation not to be neglected.

The outcome of this was that[64]

On one hand, the definition of a state of a physical system, as ordinarily understood, claims the elimination of all external disturbances. But in that case . . . any observation will be impossible, and, above all, the concepts of space and time lose their immediate sense. On the other hand, if in order to make observation possible we posit certain interactions with suitable agencies of measurement, not belonging to the system, an unambiguous definition of the state of the system is naturally no longer possible, and there can be no question of causality in the ordinary sense of the word.

In one sense at least Bohr's argument was not wholly satisfactory, for while he claimed that the quantum postulate "forced" the impossibility of joint observation and definition this is far from self-evident.[65] But he did place Heisenberg's analysis on much sounder and more general grounds. Heisenberg had adopted Pauli's operationalist creed that observability and defineability should be equated in a consistent theory. But unable to create such a theory he carried this creed into his analysis of a theory that was still, in terms of its conceptual foundations, inconsistent. Bohr's analysis showed that this inconsistency lay in the very fact that an operational definition of the kinematic concepts needed was impossible. The ideals of observation and definition, both necessary to any physical theory, were in fact incompatible. Bohr defined this combination of joint necessity and mutual incompatibility through the notion of "complementarity", and from the complementarity of observation and definition he derived that of space-time description and causality, and that of the wave and particle pictures.[66] He also gave two derivations of the uncertainty principle, one from the wave-particle duality that reflected his earlier arguments with Heisenberg, and one directly from complementarity.[67]

Pauli, as may be expected, accepted Bohr's new ideas enthusiastically, criticising the presentation on a few minor points but declaring that "in general I am *very* much in agreement, both with the overall thrust of your paper and with most of its details."[68] The main force of the complementarity principle lay in its demonstration that for quantum phenomena *any* operationally defined system of concepts was impossible, the processes of operation and definition being themselves incompatible, and in this sense the principle represented a victory for Bohr over Pauli. But Bohr had also now admitted that the classical conceptions in particular were incapable of consistent application, and that this limitation arose, as Pauli had always insisted, from their being operationally ill-defined. In a few years the dialogue between the two men was to be reopened in the new context of nuclear physics, but for the moment they were in agreement, a situation that must have given great satisfaction to them both. Of the other physicists, Jordan later became one of the chief advocates of complementarity,[69] and for the present he, Heisenberg, Dirac and Born at least refrained from criticism.

A COMMON INTERPRETATION

By the Summer of 1927 there was still a large measure of disagreement between the quantum physicists. Bohr and Heisenberg still disagreed as to how much emphasis should be placed on the particulate nature of measurement, and Pauli could still write to Bohr in August in condemnation of Jordan and Klein's continued use of the "phenomenological" wave-mechanical approach.[70] But Bohr could now write back in agreement with Pauli, and at the fifth Solvay congress in October Born and Heisenberg could sink their differences of the past few years in a joint paper.[71] By December, another joint paper had bridged the gap between Pauli and Jordan.[72]

In the papers and comments delivered to the Solvay congress by Born and Heisenberg, Bohr, Pauli and Dirac, there were still significant differences of emphasis. Bohr reiterated his notion of complementarity, Born and Heisenberg concentrated on the interpretation of the theory within the classical concepts, and Dirac, in discussion, insisted on the restriction of the theory's application to our knowledge of a system, and on its lack of ontological content.[73] But these differences were now treated more as matters of personal taste and metaphysics than as anything physically serious, and it was clear that all the physicists concerned, together with Pauli who was also present, effectively shared a common interpretation. While their opponents, Einstein, Schrödinger, Lorentz and, for the time being, de Broglie, continued to argue heatedly about the interpretation,[74] they treated it rather as a problem of the past, and one that had now been more or less settled. What mattered now was the problem of extending quantum mechanics to general electromagnetic phenomena, of developing a relativistic quantum electrodynamics.[75] This problem had already dominated the thoughts of Jordan, Pauli and Dirac since the previous February,[76] and before long Heisenberg was also fully engaged upon it.[77]

Opposition to what became known as the Copenhagen interpretation has never ceased, and this interpretation has developed, particularly through analysis of the measurement problem, since the 1927 Solvay congress.[78] But as the least common denominator of the leading quantum physicists' views it was already established at that congress as a commonly held – and, since its opponents were unable to agree among themselves, *the* only commonly held – interpretation. On this interpretation it was agreed that, as Dirac explained, the wave functions represented our knowledge of the system, and the reduced wave packets our more precise knowledge after measurement. Within the classical conceptions of waves and particles the waves could be interpreted, as Born and Heisenberg wrote, as probability functions, corresponding to our knowledge of the system, of localised measurements. These measurements corresponded to kinematic and mechanical concepts associated with particles, and effectively created a situation that corresponded, within the limits of uncertainty, to the particle concept. Similarly, successive measurements could create a situation corresponding to the notion "path of a particle", just as Heisenberg had earlier explained. Within the classical conceptions, causality was effectively abandoned as irrelevant (after some toing and froing in individual papers by Born and Heisenberg, in which each advocated the position earlier adopted by the other, it was replaced by the concept of weak causality or causation).[79] And it was agreed that the quantum-mechanical theory corresponded precisely to the limits of observation, that within the context of the theory anything beyond these limits was practically irrelevant, and that in this sense the theory was "complete" and, given the interpretation, "without contradictions".[80] Finally, the limits of the classical concepts could be explained, if desired, through Bohr's notion of complementarity.

CHAPTER 10

CONCLUDING REMARKS

Bohr's principle of complementarity and the ensuing Copenhagen interpretation brought both the creation of quantum mechanics and, for the time being, the Bohr-Pauli dialogue to an end. The long sought for new system of operationally defined concepts upon which quantum mechanics was to have been built upwards had remained elusive, and in this sense Bohr's views had finally prevailed. But the inadequacy of the existing concepts, and in particular of visualisable models, had now been established and given a foundation. As Pauli had always maintained this foundation lay in their operational inadequacy, expressed in terms of a complementarity between definition and observation which effectively prohibited any operationally based definition. In philosophical terms the compromise was somewhat unfortunate. By stopping short of any assertions as to the nature and location of any reality underlying the defined limits of knowledge the Copenhagen interpretation masked the very real differences of philosophical viewpoint between its main creators and proponents, and in fact ran generally counter to their views. For all their insistence on the role of observation, both Pauli and Heisenberg were for example, and were to remain, philosophical realists. Scientifically, however, this limitation of the Copenhagen interpretation provided it with a great strength, wedding it closely to the theory and rendering it effectively immune to changes of philosophical opinion. Only when identified with an anti-realist philosophy has it ever come under serious attack, and such an identification is, as our analysis has shown, a misleading one.

The close relationship between the formalism of quantum mechanics and its Copenhagen interpretation suggests that their roots too were closely related, and our analysis has confirmed that this was in fact so. The analysis is of course far from complete. In portraying the creation and development of quantum mechanics primarily in terms of the Bohr-Pauli dialogue, we have inevitably overstressed some and understressed other aspects of the history. The most significant omission has, of course, been that of the detailed background to Schrödinger's wave mechanics, especially in respect of its relationship to Einstein's work on quantum gas statistics.[1] Other areas in which we have done less than full justice to the available historical literature include the technical details of Heisenberg's work in the early middle 1920s;[2] the contributions to quantum mechanics of Dirac;[3] and the technical details of Bohr's struggles with the problems of atomic structure in the early 1920s.[4] In stressing the importance of Pauli's views and their opposition to those of Bohr we may also have painted rather too brilliant a picture of the young Pauli. Brilliant he certainly was, and arguably one of the most brilliant physicists of this century; but he was also young. His youthful criticisms of Weyl, made we must remember when he was still in his teens, were certainly pointed. But he had not yet acquired the impressive soundness of his mature years and the criticisms were not altogether coherent in their details. Moreover, the maturing process took place largely under Bohr's supervision during the year that Pauli spent in Copenhagen, and to the relationship between these two physicists that we have portrayed, that of friendly adversaries, must be added that of pupil and teacher.[5]

Pauli's prominence in our story must also not be mistaken for dominance, which characteristic belongs rather to Bohr. But despite this qualification our analysis has indicated that Pauli's creative influence was a crucial factor in the origins and development of quantum mechanics. This is not to take away anything from Heisenberg's abilities and achievements. Whereas Heisenberg's ability to get to the technical heart of a problem might still conceivably have led him in due course to something like the new kinematics, even without Pauli's aid, it is most unlikely that Pauli on his own would have got there. It is however clear that the creation of the new kinematics, though Heisenberg's in the end, was the result of a very close collaboration between the two young physicists and of a methodological programme, based on a demand for operational consistency and for an operationally defined system of concepts, laid down by Pauli.

Elsewhere too Pauli's views, standing in opposition to Bohr's insistence that the classical concepts were somehow rooted in our perceptions and thus irreplaceable, again provided the starting point for Heisenberg's brilliant analyses. How far Bohr and Pauli engaged in an active and conscious dialogue concerning their beliefs is far from clear, but looking at the history of the creation of quantum mechanics in terms of such a dialogue we find that the key events fall naturally into place. The peculiar history of the virtual oscillator theory, the reactions to which cut right across the traditional quantum-theoretical camps, finds a straight-forward explanation and interpretation. Heisenberg's new kinematics appears not only as a development of his own work on fluorescent polarisation, in terms of which alone its pedigree seems incomplete, but also as a direct consequence of his acceptance, as a result of his own studies, of Pauli's arguments as to what the long sought for new mechanics should set out to achieve. Continuing analysis of the applicability of the classical kinematic concepts also prepared the way for the development of the transformation theory, and led directly to the consideration of position and momentum measurement and thus to Heisenberg's uncertainty principle. Finally, Bohr's principle of complementarity also fits naturally into the story as a further development of Heisenberg's analysis and as a considered response to Pauli, producing an agreed conclusion, for the time being at least, to their long debate.

This debate was clearly not the only focal point in the development of quantum mechanics. Other physicists viewed the problems from other perspectives, and the polarisation between Bohr and Pauli, conditioned by their teacher-pupil relationship and by their common stance on topics such as the core theory of the atom, was not always in emphasis. But the connections we have made are nevertheless genuine ones, and although the presentation of the historical creation of quantum mechanics in terms of the Bohr-Pauli dialogue may well be somewhat artificial in purely historical terms it does offer an insight to the underlying themes and currents of thought, and does add to our understanding of the events that took place. Heisenberg did set out in 1925 after discussions with Pauli to do precisely what Pauli had prescribed a few months earlier. Bohr did present his principle of complementarity primarily in terms of the relationship between observation and definition, and not in terms of what was in fact the deduced relationship between wave and particle pictures. And there can be little doubt that throughout the period we have discussed, and throughout the creation of quantum mechanics, the relationship between observation and definition was in fact crucial to the creative enterprise.

Another feature to have emerged from our treatment, and one that seems to be of some historical importance, concerns the relationship between different branches of physics. Perhaps because historians of physics have often themselves been specialists in one branch of physics or another, the way in which the fundamental problems transcended the boundaries within the discipline seems to have been rather overlooked. Some authors have tried to relate different areas of speciality with differing views on the nature of the quantum problem and the interpretation of quantum mechanics.[6] But while such lines may perhaps be drawn for Einstein and Bohr, the most remarkable feature of the work of the key creative figures, Pauli and Schrödinger, is its lack of sub-disciplinary barriers. Both were not only informed but also active in all the key areas of theoretical physics of the period, including the theoretical spectroscopy of the Bohr atomic theory, quantum statistical thermodynamics, and general relativity theory. Other key figures such as Born and Dirac shared this catholicism, and while it is difficult to assess just how crucial the cross-fertilisation of ideas between the different subject areas may have been, there can be no doubt that the fundamental problems of physics not only cut across the internal boundaries but were also widely perceived as so doing. In purely technical terms, the new kinematics emerged from the problem complex of the Bohr atomic theory, but its emergence was founded on a crystallisation of ideas derived from much wider considerations. In particular our treatment has suggested that, as one would indeed expect, the two great physical achievements of the inter-war period, quantum mechanics and general relativity, were far from being the isolated and independent developments they have previously been portrayed as. In this case as in others the scientific creativity operated primarily through the consideration of fundamental physical, methodological and epistemological problems, and only secondarily through that of their manifestations in any specific area.

Once our attention has been drawn to the organic relationship between developments in different areas of physics it is natural to enquire how the creation of quantum mechanics fits in to the much wider context of Western intellectual history. That some connection exists is not to be doubted, but the partially progressive nature of physical science and the technical language in which it is expressed make it no easy matter to determine what that connection might be. The fact that quantum mechanics emerged from a debate between scientists of very different philosophical persuasions, with the Copenhagen interpretation itself being something of a philosophically neutral compromise, complicates the issue further. The perception of quantum mechanics by those not responsible for its creation, in terms of indeterminism and anti-realist instrumentalism, may certainly be linked with other, non-scientific, trains of thought. Paul Forman has demonstrated the romantic and anti-determinist nature of the prevailing intellectual milieu of the period, and has shown that in Germany at least physicists recognised and were to some extent influenced by this milieu.[7] Stephen Brush has connected the instrumentalism associated with quantum mechanics with a general tendency towards romanticism and anti-realism in inter-war intellectual and artistic activity.[8] But causality does not seem to have been a central issue in the creation of quantum mechanics, and with a few exceptions the physicists responsible for its creation and interpretation were not themselves instrumentalists. The line they did take was indeed much closer to the philosophy of Kant,[9] and their considerations were in many ways closer to those of their mid-nineteenth century predecessors, Helmholtz and Maxwell, than they were to the philosophies of the twentieth century, by when Kantian epistemology was distinctly out of fashion.[10] The question as to whether the quantum mechanics itself, rather than just the way in which it was perceived, may be directly related to other cultural, philosophical or artistic developments thus remains an open one. Perhaps it may, but if any worthwhile attempt to portray such a relationship is to be made it will be necessary first to reach a very clear picture indeed of the very complex origins of quantum mechanics, and it is towards this more proximate end that our analysis here has been conducted.

NOTES

Short titles are used for works cited in the bibliography, or for repeat citations within the same chapter. The abbreviations *SHQP, BSC, Bohr MSS* and *ETHZ* are used for the primary source archives given in Section A of the bibliography, *PB* for the Pauli *Briefwechsel*, Bd. 1 (Section B of the bibliography) and *SQM* for Waerden's *Sources* (Section C of the bibliography).

CHAPTER 2. WOLFGANG PAULI AND THE SEARCH FOR A UNIFIED THEORY

1 Stuewer, *Compton Effect*: but see Chapter 4 below. The Compton effect seems to have been a turning point for American physicists, but less important to the Europeans. In Cambridge, for example, a talk on Compton's theory was given to the Kapitza Club in August 1923 by Herbert Skinner, and the members present signed a note in the minute book: "Compton is wrong. PK. HWBS. PMB. DRH. J. E. Jones." (Kapitza, Skinner, Blackett, Hartree were the first four signatories.) Only after a discussion of Pauli's theory of radiative equilibrium, also based upon light-quantum collisions, was one member, Blackett, won over: "Compton right we hope. ECS. PMB. We hope wrong. P. Kapitza. DRH. EGD. MHAN. HWBS." (The additional signatory in favour of light-quanta was Stoner, those against Dymond and Newman.) See Minutes of Kapitza Club, 3 August 1923 and 29 January 1924, *SHQP*, 38, 2.

2 One possible prominent exception is Sommerfeld, for whom see Chapters 3, 4 below.

3 Hendry, 'Wave-particle duality'. We shall see that it was the rejection of the requirement that a description be visualisable that characterised the new quantum mechanics. Although the criterion of visualisability was firmly established, however, this was far from being the first time it had been challenged. In the early nineteenth century Coleridge and other opponents of Laplacian mechanics had already attacked the "despotism of the eye" that made people judge a theory by its visual properties rather than by more fundamental criteria.

4 This is reflected in their work, much of which is discussed below. See also Klein, 'First phase', and Raman and Forman, 'Why Schrödinger'.

5 See for example Jammer, *Conceptual Development*, Waerden, *Sources*, Serwer, 'Unmechanischer Zwang', and MacKinnon, 'Heisenberg'.

6 A. Einstein, 'Die Grundlagen der allgemeinen Relativitätstheorie', *Ann. der Phys. 49* (1916), 769–822, translated in Lorentz, *Principle of Relativity*, 111–164. The best account of general relativity theory is probably still W. Pauli, 'Relativitätstheorie', *Encykl. Math. Wiss.* **19** (1921), translated as *Theory of Relativity* (Oxford, 1958). For a recent historical account see Mehra, *Einstein and Hilbert*.

7 The covariant tensor formulation was introduced by Einstein and Grossmann in 1913, abandoned by Einstein in 1914, and reintroduced by him in 1915: A. Einstein and

M. Grossmann, 'Entwurf einer verallgemeinerten Relativitätstheorie', *Zeit. Math. u. Phys. 62* (1913), 225–261, and 'Kovarianzeigenschaften der Feldgleichungen der auf die verallgemeinerte Relativitätstheorie gegründeten Gravitationstheorie', *ibid. 63* (1914), 215–225; A. Einstein, 'Die formale Grundlage der allgemeinen Relativitätstheorie', *S.-B. Preuss. Akad. Wiss.* (1915), 778–786 and 799–801, and 'Die Feldgleichungen der Gravitation', *ibid.*, 844–847. For a more complete set of references, see Mehra, *Einstein and Hilbert*. For the principle of equivalence, see A. Einstein, 'Über das Relativitätsprinzip und die aus demselben gezogenen Folgrungen', *Jahrb. der Radioaktivität u. Elektronik 4* (1907), 411–461, and 'Über den Einfluss der Schwerkraft auf die Ausbreitung des Lichtes', *Ann. der Phys. 35* (1911), 898–908, translated in Lorentz, *Principle of Relativity*, 99–108. The postulate of general covariance corresponds to the requirement that the tensor field equations be covariant with respect to the arbitrary coordinate transformations; it is required only that the coordinate systems be unique and continuous (Gaussian).

[8] They were also concerned with the rival theories of Nordström, Abraham, Ishiwara and especially Mie, but Hilbert's work in particular closely parallelled that of Einstein, and in 1915 he derived independently the field equations of Einstein's theory: D. Hilbert, 'Grundlagen der Physik', *Nach. königliche Ges. Wiss. Göttingen, Math.-Phys. Kl.* (1915), 395–407, and see also *ibid.* (1917), 477–480. The relationship between Einstein's work and Hilbert's is discussed by Mehra, *Einstein and Hilbert*, Earman and Glymour, 'Einstein and Hilbert', and Pyenson, 'Göttingen reception'. Born worked with Hilbert from 1908 to 1914, when he published a work based on Mie's theory: M. Born, 'Der Impuls-Energie-Satz in der Elektrodynamik von Gustav Mie', *Nach. königliche Ges. Wiss. Göttingen, Math.-Phys. Kl* (1914), 23–37. Weyl worked with Hilbert from 1906 to 1913. His work on Mie's theory was published in his book: H. Weyl, *Raum-Zeit-Materie* (Berlin, 1918), Section 26. The fourth edition (1920) of this book was translated as *Space-Time-Matter* (London, 1922).

[9] Einstein to Weyl, 23 December 1916, *ETHZ* 91, 536. See J. Dieudonné, 'Weyl', *Dict. Sci. Biog. 14* (1976), 281–285. The Einstein-Weyl correspondence deserves publication at length, but permission to do this has so far been refused, no reason having been given, by the Einstein Estate.

[10] Weyl, *Raum-Zeit-Materie*, Sections 34 to 36; H. Weyl, 'Gravitation und Elektrizität', *S.-B. Preuss. Akad. Wiss.* (1918), 465–480, translated in part in Lorentz, *Principle of Relativity*, 201–216; 'Reine Infinitesimalgeometrie', *Math. Zeit. 2* (1918), 384–411; 'Eine neue Erweiterung der Relativitätstheorie', *Ann. der Phys. 59* (1919), 101–133.

[11] Weyl, 'Grav. u. Elek', 477–478, translation, 215–216.

[12] Mie to Weyl, 26 October 1918, *ETHZ* 91, 674.

[13] Sommerfeld to Weyl, 3 July 1918, *ETHZ* 91, 751. Sommerfeld also criticised the theory, however, and his criticisms are to be found in Sommerfeld to Weyl, 7 November 1919, 11 December 1919 and 6 January 1920, *ETHZ* 91, 752–754.

[14] Eddington to Weyl, 18 August 1920, *ETHZ* 91, 523. Eddington's first enthusiastic response was in a letter Eddington to Weyl, 16 December 1918, *ETHZ* 91, 522.

[15] Einstein to Weyl, 8 March 1918, *ETHZ* 91, 539, including two pages of extravagant praise.

[16] W. Pauli, 'Zur Theorie der Gravitation und der Elektrizität von Hermann Weyl', *Phys. Zeit. 20* (1919), 457–467, and 'Mehrkurperihelbewegung und Strahlenableitung in Weyl's Gravitationstheorie', *Verh. Deut. Phys. Ges. 21* (1919), 742–750.

[17] E. T. Whittaker, *A History of the Theories of Aether and Electricity* 2, (London, 1953), 214–217.
[18] Einstein to Weyl, 23 December 1916, *ETHZ* 91, 536.
[19] Einstein to Weyl, 8 March 1918, in response to Weyl to Einstein, 1 March 1918, *ETHZ* 91, 539, 538a.
[20] Einstein to Weyl, 8 April 1918, *ETHZ* 91, 540.
[21] Einstein to Weyl, 15 April 1918, *ETHZ* 91, 541: "So schön Ihre Gedanke ist, muss ich doch offen sagen, dass er nach meiner Ansicht ausgeschlossen ist, dass die Theorie der Natur entspricht." Einstein had already communicated this paper, though Weyl had meanwhile become dissatisfied with it, feeling that it did not go far enough: Weyl to Einstein, 15 April 1918, *ETHZ* 91, 540a.
[22] Einstein to Weyl, 15 April 1918, *ETHZ* 91, 541, and see also Einstein to Besso, 20 August 1918 and 26 July 1920 in Einstein and Besso, *Correspondance*, 132–134, 155–158 (Items 46, 52.1).
[23] Einstein to Weyl, 19 April 1918, *ETHZ* 91, 543, with repercussions in Weyl to Einstein, 27 and 28 April 1918, and Einstein to Weyl, 1 May 1918, *ETHZ* 91, 543a, 543b, 544; Weyl, 'Grav. u. Elek.', 478–480, not in translation.
[24] Weyl, *ibid.*
[25] Weyl, *Raum-Zeit-Materie*, 1st edition, 226–227, not in 4th edition or translation.
[26] Weyl to Einstein, 19 May 1918, *ETHZ* 91, 545a: "Weiss ich doch nur zu gut, in einem wie viel Lauteren Verhältnis Sie zur Wirchlichkeit stehen als ich."
[27] *Ibid.*: "Behalten Sie für die wirkliche Welt recht, so bedaure ich, den liebern Gott einer mathematische Inkonsequenz zeihen zu müssen."
[28] Einstein to Weyl, 31 June 1918, *ETHZ* 91, 546. Writing again on 27 September 1918, *ETHZ* 91, 548, Einstein lamented that God had not made it easy for them. This resort to the highest authority was partly light-hearted banter (all the protagonists in this particular controversy remained on the friendliest of terms), but it also reflects a traditional Platonic approach to theoretical physics, namely that of asking how a creator might reasonably have designed things.
[29] Einstein to Weyl, 27 September 1918, *ETHZ* 91, 548, and see also Einstein to Besso, 26 July 1920, Einstein and Besso, *Correspondance*, 155–158 (Item 52.1). Einstein and Weyl repeated their respective positions in papers of 1920 and in a number of letters between 1918 and 1923, when the extant correspondence breaks off for a few years: H. Weyl, 'Elektrizität und Gravitation', *Phys. Zeit. 21* (1920), 649–650; A. Einstein, *ibid.*, 651; correspondence between Weyl and Einstein, *ETHZ* 91, 548a–556.
[30] For Einstein's continuing conviction see Einstein, *ibid.*, but in September 1918 he wrote that he was certain both of them had no other aim than to find the truth, and suggested that this would be established one way or the other within a couple of years: Einstein to Weyl, 27 September 1918, *ETHZ* 91, 548.
[31] Pauli, *Relativity*, 196.
[32] A. S. Eddington, 'Relativity of field and matter', *Phil. Mag. 42* (1921), 800–806, and 'A generalisation of Weyl's theory of the electromagnetic and gravitational fields', *Proc. Roy. Soc. A99* (1921), 104–122. See also A. S. Eddington, *Space, Time and Gravitation* (Cambridge, 1920), Chapter 11, where Weyl's theory is enthusiastically reviewed, and *Mathematical Theory of Relativity* (Cambridge, 1923).
[33] Eddington to Weyl, 10 July 1921, *ETHZ* 91, 525. Eddington wrote here in a reply to a lost letter from Weyl that his own theory was so directly inspired by Weyl's that

he had not seen it as a rival until he had received Weyl's objections. Then he had recognised that they had started from opposite ends, he from the pure geometry and Weyl from the action principle.

34 *Ibid.*

35 *Ibid.*; Einstein saw Eddington's theory as even further removed from reality than Weyl's, describing it as "beautiful but physically meaningless": Einstein to Weyl, 5 September 1921, *ETHZ* 91, 551, and see also his letter 6 June 1922, *ETHZ* 91, 554.

36 G. Mie, 'Grundlagen einer Theorie der Materie', *Ann. der Phys. 37* (1912), 511–534, *39* (1912), 1–40, *40* (1913), 1–66. See Mehra, *Einstein and Hilbert*, Pyenson, 'Göttingen reception', and L. Pyenson, 'Mathematics, education, and the Göttingen approach to physical reality, 1890–1914', *Europa 2* (1979), 91–127.

37 For the background to this electromagnetic world view see R. McCormmach, 'H. A. Lorentz and the electromagnetic view of nature', *Isis 61*(1970), 459–497.

38 Hilbert, 'Grundlagen'.

39 See Note 8 above and Einstein to Weyl, 23 December 1916 and 3 January 1917, *ETHZ* 91, 536, 537.

40 Mie, 'Grundlagen'.

41 Weyl, *Raum-Zeit-Materie*, 1st edition, Section 35, and 'Neue Erweiterung'.

42 Pauli, 'Grav. u. Elek.' and *Relativity*, 202, 205–206. See also A. Einstein, 'Spielen Gravitationsfelder in Aufbau der materiellen Elementarteilchen eine wesentliche Rolle?', *S.-B. Preuss. Akad. Wiss.* (1919), 348–356, translated in Lorentz, *Principle of Relativity*, 191–198.

43 The positron was yet to be discovered.

44 Pauli, *Relativity*, 206.

45 I do not use terms such as 'operationalism' to define precise philosophical systems, for the physicists in general and Pauli in particular did not work out or set down such precise systems. The terms are rather used in a sense consistent with the views they are used to describe, as indicating general classes and tendencies of belief only.

46 Mehra, *Einstein and Hilbert*, 56, dates Einstein's serious attempts at a unified theory from 1928; but the general aim is clearly apparent in his work from 1907, as he tried to place both gravitation and light-quanta upon a field-theoretical basis. His desire for a unified theory is explicit in his letter to Weyl of 27 September 1918, *ETHZ* 91, 548.

47 Einstein to Born, 27 January 1920, in Born and Einstein, *Letters*, 20–23 (Item 13).

48 Einstein, 'Spielen Grav.'.

49 *Ibid.*, 356.

50 Einstein to Born, 27 January 1920, and see also Einstein to Born, 3 March 1920, Born and Einstein, *Letters*, 20–26 (Items 13, 14).

51 Einstein's programme was presented in A. Einstein, 'Bietet die Feldtheorie Möglichen für die Lösung des Quantenproblems?', *S.-B. Preuss. Akad. Wiss.* (1923), 359–364. The programme was also outlined in Einstein to Besso, 4 January 1924, Einstein and Besso, *Correspondance*, 197–199 (Item 72), partly translated in Mehra, *Einstein and Hilbert*, 80; and in Einstein to Lorentz, 25 December 1923, quoted by Forman, 'Weimar culture', 96. He wrote to Besso that the programme represented a "logical possibility", but that the mathematics was too difficult for him.

52 See for example Besso to Einstein, 25 December 1923, Einstein and Besso, *Correspondance*, 192–194 (Item 71). When in late 1921 Einstein thought, wrongly, that some

experiments by Geiger on canal rays offered conclusive proof of the particular nature of light, he wrote gleefully on this point to Ehrenfest; but writing to Weyl at the same time he was more concerned with the problems the experiment seemed to propose for the field-theoretical programme. See Klein, 'First phase', and Einstein to Weyl, 16 and 22 December 1921, *ETHZ* 91, 552, 553.

53 Eddington, 'Field and matter' and 'Generalisation of Weyl's theory'.

54 *Ibid.*

55 *Ibid.*; writing to Weyl, however, he accepted that this was unlikely: Eddington to Weyl, 10 July 1921, *ETHZ* 91, 525.

56 Eddington to Weyl, 10 July 1921, *ETHZ* 91, 525.

57 Weyl to Pauli, 10 May 1919, *PB*, 3–5 (Item 1).

58 *Ibid.*

59 Weyl to Pauli, 9 December 1919, *PB*, 5–8 (Item 2). This was the basis of Weyl's acausal manifesto: H. Weyl, 'Das Verhältnis der kausalen zur statistischen Betrachtungsweise in der Physik', *Schweizerische Medizinische Wochenschrift 1* (1920), 737–741. In the letter Weyl, having discussed the connection between choice of sign of electric charge and the direction of time, concluded: "In contrast with most physicists, I hold the essential distinction [between past and future] to be a fact of even more fundamental significance than that between positive and negative electricity. Nevertheless, modern physics may be right in finding no place for 'lawful' or 'field' physics. For I am completely convinced that the statistics are in principle somewhat independent, compared with the causality, which is 'lawful'; because it is in general paradoxical to introduce a continuum as something ready made. I think that field physics actually plays only the role of the 'world geometry'; in matter there is still something else, something of reality, which is not causal but which is perhaps to be thought of in terms of 'independent decisions', which we take account of in physics through the medium of statistics.

60 Weyl to Pauli, 9 December 1919, *PB*, 5–8 (Item 2). This idea was explored in H. Weyl, 'Feld und Materie', *Ann. der Phys. 65* (1921), 541–563, and 'Was ist Materie?', *Naturwissenschaften 12* (1924), 561–568, 585–593, 604–611.

61 Weyl, *Space-Time-Matter*, 311.

62 Weyl to Pauli, 9 December 1919, *PB*, 5–8 (Item 2). The context of this observation was Weyl's reply to Pauli's criticism of the unobservability of the field within an electron. Weyl argued that internal motions of the electron might well have measurable consequences elsewhere. This did not, however, answer Pauli's point, which was concerned with the problem of the definition of the field concepts.

63 See Forman, 'Weimar culture' and Hendry, 'Weimar culture'.

64 Weyl, 'Das Verhältnis'.

65 See Weyl, *Space-Time-Matter*, 212–213, and Pauli, *Relativity*, 192.

66 Pauli, 'Grav. u. Elek.', and *Relativity*, 205–206.

67 Einstein had also expressed doubts on the maintenance of energy conservation in Weyl's theory: Einstein to Weyl, 27 September 1918, *ETHZ* 91, 548.

68 A. Einstein, 'Rapport sur l'état actuel du problème des chaleurs spécifiques', in P. Langevin and M. de Broglie (eds.), *La Théorie du Rayonnement et les Quanta* (Paris, 1912), 407–435, and 'Zur Quantentheorie der Strahlung', *Phys. Zeit. 18* (1917), 121–128.

69 Hendry, 'Weimar culture'. Einstein's derivation of Planck's law also appeared on the

surface to rest on the particulate nature of light, but it actually entailed several assumptions (including that of the existence of stimulated or induced emission) that could be understood only in terms of the wave theory. See A. S. Eddington, 'On the derivation of Planck's law from Einstein's equation', *Phil. Mag. 50* (1925), 803–808, and J. Hendry, 'An investigation of the mathematical formulation of quantum theory and its physical interpretation, 1900–1927', Ph.D. thesis, London University, 1978, Appendix F.

70 Hendry, 'Weimar culture', but see also Forman, 'Weimar culture'.

71 W. Schottky, 'Das Kausalproblem der Quantentheorie als eine Grundfrage der modernen Naturforschung überhaupt', *Naturwissenschaften 9* (1921), 492–496, 506–511.

72 They were not at the same institution, but this did not stop them working in close liaison, as did Weyl and Pauli in later years.

73 K. Försterling, 'Bohrsches Atommodell und Relativitätstheorie', *Zeit. Phys. 3* (1920), 404–407, reviewed by Pauli in *Phys. Ber. 2* (1921), 489.

74 E. Schrödinger, 'Dopplerprinzip und Bohrsche Frequenzbedingung', *Phys. Zeit. 23* (1922), 301–303.

75 Schrödinger to Pauli, 8 November 1922, *PB*, 69–73 (Item 29).

76 *Ibid.*

77 E. Schrödinger, 'Was ist ein Naturgesetz?', *Naturwissenschaften 17* (1929), 9–11, inaugural lecture at Zürich, December 1922.

78 Schrödinger to Pauli, 8 November 1922, *PB*, 69–73 (Item 29).

79 Schrödinger, 'Naturgesetz'.

80 This decision of Schrödinger's remains obscure. He seems to have been working at the time on both quantum atomic theory and classical relativity theory: E. Schrödinger, 'Die Wasserstoffahnlichen Spektren vom Standpunkte der Polarisierbarkeit des Atomrumpfes', *Ann. der Phys. 72* (1925), 43–70, and 'Die Erfüllbarkeit der Relativitätsforderung in der klassischen Mechanik', *Ann. der Phys. 77* (1925), 325–336. The latter paper was once again connected with Weyl's work, to which it referred. Schrödinger's theme was that the cosmological problem might be solved through the reconciliation of general relativity with the Mach principle.

81 E. Schrödinger, 'Quantisierung als Eigenwertproblem', *Ann. der Phys. 79* (1926), 489–527, esp. 489. The words were however omitted from the English translation of the paper, perhaps as unsuitable for an English audience.

82 E. Schrödinger, 'Über eine bemerkenswerte Eigenschaft der Quantenbahnen eines einzelnen Elektrons', *Zeit. Phys. 12* (1922), 13–23. See also Raman and Forman, 'Why Schrödinger', 305.

83 Schrödinger, *ibid.*, and see Raman and Forman, 'Why Schrödinger', 303–310.

84 Raman and Forman, *ibid.*

85 Eddington, *Mathematical Theory*. This was itself an extended version of the mathematical appendix to the French edition of his more popular *Space, Time and Gravitation*, the first edition of which contained the first extended presentation of his philosophical ideas.

86 Pauli, *Relativity*, 206.

87 Pauli to Eddington, 20 September 1923, *PB*, 115–119 (Item 45). This letter was sent in response to a copy of Eddington's *Mathematical Theory*.

CHAPTER 3. NIELS BOHR AND THE PROBLEMS OF ATOMIC THEORY

1 Pauli's work on atomic theory had been planned for some time. See Schrödinger to Pauli, 12 July 1920, 13 February 1921, *PB*, 24–26 (Items 8, 9), and for the context *ibid*., pp. 23–24.

2 Serwer, 'Unmechanischer Zwang', 23–24.

3 Interview with J. Franck, *SHQP*. See Jammer, *Conceptual Development*, 85, for the work of Franck and Hertz. Recalled by Heisenberg (interview with Heisenberg, *SHQP*) as a "mathematical methods" man, Born seemed more interested in the existence or otherwise of solutions than in the solutions themselves, and was as happy with negative as with positive results. He was the ideal man for the proposed investigation.

4 Born to Pauli, 23 December 1919, *PB*, 9–11 (Item 4): "I was especially interested by your remark at the end, that you hold the application of the continuum theory to the inside of the electron to be meaningless, because it is a case there of in principle non-observable things. I have followed just this line of thought for some time, but so far without any positive result, namely, that the way out of all the quantum difficulties over wholly fundamental points must be sought thus: one must not carry over the concept of space and time as a four-dimensional continuum from the macroscopic world of experience to the atomic world."

5 Fierz, 'Pauli'.

6 In general Bohr used his assistants as sounding boards for his new ideas, but did not actually work with them; Pauli was in this an exception. See Serwer, 'Unmechanischer Zwang', 225.

7 N. Bohr, 'On the constitution of atoms and molecules', *Phil. Mag. 26* (1913), 1–25, 476–502, 857–875.

8 See Jammer, *Conceptual Development*, and Heilbron, 'Kossel-Sommerfeld theory'.

9 **A. Einstein 'Zur Quantentheorie der Strahlung', *Mitt. Phys. Ges. Zürich 18* (1916),** 47–62; *Phys. Zeit. 18* (1917), 121–128, translated in *SQM*, 63–78.

10 N. Bohr, *On the Quantum Theory of Line Spectra* (Copenhagen, 1918). The terminology was introduced in N. Bohr, 'Über die Linienspektren der Elemente', *Zeit. Phys. 2* (1920), 423–469, translated in Bohr, *Theory of Spectra*, 20–60. It is sometimes aruged that the correspondence principle should be dated to 1913, but this is to miss its whole point. Bohr did talk of a correspondence with the classical theory in 1913, but this was no more than that correspondence which relates almost any theory to its predecessors, in the domain in which they represent valid approximations. It was not until 1918 that Bohr talked of a correspondence between quantum and classical theories in which the latter formed an integral part of the foundations of the former, as discussed below.

11 Bohr considered a system with one degree of freedom, and a transition between two states of this system with high quantum numbers and close frequencies. For two states (n', ω') and (n'', ω'') whose energies were $E' = hn'\omega'$, $E'' = hn''\omega''$ (for the Planck oscillator, $\omega' = \omega''$, but this does not hold in general), he considered the limit "where n is very large, and where the ratio between the frequencies of the motion in successive stationary states differs very little from unity", in fact requiring $n' - n'' << n'$, with $\omega'/\omega'' \rightarrow 1$ faster than $n \rightarrow$ infinity. In this case the frequency of emitted or absorbed radiation was given by $\nu = (n' - n'')\omega$, where $\omega \approx \omega', \omega''$, and Bohr compared this

with the classical Fourier expression for particle displacement, $\Sigma_{\tau} C_{\tau} \cos 2\pi(\tau\omega t + c_{\tau})$. He linked the classical harmonic given by τ_0 with the transition $\tau_0 = (n' - n'')$, and reasonably expected the probability of a quantum emission of frequency $\tau_0 \omega$ to be the same as the intensity of a classical emission of the same frequency, namely C_{τ_0}. Finally, he suggested that this relationship should hold in some (vague) way for small quantum numbers too, despite the fact that the frequencies could no longer be expected to correlate there ($\omega' \neq \omega''$) as they had done, in practice, for the higher states.

12 See A. S. Eddington, 'On the derivation of Planck's law from Einstein's equation', *Phil. Mag. 50* (1925), 803–808.

13 Heilbron and Kuhn, 'Genesis'.

14 See Note 10 above.

15 Bohr, 'Linienspektren'.

16 Bohr to Darwin, catalogued July 1919, more likely 1920, *BSC* 1, 4.

17 N. Bohr, 'L'application de la théorie des quanta aux problèmes atomiques', in Institut International de Physique Solvay, *Atomes et Électrons* (Paris, 1921), 228–247; original English draft in Bohr, *Works*, Vol. 3, 364–380.

18 N. Bohr, 'On the application of the quantum theory to atomic structure: Part I. The fundamental postulates', *Supplement to Proc. Camb. Phil. Soc.* (1924), p. 35. This paper is a translation from *Zeit. Phys.* 13 (1923), 117–165, submitted November 1922.

19 Sommerfeld to Bohr, 10 May 1918, *BSC* 7, 3.

20 See for example Stuewer, *Compton Effect*, for some strong views on duality expressed by physicists not closely concerned with the problem. See also Hendry, 'Wave-particle duality'.

21 Bohr, 'Fundamental postulates', 35, 1.

22 Bohr to Høffding, 22 December 1923, *BSC*, as quoted by Honner, 'Transcendental philosophy', 7.

23 A. Einstein, 'Rapport sur l'état actuel du problème des chaleurs spécifiques', in P. Langevin and M. de Broglie (eds.), *La Théorie du Rayonnement et les Quanta* (Paris, 1912), 407–435, esp. 429.

24 *Ibid.*, 429.

25 Einstein to Ehrenfest, 31 May 1924 and see also 12 July 1924, and Einstein to Born, 29 April 1924, all of which are discussed with extracts in Klein, 'First phase', 32–35.

26 Einstein, 'Rapport', 443.

27 See Hendry, 'Weimar culture'.

28 Darwin to Bohr, 20 July 1919, *BSC* 1, 4.

29 Bohr to Darwin, catalogued July 1919, *BSC* 1, 4.

30 *Ibid.*

31 Bohr, 'L'application', 374.

32 N. Bohr, 'Application of the quantum theory to atomic problems in general', in his *Works*, Vol. 3, 397–414, esp. 413.

33 Bohr, 'Fundamental postulates', 40.

34 See Klein, 'First phase', and Chapter 5 below. Einstein almost always linked the two issues together.

35 M. Planck, 'Über die Begründung des Gesetzes der schwarzen Strahlung', *Ann. der Phys. 37* (1912), 642–656, esp. 644.

[36] Forman, 'Weimar culture'.
[37] Bohr, *Atomic Theory and the Description of Nature*, 91. For a wide-ranging discussion see Meyer-Abich, *Korrespondenz.*
[38] See Jammer, *Conceptual Development*, 84.
[39] For example, H. Høffding, *History of Modern Philosophy, Vol. 2* (London, 1900), 286.
[40] See for example O. W. Richardson, *The Electron Theory of Matter* (Cambridge, 1916), 507.
[41] Bohr, 'Fundamental postulates', 21.
[42] N. Bohr, 'Problems of the atomic theory', in his *Works*, Vol. 3, 569–574.
[43] Darwin, manuscript draft of July 1919, *SHQP* 36, 3.
[44] R. von Mises, 'Über die gegenwärtige Krise der Mechanik', *Zeit. Ang. Math. u. Mech. 1* (1921), 425–431, and *Naturwissenschaft und Technik der Gegenwart* (Leipzig, 1922). See Hendry, 'Weimar culture', 81.
[45] H. A. Senftleben, 'Zur Grundlagen der Quantentheorie', *Zeit. Phys. 22* (1923), 105–156, esp. 127.
[46] Bohr, 'Fundamental postulates', 42.
[47] W. Pauli, 'Über das thermische Gleichgewicht zwischen Strahlung und freien Elektronen', *Zeit. Phys. 18* (1923) 272–286; review of K. Försterling, 'Bohrsches Atommodell und Relativitätstheorie', *Phys. Ber. 2* (1921), 489.
[48] See Schrödinger to Pauli, 8 November 1922, *PB*, 69–73 (Item 29).

CHAPTER 4. THE TECHNICAL PROBLEM COMPLEX

[1] Forman, 'Weimar culture'.
[2] Hendry, 'Weimar culture'.
[3] *Ibid.*
[4] Interview with Heisenberg, *SHQP*. The reference was to physicists at Munich (under Sommerfeld) and at Göttingen (under Born and Franck).
[5] A. Sommerfeld, *Atombau und Spektrallinien* (Braunschweig, 1922), translated as *Atomic Structure and Spectral Lines* (London, 1923), 253. See also Heisenberg to Landé, 28 November 1921, *SHQP* 6, 2.
[6] A. Einstein and P. Ehrenfest, 'Quantentheoretische Bemerkung zum Experiment von Stern und Gerlach', *Zeit. Phys. 11* (1922), 31–34.
[7] M. Born and W. Heisenberg, 'Die Elektronenbahnen im angeregten Heliumatom', *Zeit. Phys. 16* (1923), 229–243.
[8] Heisenberg to Bohr, 2 February 1923, *BSC.*
[9] Heisenberg to Pauli, 19 February 1923, *PB*, 79–81 (Item 31).
[10] A. Landé, 'Schwierigkeiten in der Quantentheorie des Atombaues, besonders magnetischer Art', *Phys. Zeit. 24* (1923), 441–444. M. Born, 'Quantentheorie und Störungstheorie', *Naturwissenschaften 11* (1923), 537–542. See also F. Paschen, 'Die spektroscopische Erforschung des Atombaus', *Phys. Zeit. 24* (1923), 401–407, who wrote that " the present contradiction must be augmented by further incomprehensible problems".
[11] Born, 'Quantentheorie', 542.
[12] W. Heisenberg, 'Zur Quantentheorie der Linienstruktur und der anomalen Zeemaneffekte', *Zeit. Phys. 8* (1922), 273–297, esp. 281.

[13] Sommerfeld to Einstein, 11 January 1922, Einstein and Sommerfeld, *Briefwechsel*, 95–97 (Item 40).

[14] Einstein did try to set up a decisive experiment, but failed: see Klein, 'First phase'.

[15] A. H. Compton, 'Secondary radiation produced by X-rays, and some of their applications to physical problems', *Bull. Nat. Res. Counc. 4* (1922), No. 20; 'A quantum theory of the scattering of X-rays by light elements', *Phys. Rev. 21* (1923), 483–502; 'Wavelength measurements of scattered X-rays', *ibid.*, 715; 'The scattering of X-rays', *J. Franklin Inst. 198* (1924), 61–72. In the Compton effect, X-ray scattering was interpreted as a collision process between individual particulate light-quanta and individual electrons in the scattering substance. This gave the correct change in wavelength as a function of scattering angle, which change could not easily be interpreted classically, on the wave theory of light. For further details see Jammer, *Conceptual Development*, 157 ff., and Stuewer, *Compton Effect*. As has been noted the Compton effect does not in general seem to have had a decisive effect on physicists' views of the nature of light, but one exception may be Sommerfeld, whose response was reported by Compton, 'The scattering of X-rays', 69: "In a recent letter to me Sommerfeld has expressed the opinion that the discovery of the change of wavelength of radiation, due to scattering, sounds the death-knell for the wave theory of radiation." It is unclear whether Compton noted the irony.

[16] P. Debye, 'Zerstreuung von Röntgenstrahlen und Quantentheorie', *Phys. Zeit. 24* (1923), 161–166; W. Duane, 'The transfer in quanta of radiation momentum to matter', *Proc. Nat. Acad. Sci. 9* (1923), 158–164; A. H. Compton, 'The quantum integral and diffraction by a crystal', *ibid.*, 359–362.

[17] W. Pauli, 'Über das thermische Gleichgewicht zwischen Strahlung und freien Elektronen', *Zeit. Phys. 18* (1923), 272–286; see also A. Einstein and P. Ehrenfest, 'Zur Quantentheorie des Strahlungsgleichgewichts', *Zeit. Phys. 19* (1923), 301–306. While disagreeing with Einstein's ultimate goal of a pure field theory, Pauli agreed with him completely on the necessity of the light-quantum concept.

[18] P. S. Epstein and P. Ehrenfest, 'The quantum theory of the Fraunhofer diffraction', *Proc. Nat. Acad. Sci. 10* (1924), 133–139, and 'Remarks on the quantum theory of diffraction', *ibid.*, *13* (1927), 400–408; A. H. Compton, 'The total reflexion of X-rays', *Phil. Mag. 45* (1923), 1121–1131, esp. 1130.

[19] Sommerfeld introduced the general conditions $\oint p_k dq_k = n_k h$: A. Sommerfeld, 'Zur Quantentheorie der Spektrallinien', *Ann. der Phys. 52* (1916), 1–94, 125–167. These conditions were then expressed in terms of the classical Hamilton-Jacobi theory, the action variable being defined as in the main text, by Epstein and Schwarzchild: K. Schwarzchild, 'Zur Quantenhypothese', *S.-B. Preuss. Akad. Wiss.* (1916), 548–568, and P. S. Epstein, 'Zur Quantentheorie', *Ann. der Phys. 52* (1916), 168–188.

[20] See N. Bohr, *On the Quantum Theory of Line Spectra* (Copenhagen, 1918); A. Sommerfeld, *Atombau und Spektrallinien* (Braunschweig, 1919–1920). Also essential to the theory was Ehrenfest's adiabatic principle: P. Ehrenfest, 'Adiabatic invariants and the theory of quanta', *Phil. Mag. 33* (1917), 500–513. See Jammer, *Conceptual Development*, 98, and Klein, *Ehrenfest*, 264ff.

[21] J. H. van Vleck, 'The normal helium atom and its relation to the quantum theory', *Phil. Mag. 44* (1923), 842–869, and 'The dilemma of the helium atom', *Phys. Rev. 19* (1922), 419–420; H. A. Kramers, 'Über das Modell des Heliumatom', *Zeit. Phys. 13* (1923), 312–341; I. Langmuir, 'The structure of the helium atom', *Phys. Rev. 17*

(1921), 339–353; P. S. Epstein, 'Problems of the quantum theory in the light of the theory of perturbations', *Phys. Rev. 19* (1922), 578–608, completed September 1921. For a detailed treatment of all this work see Small, 'The helium atom'.

22 M. Born and W. Pauli, 'Über die Quantelung gestörten mechanischer Systeme', *Zeit. Phys. 10* (1922), 137–158. See also M. Born and E. Brody, 'Über die Schwingungen einer mechanischer System mit endlicher Amplitude und ihre Quantelung', *ibid.*, *6* (1921), 140–152.

23 Born and Pauli, 'Quantelung'; Born to Einstein, 21 October 1921, Born and Einstein, *Letters*, 57–59 (Item 33).

24 W. Pauli, 'Über das Modell des Wasserstoffmolekulions', *Ann. der Phys. 68* (1922), 177–240.

25 W. Heisenberg, 'Linienstruktur u. anomalen Zeemaneffekte'; A. Landé, 'Zur Theorie der anomalen Zeeman- und magnetomechanischen Effekte', *Zeit. Phys. 11* (1922), 353–363.

26 See Forman, 'Landé', and Cassidy, 'Werner Heisenberg'.

27 *Ibid.*, and see also Serwer, 'Unmechanischer Zwang'. To comprehend the magnitude of the problem faced we should remember that they were working without either the spin or the parity concept.

28 Bohr to Landé, 15 May 1922, quoted in Serwer, 'Unmechanischer Zwang', 224–225.

29 *Ibid.*

30 M. Born and W. Heisenberg, 'Über Phasenbeziehungen bei den Bohrschen Modellen von Atomen und Molekeln', *Zeit. Phys. 14* (1923), 44–55.

31 Born and Heisenberg, 'Elektronenbahnen', 229.

32 Interview with Heisenberg, *SHQP*. See also Heisenberg to Pauli, 26 March 1923, *PB* 85–86 (Item 34); "Basically we [Heisenberg and Born] are now both of the conviction that all helium models so far are just as erroneous as the whole of atomic physics."

33 N. Bohr, 'On the application of the quantum theory to atomic structure. Part I. The fundamental postulates', *Supplement to Proc. Camb. Phil. Soc.* (1924), especially 32–33.

34 *Ibid.*, 33–34.

35 Bohr to Landé, 3 March 1923, *SHQP* 4, 1: "It was, as you saw, a desperate attempt to stick with integral quantum numbers, because we hoped to see, even in the paradoxes themselves, a hint of the paths upon which one might seek the solution of the anomalous Zeeman effect." ("Es war, wie Sie gesehen haben, ein Verzweiflungsversuch, den ganzen Quantenzahlen treu zu bleiben, indem wir hofften, eben in den Paradoxen einen Fingerzeig zu sehen für die Wege, auf denen man die Lösung des annomalen Zeemaneffektes suchen dürfte.")

36 Heisenberg to Landé, 13 November 1922, quoted by Serwer, 'Unmechanischer Zwang', 210–211.

37 *Ibid.*

38 *Ibid.*

39 A. Landé, 'Termstruktur und Zeemaneffekt der Multipletts', *Zeit. Phys. 15* (1923), 189–205.

40 N. Bohr, 'Der Bau der Atome und die physikalischen und chemischen Eigenschaften der Elemente', *Zeit. Phys. 9* (1922), 1–67. See Kragh, 'Bohr's second theory'.

41 Bohr, 'Fundamental postulates', 16. See Note 20 above.

42 Bohr, *Quantum Theory of Line Spectra.*

[43] Landé, 'Theorie der anomalen Zeeman- und magnetomechanischen Effekte', 361.
[44] N. Bohr, manuscript printed in his *Works*, Vol. 3, 502–531, translation 532–565, esp. 558. See also N. Bohr, 'Linien-spektren und Atombau', *Ann. der Phys. 71* (1923), 228–288.
[45] Pauli to Bohr, 11 February 1924, *PB*, 143–145 (Item 54).
[46] See for example Bohr to Landé, 15 May 1922, quoted by Serwer, 'Unmechanischer Zwang', 224–225.
[47] Pauli to Landé, 23 May 1923, *PB*, 87–90 (Item 35).
[48] Pauli to Sommerfeld, 6 June 1923, *PB*, 94–101 (Item 37).
[49] *Ibid.*
[50] Pauli to Bohr, 16 July 1923, *PB*, 102–105 (Item 39).
[51] Pauli, 'Thermische Gleichgewicht'.
[52] Pauli to Eddington, 20 September 1923, *PB*, 115–119 (Item 45).
[53] Pauli to Bohr, 21 February 1924, *PB*, 147–149 (Item 56).
[54] Heisenberg to Pauli, 19 February 1923, *PB*, 79–81 (Item 31).
[55] Heisenberg to Pauli, 9 October 1923, *PB*, 125–128 (Item 47). Heisenberg to Bohr, 22 December 1923, *BSC*. W. Heisenberg, 'Über eine Abänderung der formalen Regeln der Quantentheorie beim Problem der anomalen Zeemaneffekte', *Zeit. Phys. 26* (1924), 291–307; and see Serwer, 'Unmechanischer Zwang', 213–218. The formalism as first expressed in the letter to Pauli did not include the integration limits, but these could be inferred.
[56] Serwer, 'Unmechanischer Zwang', 215–216.
[57] Heisenberg to Pauli, 9 October 1923, *PB*, 125–128 (Item 47).
[58] *Ibid.*
[59] *Ibid.*
[60] Interview with Heisenberg, *SHQP*. The seminar was given by Courant and Siegel, and would probably have been attended, as was normal, by all those in Born's department as well as by those in Hilbert's.
[61] Landé, "Termstruktur".
[62] Bohr to Heisenberg, 31 January 1924, *BSC*.
[63] Heisenberg to Pauli, 9 October 1923, *PB*, Item 47: "The considerable negative aspect of the theory is that we now understand the quantum theory no more whatsoever. But that appeals very much to me. The proper aim must now be to reach the discrete states *unequivocally* from the Symbol-model; whether the formulae thus attained will approach a comprehensible meaning I doubt."
[64] Pauli to Landé, 14 December 1923, *PB*, 134 (Item 51).
[65] A. Landé and W. Heisenberg, 'Termstruktur der Multipletts hoherer Stufe', *Zeit. Phys. 25* (1924), 279–286. See Heisenberg to Bohr, 3 February 1924, *BSC*, for his hopes, and Serwer, 'Unmechanischer Zwang', for discussion.
[66] R. Ladenberg, 'Die quantentheoretische Zahl der Dispersionselektronen', *Zeit. Phys. 4* (1921), 451–471, translated in *SQM*, 139–158. See also N. Bohr 'L'application de la théorie des quanta aux problèmes atomiques', in Institut International de Physique Solvay, *Atomes et Électrons* (Paris, 1921), 228–247, and original English draft in Bohr, *Works*, Vol. 3, 364–380.
[67] R. Ladenburg and F. Reiche, 'Absorption, Zerstreuung und Dispersion in der Bohrsche Atomtheorie', *Naturwissenschaften 11* (1923), 584–598, esp. 597.
[68] Bohr, 'Application to problems in general', 414.

69 Bohr to Darwin, 21 December 1922, *BSC.*
70 Bohr, 'Fundamental postulates', 38.
71 H. A. Kramers, 'On the theory of X-ray absorption and the continuous X-ray spectrum', *Phil. Mag. 46* (1923), 836–871, esp. 861. This work was partly a response to Eddington's theory of X-ray capture as a means of absorption of radiation in stars. See A. S. Eddington, 'On the absorption of radiation inside a star', *Month. Not. Roy. Astr. Soc. 83* (1922), 32–46, *84* (1923), 104–123; and 'Das Strahlungsgleichgewicht der Sterne', *Zeit. Phys.* 7 (1921), 351.
72 Kramers, 'X-ray absorption', 843.
73 *Ibid.*, 852.
74 See below, this chapter.
75 N. Bohr, 'Problems of the atomic theory', in his *Works*, Vol. 3, 569–574.
76 *Ibid.*, 571.
77 Pauli to Sommerfeld, 6 June 1923, *PB*, 94–101 (Item 37). Pauli added that Epstein had already made a similar remark, but I have been unable to trace this. Pauli's remark may have been linked with Born's work at the time, for Born ('Quantentheorie') ran into considerable difficulties with high-frequency electron-electron coupling in the atom.
78 Pauli to Bohr, 21 February 1924, *PB*, 147–149 (Item 56).
79 Heisenberg, *Physics and Beyond*, 36.
80 Heisenberg to Pauli, 9 October 1923, *PB*, 125–128 (Item 47).
81 A. Sommerfeld and W. Heisenberg, 'Eine Bemerkung über relativistische Röntgendubletts und Linienscharfe', *Zeit. Phys. 10* (1922), 393–398, esp. 398.
82 Heisenberg to Landé, 29 October 1921, *SHQP* 6, 2.
83 See for example A. Sommerfeld and W. Heisenberg, 'Die Intensität der Mehrfachlinien und ihre Zeemankomponenten', *Zeit. Phys. 11* (1922), 131–154, esp. 132.
84 Pauli to Bohr, 11 February 1924, *PB*, 143–145 (Item 54).

CHAPTER 5. FROM BOHR'S VIRTUAL OSCILLATORS TO THE NEW KINEMATICS OF HEISENBERG AND PAULI

1 N. Bohr, H. A. Kramers and J. C. Slater, 'The quantum theory of radiation', *Phil. Mag. 47* (1924), 785–802, reprinted in *SQM*, 159–176.
2 This seems to me to be the true value of the virtual oscillator treatment, but Jammer, *Conceptual Development*, 181–195, emphasises the break with classical causality and conservation laws. See also Klein, 'First phase', 23–29; *SQM*, 11–14; and K. R. Popper, *Objective Knowledge* (London, 1970), 206. For a treatment similar to that given here but self-contained see Hendry, 'Bohr-Kramers-Slater'.
3 The habitual reference to "the present state of the theory" is omitted, but compare the statement given below (Note 27) with Bohr's earlier statement in N. Bohr, 'On the application of the quantum theory to atomic structure. Part I. The fundamental postulates', *Supplement to Proc. Camb. Phil. Soc.* (1924), 20: "According to this method of treatment, we do not seek a *cause* for the occurrence of radiative processes, but we simply assume that they are governed by the laws of probability."
4 See below.
5 L. de Broglie, 'Sur la dégradation du quantum dans les transformations successives des radiations de haute fréquence', *Comptes Rendus 173* (1921), 1160–1162. See also A. Einstein, 'Über einen die Erzeugung und Verwändlung des Lichtesbetreffenden

heuristischen Gesichtspunkt', *Ann. der Phys. 17* (1905), 132–148, and M. de Broglie, 'La relation $h\nu = \epsilon$ dans les phénomènes photélectriques: production de la lumière dans le choc des atomes par les électrons et production des rayons de Röntgen', in Institut International de Physique Solvay, *Atomes et électrons* (Paris, 1921), 80–119.

[6] L. de Broglie, 'Rayonnement noir et quanta des lumière', *J. de Phys. 3* (1922), 422–428, translated in de Broglie and Brillouin, *Selected Papers*, 1–8.

[7] *Ibid.*, 1. See also L. de Broglie, 'Ondes et quanta', *Comptes Rendus 177* (1923), 507–510; 'Quanta de lumière', *ibid.*, 548–550; 'Les quanta, la théorie cinétique des gaz et le principe de Fermat', *ibid.*, 630–632; 'A tentative theory of light quanta', *Phil. Mag. 47* (1924), 446–458.

[8] The reception of de Broglie's ideas is analysed in Raman and Forman, 'Why Schrödinger' but see also discussion of Born's response below.

[9] de Broglie, 'Tentative theory'.

[10] Slater to Kramers, 8 December 1923, *SHQP* 8, 10.

[11] Interview with J. C. Slater, *SHQP*. In the classical theory the sharpness of a spectral line was related to the period over which a wave was emitted, the wave undergoing a gradual change of frequency as the electron emitting it spiralled inwards.

[12] *Ibid.*

[13] Slater to Kramers, *SHQP* 8, 10.

[14] *Ibid.*

[15] See Chapter 4 above, and Stuewer, *Compton Effect.*

[16] H. A. Lorentz, lectures given in California in 1923 and printed in his *Problems of Modern Physics* (New York, 1927), especially pp. 150 ff.

[17] Slater's conception is closest to that of de Broglie, who was the only writer of any importance to suggest quanta moving at less than the traditional speed of light. de Broglie's work appeared in the *Comptes Rendus* over 1923, and the English version was communicated to the *Philosophical Magazine*, for publication in February 1924, by Slater's host Fowler, so that some influence on Slater would seem to be quite probable. On the other hand, the idea of light-quanta guided by a system of ghost waves was very commonplace, and Slater's conversion to the light-quantum concept could also have been related to the Compton effect results of the previous year. In his interview, given in 1963, Slater denied the influence of de Broglie, and said that he got nothing from Fowler but politeness. Recounting his idea in 1925, however, he wrote that "the theory in this form was developed in England, under the guidance of Mr R. H. Fowler, to whom my sincerest thanks are due'. (J. C. Slater, 'The nature of radiation', *Nature 116* (1925), 278); and Fowler wrote to Bohr on 14 January 1924, *BSC*, that "I thought he [Slater] was on sound lines and I encouraged him as much as I could". Slater's interview was given only reluctantly, and revealed a marked hostility towards Bohr, whose vague "handwaving" clearly repelled him. The Bohr-Kramers-Slater episode had hurt him considerably, and this may have affected his recollections. The question as to his sources if any must remain an open one.

[18] Slater to Kramers, 8 December 1923, *SHQP* 8, 10. Bohr's 'Fundamental postulates' had also been communicated, to the Cambridge Philosophical Society, by Fowler in 1923.

[19] Interview with Slater, *SHQP*.

[20] The light-quanta were emitted instantaneously, but they were governed by the waves, which were not.

[21] See Note 17 above.
[22] Kemble to Bohr, 4 January 1924, *BSC.*
[23] Kramers and Holst, *The Atom*, 175. Kramers's views and abilities are difficult to assess. He was obviously a first rate physicist, and a philosophical one at that, but on matters of interpretation he often comes across, perhaps unfairly, as naive. See forthcoming papers by H. Radder in *History of Science* and *Janus* for a more generous view of Kramers than that taken here.
[24] J. C. Slater, 'Radiation and atoms', *Nature 113* (1924), 307: "But when the idea with this [the light-quantum] interpretation was described to Dr. Kramers, he pointed out that it scarcely suggested the definite coupling between emission and absorption processes which light-quanta would provide." The same account is repeated in Slater, 'Nature of radiation'.
[25] Slater, 'Radiation and atoms', 307.
[26] Interview with Slater, *SHQP*; Slater to van Vleck, 27 July 1924, *SHQP* 49, 14, in which the Bohr-Kramers-Slater theory is clearly rejected.
[27] Bohr, Kramers and Slater, 'Quantum theory of radiation', 796.
[28] Interview with Slater, *SHQP.*
[29] A. H. Compton, 'The scattering of X-rays', *J. Franklin Inst. 198* (1924), 61–72, esp. 69, where Sommerfeld's views are also quoted.
[30] Bohr, Kramers and Slater, 'Quantum theory of radiation', 799.
[31] Note 29 above.
[32] E. C. Stoner, 'The structure of radiation', *Proc. Camb. Phil. Soc. 22* (1925), 577–594, esp. 592, and see also Note 1 to Chapter 2 above.
[33] Einstein to Ehrenfest, 31 May 1924 and see also 12 July 1924 and Einstein to Born, 29 April 1924, all discussed with quotations in Klein, 'First phase', 32–35.
[34] Ehrenfest to Einstein, 9 January 1925, *ibid.*, 31.
[35] Pauli to Bohr, 2 October 1924, *PB*, 163–167 (Item 66).
[36] Pauli to Bohr, 21 February 1924, *PB*, 147–149 (Item 56). See also Heisenberg to Bohr, 8 January 1925, *BSC*: "[Pauli] does not believe in the virtual oscillators and denounces the virtualisation of physics. It is not clear to me what he means by this." "Er glaube . . . nicht aber an virtuelle Oszillatoren und schimpft über die 'Virtualisierung' der Physik. Mir ist nicht klar, was es damit meint.")
[37] Pauli to Bohr, 2 October 1924, *PB*, 163–167 (Item 66). See also Pauli to Sommerfeld, November 1924, *PB*, 173–176 (Item 70).
[38] Schrödinger's position is clearest in his letter to Pauli on 8 November 1922, *PB* 69–73 (Item 29), for which see Chapter 2 above. For his reaction to Bohr-Kramers-Slater see Schrödinger to Bohr, 24 May 1924, *BSC*, and E. Schrödinger, 'Bohr's neue Strahlungshypothese und der Energiesatz', *Naturwissenschaften 12* (1924), 720–724.
[39] H. A. Kramers, 'The law of dispersion and Bohr's theory of spectra', *Nature 113* (1924), 673–674; 'The quantum theory of dispersion', *Nature 114* (1924), 310 (in response to Breit's criticisms); H. A. Kramers and W. Heisenberg, 'Über die Streuung von Strahlen durch Atome', *Zeit. Phys. 31* (1925), 681–707.
[40] Jordan's first published research was a critique of Einstein's 1916 demonstration of light-quanta (P. Jordan, 'Zur Theorie der Quantenstrahlung', *Zeit. Phys. 30* (1924), 297–319, refuted by A. Einstein, 'Bemerkung zu P. Jordans Abhandlung . . . ', *ibid.*, *31* (1925), 784–785). During the remainder of the decade his work showed a strong preference for wave rather than particle formulations, but his explicit discussions of physical issues remained strictly neutral.

[41] R. H. Fowler, manuscript notes taken by Dirac on a lecture course 'Recent Developments' (1925), *SHQP* 36, 8. R. Becker, 'Über Absorption und Dispersion in Bohr's Quantentheorie', *Zeit. Phys. 27* (1924), 173–188.
[42] M. Born, 'Über Quantenmechanik', *Zeit. Phys. 26* (1924), 379–395, translated in *SQM*, 181–198, esp. 189.
[43] Heisenberg to Pauli, 8 June 1924, *PB*, 154–156 (Item 62).
[44] Heisenberg to Pauli, 4 March 1924, *PB*, 149–150 (Item 57).
[45] Ladenburg to Kramers, 31 May 1924, *SHQP* 8, 9.
[46] In a further letter of 8 June 1924, *SHQP* 8, 9, Ladenburg wrote to Kramers of Einstein's reaction to the oscillator theory that "Seine Meinung was entschrieden nicht ungünstig". If anything this would support the view that Ladenburg was again talking of the technique rather than of the interpretation, but it is much more likely to be a slip of the pen: there is already one slip in "entschrieden" for "entschieden", and the inclusion of "nicht" may well be another. In neither letter is the interpretation referred to explicitly.
[47] C. Ramsauer, 'Über der Wirkungsquerschnitt der Gasmolekule gegenüber langsamen Elektronen', *Ann. der Phys. 64* (1921), 513–540, *66* (1922), 546–558, and *72* (1923), 345–352.
[48] Geiger to Bohr, 17 April 1925, *BSC*; W. Bothe and H. Geiger, 'Über das Wesen des Comptoneffektes; ein experimenteller Beitrag zur Theorie der Strahlung', *Zeit. Phys. 32* (1925), 639–663, and 'Experimentelles zur Theorie von Bohr, Kramers und Slater', *Naturwissenschaften 13* (1925), 440–441. See also A. H. Compton and A. W. Simon, 'Directed quanta of scattered X-rays', *Phys. Rev. 26* (1925), 289–299.
[49] Bohr to Heisenberg, 18 April 1925, *BSC*, written before the letter from Geiger had been recieved. Bohr wrote that he was preparing himself for Geiger's results, which he clearly thought would go against his interpretation. Writing to Franck on 21 April 1925, *BSC*, he wrote that he had long been worried by Ramsauer's results, and he referred to a much earlier letter, now lost, in which he had discussed these.
[50] Born to Bohr, 24 April 1925, *BSC*.
[51] Bohr to Geiger, 21 April 1925, Bohr to Franck, 21 April 1925, and Bohr to Born, 1 May 1925, *BSC*. See also Stuewer, *Compton Effect*, 301.
[52] Pauli to Kramers, 27 July 1925, *PB*, 232–235 (Item 97).
[53] Interview with J. C. Slater, *SHQP*.
[54] Kramers, 'Law of dispersion'.
[55] Interview with W. Heisenberg, *SHQP*.
[56] Kramers, 'Quantum theory of dispersion' and Born, 'Quantenmechanik'. Kramers wrote that his formula was derived by substituting a quantum expression for the classical dispersion form, but the substitution given was far from obvious. It could only have been derived from physical considerations or from Born's general rules, and even in the former case a knowledge of Born's work would seem to have been necessary to get the exact form given by Kramers. The timing of the papers allows plenty of scope for this, and it would indeed be remarkable if Kramers were not aware of Born's work when he completed his own.
[57] See Chapter 4 above.
[58] Heisenberg to Pauli, 9 October 1923, *PB*, 125–128 (Item 47): see Note 63 to Chapter 4 above.
[59] That the theory had such implications was recognised by Heisenberg in his letter to Bohr of 22 December 1923, *BSC*.

60 Heisenberg to Pauli, 8 June 1924, *PB*, 154–156 (Item 62).
61 Interview with W. Heisenberg, *SHQP*, and Born, *My Life*, 216.
62 Born, 'Quantenmechanik'; W. Heisenberg, 'Über eine Abänderung der formalen Regeln der Quantentheorie beim Problem der anomalen Zeemaneffekte', *Zeit. Phys. 26* (1924), 291–307.
63 Born, 'Quantenmechanik': *SQM*, 182.
64 It is an historian's job to speculate to some extent, but when a new theory arises from group discussions, or from ideas already 'in the air', such speculation tends to distort the reality. This seems to have happened, for example, in respect of the history of the spin concept, when van der Waerden wrote an excellent detailed study, only to be censured, also with good reason, by two of the main participants. Goudsmit and Uhlenbeck argued that the history had been distorted by overemphasis on "micro-history" and consequent loss of its "irrational" part, that the ideas were in the air and that it did not really matter who published what or when. See S. Goudsmit and G. Uhlenbeck, manuscripts on the advent of spin, *SHQP*, and Waerden, 'Exclusion principle'. The present situation was described by Born as a groping towards the new theory by his whole department jointly.
65 All Born's work on the quantum theory had been conducted through the medium of perturbation theory, and he took the same approach to the new quantum mechanics in 1925–1926.
66 Born, 'Quantenmechanik': *SQM*, 181.
67 *Ibid.*, 182.
68 M. Born, 'Quantentheorie und Störungsrechnung', *Naturwissenschaften 11* (1923), 537–542.
69 Born, 'Quantenmechanik': *SQM*, 193; he also wrote that "the two kinds of resonator have a different behaviour".
70 *Ibid.*, 190.
71 Kramers and Heisenberg, 'Streuung von Strahlen'.
72 Interview with W. Heisenberg, *SHQP*. Smekal's treatment was less general than that of Kramers and Heisenberg, but more so than Kramers's earlier one, and it is somewhat surprising that Kramers had not apparently read it, or chose not to cite it, when he published the latter: Kramers, 'Law of dispersion', and A. Smekal, 'Zur Quantentheorie der Dispersion', *Naturwissenschaften 11* (1923), 873–875.
73 Interview with W. Heisenberg, *SHQP*.
74 The work is discussed thoroughly by MacKinnon, 'Heisenberg', but he seems to me to miss the point rather in treating the virtual oscillator approach as here entailing a physical model. Given that it did not do so then the otherwise contradictory treatments of MacKinnon and Serwer ('Unmechanischer Zwang', especially 221) become more or less consistent.
75 R. Wood and A. Ellett, 'On the influence of magnetic fields on the polarisation of resonance radiation', *Proc. Roy. Soc. A103* (1923), 396–403; P. D. Foot, A. E. Ruark and F. L. Mohler, 'The D_2 Zeeman pattern for resonance radiation', *J. Opt. Soc. America 7* (1923), 415–418.
76 N. Bohr, 'Zur Polarisation des Fluorescenzlichtes', *Naturwissenschaften 12* (1924), 1115–1117; W. Heisenberg, 'Über eine Anwendung des Korrespondenzprinzips auf die Frage nach der Polarisation des Fluorescenzlichtes', *Zeit. Phys. 31* (1925), 617–626.
77 Bohr, 'Polarisation', 1115.
78 H. Burger and H. Dorgelo, 'Beziehung zwischen inneren Quantenzahlen und Inten-

sitäten von Mehrfachlinien', *Zeit. Phys.* *23* (1924), 258–266; L. Ornstein and H. Burger, 'Strahlungsgesetz und Intensität der Komponenten im Zeemaneffekt', *ibid.*, *28* (1924), 135–141, and *29* (1924), 241–242.

79 Interview with W. Heisenberg, *SHQP*, and see MacKinnon, 'Heisenberg', 154.

80 Heisenberg, 'Polarisation', 621.

81 *Ibid.*, 617.

82 M. Born and W. Heisenberg, 'Über den Einfluss der Deformierbarkeit der Ionen auf optische und chemische Konstanter, I', *Zeit. Phys.* *23* (1924), 388–410. A. Landé, 'Zur Struktur des Neonspektrums', *ibid.*, *17* (1923), 292–294, and 'Das Wesen der relativistisch Röntgendubletts', *ibid.*, *24* (1924), 88–97; A. Landé and W. Heisenberg, 'Termstruktur der Multipletts höherer Stufe', *ibid.*, *25* (1924), 279–286.

83 W. Pauli, 'Über den Einfluss der Geschwindigkeitsabhängigkeit der Elektronenmasse auf den Zeemaneffekt', *ibid.*, *31* (1925), 373–385.

84 W. Pauli, 'Über den Zusammenhang des Abschlusses der Elektronengruppen in Atom mit der Komplexstruktur der Spektren', *ibid.*, *32* (1925), 765–783; the *Zweideutigkeit* is introduced in this sense in Pauli, 'Einfluss der Geschwindigkeitsabhängigkeit', 385.

85 E. C. Stoner, 'On the distribution of electrons among atomic levels', *Phil. Mag.* *48* (1924), 719–736.

86 Pauli, 'Zusammenhang'.

87 Pauli to Bohr, 12 and 31 December 1924, *PB*, 186–189, 197–199 (Items 74, 79).

88 Pauli to Bohr, 31 December 1924, *PB*, 197–199 (Item 79). He suggested that Kramers could make this sound very convincing in his popular lectures.

89 Pauli to Bohr, 12 December 1924, *PB*, 186–189 (Item 74).

90 Pauli to Bohr, 31 December 1924, *PB*, 197–199 (Item 79).

91 *Ibid.*; Serwer, 'Unmechanischer Zwang', 237ff.; Pauli to Heisenberg, 28 February 1925, and Pauli to Bohr, 30 April 1925, *PB*, 211–214 (Items 86, 87).

92 A. Sommerfeld, *Atombau und Spektrallinien* (Braunschweig, 1924).

93 W. Heisenberg, 'Zur Quantentheorie der Multiplettstruktur in der anomalen Zeemaneffekte', *Zeit. Phys.* *32* (1925), 841–860. Compare Serwer, 'Unmechanischer Zwang', 239–245, and MacKinnon, 'Heisenberg', 159–163, for different interpretations of this paper. The account given here follows Serwer.

94 Heisenberg, 'Quantentheorie der Multiplettstruktur', 856.

95 *Ibid.*, 842.

96 Interview with W. Heisenberg, *SHQP*; Heisenberg, 'Errinerungen'; Kronig, 'Turning point'.

97 W. Heisenberg, 'Über quantentheoretische Umdeutung kinematischer und mechanischer Beziehungen', *Zeit. Phys.* *33* (1925), 879–893. I have not gone into the details of Heisenberg's work here, largely because it has been treated in great depth in several other places. See especially *SQM*, Heisenberg, "Errinerungen'; Kronig, 'Turning point'; Cassidy, 'Werner Heisenberg'; and MacKinnon, 'Heisenberg'.

98 Heisenberg to Pauli, 21 June 1925, *PB*, 219–221 (Item 91).

99 See Note 9 to Chapter 6 below.

CHAPTER 6. THE NEW KINEMATICS AND ITS EXPLORATION

1 W. Heisenberg, 'Über quantentheoretische Umdeutung kinematischer und mechanischer Beziehungen', *Zeit. Phys.* *33* (1925), 879–893, translated in *SQM*, 261–276, esp. 262.

[2] In the interests of clarity and brevity I have here abandoned Heisenberg's notations. The conflicting notations he used, their origins and their implications for the genesis of his paper are examined by van der Waerden, *SQM*, 30–34. A detailed if not altogether convincing reconstruction of Heisenberg's thought processes is given in Mackinnon, 'Heisenberg', 164–184.

[3] See Chapter 5 above.

[4] The differentiation with respect to *n* could be theoretically justified only in the correspondence principle limit of large *n*. Practically, however, it was justified by the result to which it led, and that result was presumably the reason for its introduction. It provided the only means of translating from the formula obtained by substitution into (5) to a form that could be quantised as in Born's virtual oscillator mechanics of the previous year: M. Born, "Über Quantenmechanik', *Zeit. Phys. 26* (1924), 379–395, translated in *SQM*, 181–198. The result was akin to that obtained from the virtual oscillator theory by Kuhn, working in Copenhagen that Spring: W. Kuhn, 'Über die Gesämstärke der von einem Zustande ausgehende Absorptionslinien', *Zeit. Phys. 33* (1925), 408–412, translated in *SQM*, 253–257.

[5] W. Bothe and H. Geiger, 'Über das Wesen des Comptoneffektes; ein experimenteller Beitrag zur Theorie der Strahlung', *Zeit. Phys. 32* (1925), 639–663; 'Experimentelles zur Theorie von Bohr, Kramers und Slater', *Naturwissenschaften 13* (1925), 440–441.

[6] Heisenberg did establish energy conservation for the case of the anharmonic oscillator; as his recollections confirm, he could hardly have continued, in the light of the Bothe-Geiger results, had he been unable to do this: interview with W. Heisenberg, *SHQP*, and Heisenberg, *Physics and Beyond*, 61.

[7] Heisenberg had started out by considering hydrogen, but had found it too difficult. See Heisenberg, 'Errinerungen', and see also MacKinnon, 'Heisenberg'.

[8] Heisenberg to Pauli, 24 June 1924, *PB*, 225–229 (Item 93).

[9] Pauli to Kramers, 27 July 1925, reflecting a similar statement in Heisenberg to Pauli, 9 July 1925: *PB*, 231–235 (Items 97, 96).

[10] Born, *My Life*, 216.

[11] *Ibid.*, 217.

[12] In particular, Born's assistance is acknowledged in the preface of R. Courant and D. Hilbert, *Methoden der Mathematischen Physik*, Bd. 1 (Berlin, 1924), the first chapter of which is on matrices, and the authors of which were his very close associates at Göttingen.

[13] Had Born's concern been primarily with the multiplication, then his delay in recognising the matrices would have been most extraordinary. Moreover, as indicated, below, his recent work had been very much concerned with the physical implications of possible alternative theories.

[14] Born to Pauli, 23 December 1919, *PB*, 9–11 (Item 4).

[15] Born, 'Quantenmechanik'.

[16] For a discussion of these problematic results, see Chapter 9 below. On receiving the Bothe-Geiger results from Geiger in a letter dated April 17th, Bohr immediately wrote to Born, who replied that he had already anticipated the result (most physicists were indeed expecting it) and was working on a new theory to cope with it (Born to Bohr, 4 April 1925, *BSC*): "The main thing is to keep the value of the Bohr-Kramers-Slater theory, namely the emission of radiation during the stationary states. But now there are besides these periods the jumps, which may be valuable here as momentum pro-

cesses. So if one tries generally to arrange the ordering of events in space and time, one must class the stationary states with the wave emissions, and the jumps with the light-quantum emissions." (Die Hauptsache ist, dass, man das Wertvolle der Bohr-Kramers-Slaterschen Theorie beibehält: nämlich die Emission der Wellenstrahlen während der stationaren Zustände. Nun gibt es aber neben diesen Zeitabschnitten die Sprünge, die hier als Momentanprozesse gelten mögen. Wenn man also überhaupt versucht, die Ordnung der Vorgänge in Raum und Zeit vorzuordnen, so muss man den stationaren Zustände die Wellenemission, den Sprüngen die Lichtquantenemission zuordnen.") Born continued to explain that the wave would carry the light-quantum and create interference effects by controlling the absorption of radiation by matter, This corresponded to the interpretation given by de Broglie in his definitive account of the matter wave theory: L. de Broglie, *Thèse* (Paris, 1924). de Broglie had since replaced it by a guiding wave concept, L. de Broglie, 'Sur la dynamique du quantum de lumière et les interférences', *Comptes Rendus 179* (1924), 1309–1311. But Born's aquaintance with de Broglie's work would most naturally have been through the *Thèse*, which he had been studying: Born, *My Life*, 231.

[17] Born to Einstein, 15 July 1925, Born and Einstein, *Letters* 83–88 (Item 49). This was despite Bohr's criticisms, for which see Note 20 below.

[18] Interviews with J. Franck and W. Elsasser, *SHQP*.

[19] M. Born, and P. Jordan, 'Zur Quantentheorie aperiodischer Vorgänge', *Zeit. Phys. 33* (1925), 479–505, submitted in June. That they should have been pursuing the virtual oscillator theory at this late stage is further evidence of the heuristic attitude adopted towards that theory, which was now physically disproven. It is possible that Born hoped to reinterpret the formal analysis in de Broglie's sense.

[20] Note 11 above. Responding to Born's advocacy of the matter wave theory, Bohr had offered both general criticisms, for which see below, and specific ones: "The content of your note has naturally interested us very much, but I must confess that I do not believe that a contradiction-free description of the phenomena can be reached in the way proposed. It seems to me that according to your picture the binding of the light-quanta with the waves is not tight enough. On one hand, I do not understand how according to your treatment it can be achieved that the paths of the light-quanta coincide with sufficient accuracy with the propagation of the waves. If the interaction between a quantum and a scattering atom depends only on the classically calculated momentum of its virtual resonator it could indeed scarcely be avoided that the quantum, for example in reflection or refraction, should be separated from the wave train to which it was originally attached. On the other hand it seems to me that your picture can hardly reproduce the quantitative relations of light absorption, for the supposition that the probability of the absorption of a light-quantum in an atom should be proportional to the intensity of the wave indeed leads to quite different laws from those corresponding to either the wave theory or the corpuscular theory of light. The rough agreement between the two theories, so far as the rectilinear propagation of light is concerned, indeed even rests on the fact that the number of corpuscles passing through a plane surface is proportional to the intensity of the waves, and that therefore the absorption must be described through the assumption of constant effective cross-sections of the atoms." ("Der Inahlt Ihrer Note hat uns natürlich sehr interessiert, aber ich muss gestehen, dass ich nicht glaube, dass eine widerspruchsfreie Beschreibung der Phänomene sich in der vorgeschlagenen Weise erreichen lässt. Es scheint mir, dass nach Ihrem Bilde

der Verband der Lichtquanten mit den Wellen ein nicht genügend enger ist. Einerseits sehe ich nicht ein, wie es sich nach Ihre Vorstellung erreichen lässt, dass die Bahnen der Lichtquanten mit hinreichender Genauigkeit mit der Fortpflanzung der Wellen zusammenfallen. Wenn die Wechselwirkung zwischen einem Quant und einem streuenden Atom nur von dem klassisch berechneten Moment seiner virtuellen Resonatoren abhängt, dürfte es ja kaum zu vermeiden sein, dass das Quant z.b. bei Spiegelung oder Brechung vollkommen von dem ihm ursprünglich zugeordneten Wellenzug getrennt wird. Anderseits scheint mir, dass Ihr Bild kaum die quantitativen Verhältnisse der Lichtabsorption wiedergeben kann, denn die Annahme, dass die Wahrscheinlichkeit der Auffangung eines Lichtquants durch ein Atom mit der Intensität der Wellen proportional sein sollte, führt ja zu ganz anderen Gesetzmässigkeiten als es so wie der Wellentheorie wie der korpuskularen Theorie des Lichtes entsprechen würde. Die grobe Übereinstimmung dieser zwei Theorien, soweit es die gradlinige Ausbreitung des Lichtes betrifft, beruht ja eben darauf, dass die Anzahl der Korpuskeln, die durch die Flächeneinheit geht überall der Intensität der Wellen proportional ist, und dass also die Absorption durch die Annahme eines konstanten wirksamen Querschnitt der Atome beschreiben werden muss." (Bohr to Born, 1 May 1925, *BSC*.))

21 Born, *My Life*, 218.

22 *Ibid.*, 218–219. In the thick of quantum theoretical developments, Born had already been overworking for some years, and continued to do so, driving himself to a nervous breakdown in 1928–1929. It should be noted that Born's recollections do not tie in with the submission dates of the papers he refers to, and I have given the latter the greater weight. Quotations from the recollections in Jammer, *Conceptual Development*, describing Born's holiday as a health cure and describing his meeting Jordan – whom he already knew well of course – on a train, appear to be spurious.

23 M. Born and P. Jordan, 'Zur Quantenmechanik', *Zeit. Phys. 34* (1925), 858–888, abridged translation in *SQM*, 277–306.

24 For details of the redaction of the Born-Jordan paper see *SQM*, 38–40.

25 Born and Jordan, 'Quantenmechanik', 883 (not in translation).

26 Heisenberg to Pauli, 18 September 1925, *PB*, 236–241 (Item 98); *SQM*, 44–48.

27 Heisenberg to Jordan, 13 September 1925, in *SQM*, 43–44.

28 Heisenberg to Pauli, 18 September 1925, *PB*, 236–241 (Item 98).

29 *SQM*, 49.

30 Born and Jordan, 'Quantenmechanik': *SQM*, 286.

31 M. Born, W. Heisenberg and P. Jordan, 'Zur Quantenmechanik II', *Zeit. Phys. 35* (1925), 557–615, translated in *SQM*, 321–386.

32 My own phraseology.

33 Born, *My Life*, 218.

34 M. Born and W. Pauli, 'Über die Quantelung gestörter mechanischer Systeme', *Zeit. Phys. 10* (1922), 137–158.

35 W. Pauli, review of Born's *Vorlesungen über Atommechanik* in *Naturwissenschaften 13* (1925), 487.

36 Pauli to Kronig, 9 October 1925, *PB*, 242–249 (Item 100); see also *SQM*, 37.

37 Heisenberg to Pauli, 16 November 1925, *PB*, 255–257 (Item 105); see also *SQM*, 56.

38 Interview with J. Franck, *SHQP*.

39 See Chapter 4 above.

40 Heisenberg to Pauli, 16 November 1925, *PB*, 255–257 (Item 105).

[41] *Ibid.*
[42] *Ibid.*
[43] When still in his teens his mathematics had been sufficient to argue on equal terms with Weyl (and a few years later with Eddington) in the context of general relativity theory: see Chaper 2 above.
[44] Born, Heisenberg and Jordan, 'Quantenmechanik II'; for the redaction see *SQM*, 55–56.
[45] W. Pauli, 'Über das Wasserstoffspektrum vom Standpunkt der neuen Quantenmechanik', *Zeit. Phys. 36* (1926), 336–363, translated in *SQM*, 387–416.
[46] P. Jordan, 'Über kanonische Transformationen in der Quantenmechanik', *Zeit. Phys. 37* (1926), 383–386, and *38* (1926), 513–517.
[47] Heisenberg to Pauli, 16 November 1926, *PB*, 255–257 (Item 105).
[48] Heisenberg to Born, 5 October 1925, and to Jordan, 7 October 1925, *SHQP* 18, 2; see *SQM*, 55, and Born, Heisenberg and Jordan, 'Quantenmechanik II': *SQM*, 325–326.
[49] *Ibid.*, 321–325; see also W. Heisenberg, 'Über quantentheoretischer Kinematik und Mechanik', *Math. Ann. 95* (1926), 694–705, where, despite the mathematical readership at which the exposition was aimed, there is again no use of the term 'matrix mechanics'.
[50] Pauli to Bohr, 17 November 1925, *PB*, 257–261 (Item 106).
[51] *Ibid.*
[52] See Chapter 2 above.
[53] Or he may have received a very similar one.
[54] Heisenberg to Pauli, 24 November 1925, *PB*, 262–266 (Item 108). I have translated Heisenberg's "grobe" by both "rough" and "coarse", depending on the context.
[55] Born and Jordan, 'Zur Quantentheorie aperiodischer Vorgänge'.
[56] P. Jordan, 'Über das thermische Gleichgewicht zwischen Quantenatomen und Hohlraumstrahlung', *Zeit. Phys. 33* (1925), 649–655.
[57] Born, Heisenberg and Jordan, 'Quantenmechanik II': *SQM*, 375–385.
[58] S. N. Bose, 'Planck's Gesetz und Lichtquantenhypothese', *Zeit. Phys. 26* (1924), 178–181.
[59] De Broglie, *Thèse.*
[60] *Ibid.*, 79.
[61] See also P. A. M. Dirac, manuscript draft of a critical presentation of Einstein-Bose statistical mechanics, *SHQP* 36, 9: "It is a disadvantage of [Bose's] theory that the cells play so important a part in it. One assumes that the whole of phase space is divided into a number of compartments, and each atom or light-quantum as the case may be is definitely in one compartment. One can get over this difficulty by adopting the point of view, first proposed by de Broglie, that each particle is associated with a wave, and letting the waves play the part of the cells in the previous theory. Several particles may be associated with the same wave. This point of view is possible only because it turns out that the number of waves associated with a given region in phase space is equal to the number of cells into which that region of phase space was divided in the previous theory. The results in the two theories become mathematically equivalent."
[62] Born, Heisenberg and Jordan, 'Quantenmechanik II': *SQM*, 377.
[63] *Ibid.*, 378, and see Heisenberg to Pauli, 23 October 1925 and 16 November 1925, *PB*, 251–252, 255–257 (Items 102, 105).
[64] Heisenberg to Pauli, 16 November 1925, *PB*, 255–257 (Item 105).
[65] Heisenberg to Pauli, 23 October 1925, *PB*, 251–252 (Item 102).

[66] W. Heisenberg, 'Mehrkörperproblem und Resonanz in der Quantenmechanik', *Zeit. Phys. 38* (1926), 411–427. For a discussion of the earlier history of this problem see Serwer, 'Unmechanischer Zwang', and Chapter 4 above.
[67] Heisenberg, 'Mehrkörperproblem und Resonanz', 422.
[68] *Ibid.*, 423.
[69] E. Fermi, 'Zur Quantelung des idealen einatomigen Gases', *Zeit. Phys. 36* (1926), 902–912. Fermi assumed that the molecules of the gas behaved according to Pauli's exclusion principle, showed that this led to results in accordance with the experimental Stern-Tetrode values for entropy at high temperatures, and derived the "Fermi-Dirac" statistics in the form

$$N_s \propto \alpha e^{-\beta s} / (1 + \alpha e^{-\beta s})$$

for the number of molecules with energy $sh\nu$, α and β being constants.
[70] Heisenberg to Pauli, 15 November 1926, *PB*, 354–356 (Item 146).
[71] We may reconstruct the probable source of his confusion. Given two parts of a system, I and II, Heisenberg considered two possible solutions: either I was in state X, II in state Y, say (X, Y), or *vice versa*, (Y, X). Bose had treated these two possibilities as a single state, so reducing the statistical weight from 2 to 1 as Heisenberg said. The exclusion principle forbad the joint existence of (X, Y) and (Y, X), but Heisenberg took it as forbidding one of the two states altogether.
[72] Born, Heisenberg and Jordan, 'Quantenmechanik II': *SQM*, 348–364; Jammer, *Conceptual Development*, 218.
[73] Born, Heisenberg and Jordan, 'Quantenmechanik II': *SQM*, 358, but see Note 78 below.
[74] Wiener, *I am a Mathematician*, 108.
[75] N. Wiener, 'The operational calculus', *Math. Ann. 95* (1926), 557–584; for the mathematical background see Jammer, *Conceptual Development*, 223–228.
[76] M. Born and N. Wiener, 'A new formulation of the laws of quantisation of periodic and aperiodic phenomena', *J. Math. and Phys. 5* (1926), 84–98; 'Eine neue Formulierung der Quantengesetze für periodische und nicht-periodische Vorgänge', *Zeit. Phys. 36* (1926), 174–187.
[77] *Ibid.*, 84. Although references will be given here to the English version of this paper, it should be noted that as a whole the German is somewhat clearer.
[78] The operator theory had been developed mainly by Hilbert and his students in connection with integral equation theory, to which the method of infinite matrices had been applied by Fredholm. Infinite matrices were primary to this research, operators only secondary; but Born had already encountered the fact that while infinite matrix theory and operator theory were equivalent for finite-dimensional spaces and for bounded forms on more general spaces this did not seem to extend to unbounded forms, i.e., to the context of quantum mechanics. As we have seen, he had glossed over this problem in the three-man-paper, but its existence provided strong mathematical grounds for preferring the operator to the matrix formulation. An exact account of the situation, and with it a mathematically rigorous formulation of quantum mechanics, was provided only after von Neumann had taken up his general study of Hilbert spaces in 1927–1928: see especially J. von Neumann, 'Allgemeine Eigenwertetheorie Hermetischer Funktionsoperatoren', *Math. Ann. 102* (1929), 49–131.

[79] The work of Born and Wiener also raised an interesting point concerning an aspect of quantum mechanics that is not often treated, namely the intrusion of the imaginary constant *i*. They noted that their theory seemed to attribute a motion to a particular state of the form $q_k(t) = \Sigma\, q_{mk}\, e^{2\pi i \nu(m,\,k)t}$, and that this was complex even when the matrix (q_{mn}) was Hermitian. From this they deduced that "there are then two real motions belonging to every state, corresponding respectively to the real and pure imaginary parts of the line of the matrix" (Born and Wiener, 'New formulation', 86). Unable to make anything of this they did not pursue it, but Dirac had already noticed the property in October 1925 and had followed it slightly further, though without publishing his considerations: P. A. M. Dirac, manuscript treating virtual oscillators according to the new quantum mechanics, October 1925, *SHQP* 36, 9. In the virtual oscillator theory the emission and absorption oscillators had each been defined by the *real part* of a form $C\, e^{i\omega t}$, and although they did not carry energy they were related probabilistically to a transition process. In the new quantum mechanics, however, the restriction to the real part was dropped, and an atom described by *complex* oscillators of the form $C\, e^{i\omega t}$. Reviewing this situation, Dirac deduced that "the imaginary exponential is essential and fundamental in the new theory", and noted that one need a combination of $e^{i\omega t}$ and $e^{-i\omega t}$ oscillators in order to get any radiation at all. Taking another line of approach, Jordan later linked the presence of *i* with the absence of classical causality, for which see below.

CHAPTER 7. WAVE MECHANICS AND THE PROBLEM OF INTERPRETATION

[1] M. Born and N. Wiener, 'A new formulation of the laws of quantization of periodic and aperiodic phenomena', *J. Math. and Phys. 5* (1926), 84–98; *Zeit. Phys. 36* (1926), 174–187.
[2] E. Schrödinger, 'Quantisierung als Eigenwertproblem', *Ann. der Phys. 79* (1926), 362–376; 489–527; *80* (1926), 437–490; *81* (1926), 109–139. The first, second and fourth communications are translated as 'Quantisation as an eigenvalue problem', in Ludwig, *Wave Mechanics*, 94–105, 106–126, 151–167. For the background to the theory see Jammer, *Conceptual Development*; Klein, 'Einstein and duality'; Gerber, 'Gesichte der Wellenmechanik'; Raman and Forman, 'Why Schrödinger'; Hanle, 'The coming of age' and 'Schrödinger's reaction'; Wessels, 'Schrödinger's route'; Kragh, 'Schrödinger and the wave equation'.
[3] E. Schrödinger, 'Über das Verhältnis der Heisenberg-Born-Jordanschen Quantenmechanik zu der meinen', *Ann. der Phys. 79* (1926), 743–756, translated in Ludwig, *Wave Mechanics*, 127–150. For the extent and limitations of Schrödinger's demonstration see Waerden, From matrix mechanics'.
[4] Schrödinger, 'Quantisation', 163.
[5] *Ibid.*
[6] L. de Broglie, *Thèse* (Paris, 1924), 69, and 'Sur la dynamique du quantum', *Comptes Rendus 179* (1924), 1309, emphasises the priority of the wave picture, although his whole theory had been based upon the primacy of the particle concept for light.
[7] Schrödinger, 'Quantisation', 163–164.
[8] Schrödinger, 'Verhältnis', 128, and see Przibram, *Letters on Wave Mechanics*, for the responses of Einstein, Lorentz, Wien and Planck.

[9] Lorentz to Schrödinger, 27 May 1926, and Einstein to Schrödinger, 31 May 1928, in Przibram, *Letters on Wave Mechanics*, 43–54, 30–31 (Items 19, 14). L. de Broglie, 'La mécanique ondulatoire et la structure atomique de la matière et du rayonnement', *Comptes Rendus 184* (1927), 273–274, was also very critical.
[10] Interview with M. Born, *SHQP*, quoted in Jammer, *Conceptual Development*, 223.
[11] Pauli to Jordan, 12 April 1926, *PB*, 315–320 (Item 131), translated in Waerden, 'From matrix mechanics'.
[12] For a full analysis of Pauli's work and comparison with Schrödinger's, see Waerden, 'From matrix mechanics'. For Pauli's use of the operator formulation see Pauli to Heisenberg, 31 January 1926, *PB*, 283–288 (Item 118).
[13] Pauli to Jordan, 12 April 1926, *PB*, 315–320 (Item 131)'
[14] Heisenberg to Jordan, 28 July 1926, *SHQP* 18, 2.
[15] See Chapter 2 above.
[16] Pauli to Schrödinger, 24 May 1926, *PB*, 324–327 (Item 134).
[17] Pauli to Schrödinger, 22 November 1926, *PB*, 356–357 (Item 147).
[18] Pauli to Schrödinger, 12 December 1926, *PB*, 364–366 (Item 150).
[19] Heisenberg to Dirac, 26 May 1926, *SHQP* 59, 2.
[20] W. Heisenberg, 'Mehrkörperproblem und Resonanz in der Quantenmechanik', *Zeit. Phys. 38* (1926), 411–427, esp. 422; Heisenberg to Pauli, 8 June 1926, *PB*, 328–329 (Item 136).
[21] Heisenberg to Pauli, 8 June 1926, *PB*, 328–329 (Item 136): literally "dung" or "manure".
[22] *Ibid.*
[23] Heisenberg to Pauli, 28 July 1926, *PB*, 337–340 (Item 142).
[24] Sommerfeld to Pauli, 26 July 1926, *PB*, 337 (Item 141).
[25] Bohr to Kronig, 28 October 1926, *SHQP* 16, 1.
[26] Born to Schrödinger, 16 May 1927, *SHQP* 41, 7.
[27] Jordan to Schrödinger, catalogued as May 1926 but date uncertain, *SHQP* 41, 8. It should be said that Schrödinger does not appear to have been so naive as his opponents made out. His emphasis on the wave interpretation was justified in his 'Quantisation', 163–164, partly in terms of the danger of reverting to the untenable concept of three-dimensional spatial representations if the particle picture were adopted. His own interpretation centred on the proposition that the energy concept was inapplicable on the microscopic scale: E. Schrödinger, 'Energieaustausch nach der Wellenmechanik', *Ann. der Phys. 83* (1927), 956–968; Schrödinger to Planck, 31 May 1926, and to Lorentz, 6 June 1926, in Przibram, *Letters on Wave Mechanics*, 8–11 and 55–66 (Items 4, 20). See also L. Wessels, 'Schrödinger's interpretations of wave mechanics', Ph.D. dissertation, Indiana University, 1975.
[28] Born to Schrödinger, 16 May 1927, *SHQP* 41, 7: "Heisenberg war von vornherein nicht meiner Meinung, dass Ihre Wellenmechanik physikalisch mehr bedeute, als unsere Quantenmechanik .... Inzwischen habe ich mich aber wieder zu Heisenbergs Standpunkt zurückgefunden."
[29] *Ibid.*: " ... die einfache Art, aperiodische Vorgänge (Stösse) zu behandeln, bracht mir zunächst zu dem Glauben der Überlegenheit Ihre Anschauungsweise."
[30] M. Born, 'Quantenmechanik der Stossvorgänge', *Zeit. Phys. 38* (1926), 803–827, translated in Ludwig, *Wave Mechanics*, 206–225, esp. 224.
[31] M. Born, 'Physical aspects of quantum mechanics', *Nature 119* (1926), 354.

[32] M. Born, 'Zur Quantenmechanik der Stossvorgänge', *Zeit. Phys. 37* (1926), 863–867.
[33] *Ibid.*, 864.
[34] *Ibid.*, 865–866.
[35] Born, 'Quantenmechanik der Stossvorgänge': Ludwig, *Wave Mechanics*, 225.
[36] Born, 'Zur Quantenmechanik der Stossvorgänge'.
[37] Born, 'Quantenmechanik der Stossvorgänge': Ludwig, *Wave Mechanics*, 207.
[38] *Ibid.*, 207.
[39] *Ibid.*, 207.
[40] Einstein to Born, 4 December 1926, Born and Einstein, *Letters*, 90–91 (Item 52).
[41] *Ibid.*, 91.
[42] Born, *Experiment and Theory*, 23, and 'Bedeutung zur statistischen Deutung der Quantenmechanik', in Bopp, *Werner Heisenberg*, 103–118, esp. 103.
[43] Chapter 6 above, and Born, *My Life*, 231.
[44] This possibility had long been discounted: see for example Hendry, 'Wave-particle duality'.
[45] H. A. Lorentz, *Problems of Modern Physics* (New York, 1927 and 1957), 156.
[46] Hendry, 'Weimer culture'.
[47] Einstein to Born, 27 January 1920, and 3 March 1920, Born and Einstein, *Letters*, 20–26 (Items 13, 14).
[48] His assistance in the solution of Schrödinger's eigenvalue problems was acknowledged by Schrödinger, 'Quantisation': Ludwig, *Wave Mechanics*, 97. His assistance on aspects of the matrix mechanics was acknowledged by Born: Born to Weyl, 3 October 1925, *ETHZ* 91, 488.
[49] M. Born, 'Das Adiabatenprinzip in der Quantenmechanik', *Zeit. Phys. 40* (1926), 167–192, esp. 167.
[50] Hendry, 'Weimar culture'.
[51] Heisenberg to Pauli, 24 November 1925, *PB*, 262–266 (Item 108).
[52] Born, 'Zur Quantenmechanik der Stossvorgänge', 863.
[53] Note 3 of Chapter 3 above. Born's training had been with Hilbert and Minkowski.
[54] For Pauli's reaction to the virtual oscillator theory see Chapter 5 above.
[55] W. Pauli, 'Quantentheorie', *Handbuch der Physik 23* (1926), 1–278. The correspondence is not complete and Pauli may well have made comments now lost.
[56] Interview with Heisenberg, *SHQP*.
[57] Heisenberg to Pauli, 28 July 1926, *PB*, 337–340 (Item 142).
[58] K. Lanczos, 'Über eine feldmässige Darstellung der neuen Quantenmechanik', *Zeit. Phys. 35* (1926), 812–830.
[59] P. A. M. Dirac, 'The fundamental equations of quantum mechanics', *Proc. Roy. Soc. A109* (1925), 642–653, reprinted in *SQM*, 307–320.
[60] P. A. M. Dirac, 'Quantum mechanics and a preliminary investigation of the hydrogen atom', *Proc. Roy. Soc. A110* (1926), 561–579, reprinted in *SQM*, 417–427.
[61] *Ibid.*, 418.
[62] P. A. M. Dirac, 'On the theory of quantum mechanics', *Proc. Roy. Soc. A112* (1926), 661–667, esp. 667.
[63] *Ibid.*, 662.
[64] See Note 69 of Chapter 6 above.
[65] Dirac, 'Theory of quantum mechanics', 662, 666.

[66] Pauli to Heisenberg, 19 October 1926, *PB*, 340–349 (Item 143).
[67] M. Born, W. Heisenberg and P. Jordan, 'Zur Quantenmechanik II', *Zeit. Phys. 35* (1925), 557–615, translated in *SQM*, 321–386, esp. 385.
[68] Pauli to Heisenberg, 19 October 1926, *PB*, 340–349 (Item 143).
[69] Born, Heisenberg and Jordan, 'Zur Quantenmechanik II': *SQM*, 380.
[70] Heisenberg to Pauli, 15 November 1926, *PB*, 354–356 (Item 146).
[71] Pauli to Heisenberg, 19 October 1926, *PB*, 340–349 (Item 143).
[72] H. A. Kramers, 'Wellenmechanik und halbzahlige Quantisierung', *Zeit. Phys. 39* (1926), 828–840.
[73] W. Heisenberg, 'Schwankungserscheinungen und Quantenmechanik', *Zeit. Phys. 40* (1926), 501–506.

CHAPTER 8. TRANSFORMATION THEORY AND THE DEVELOPMENT OF THE PROBABILISTIC INTERPRETATION

[1] Heisenberg to Pauli, 28 October 1926, *PB*, 349–352 (Item 144).
[2] *Ibid.*
[3] Reported in Bohr to Kronig, 28 October 1926, *SHQP* 16, 1.
[4] Heisenberg to Pauli, 4 November 1926, *PB*, 352–353 (Item 145).
[5] W. Heisenberg, 'Schwankungserscheinungen und Quantenmechanik', *Zeit. Phys. 40* (1926), 501–506.
[6] *Ibid.*, 501.
[7] Heisenberg to Pauli, 4 November 1926, *PB*, 352–353 (Item 145).
[8] *Ibid.*
[9] Heisenberg to Pauli, 28 October 1926, *PB*, 349–352 (item 144).
[10] Heisenberg to Pauli, 4 November 1926, *PB*, 352–353 (Item 145); O. Klein, 'Quantentheorie und fünfdimensionale Relativitätstheorie', *Zeit. Phys. 37* (1926), 895–906; T. Kaluza, 'Zum Unitätsproblem der Physik', *S.-B. Preuss. Akad. Wiss.* (1921), 966–972.
[11] Heisenberg to Pauli, 23 November 1926, *PB*, 357–360 (Item 148).
[12] P. A. M. Dirac, 'The physical interpretation of the quantum dynamics', *Proc. Roy. Soc. A113* (1927), 621–641, esp. 621.
[13] P. Jordan, 'Über eine neue Begründung der Quantenmechanik', *Zeit. Phys. 40* (1927), 809–838.
[14] We may note also that his first research had been on a critique of Einstein's arguments in support of the necessity of light-quanta: P. Jordan, 'Zur Theorie der Quantenstrahlung', *Zeit. Phys. 30* (1924), 297–319.
[15] Jordan, 'Neue Begründung', 811.
[16] Dirac to Jordan, 24 December 1926, *SHQP* 18, 1.
[17] *Ibid.*
[18] Dirac, 'Physical interpretation', 641; Pauli to Bohr, 17 November 1925, *PB*, 257–261 (Item 106).
[19] See the discussion of Bohr's ideas in Chapter 9 below.
[20] See Chapter 2 above.
[21] D. Hilbert, J. von Neumann and L. Nordheim, 'Über die Grundlagen der Quantenmechanik', *Math. Ann. 98* (1927), 1–30.
[22] Jammer, *Conceptual Development*, 314–318.

[23] Although they left their transformations in general form the authors cannot but have been aware of the specific form required: Dirac had, after all, noted it explicitly. They did not, however, discuss the matter.
[24] Dirac, 'Physical interpretation', 625.
[25] J. von Neumann, 'Mathematische Begründung der Quantenmechanik', *Nach. königliche Ges. Wiss. Göttingen, Math.-Phys. Kl* (1927), 1–57; 'Wahrscheinlichkeitstheoretischer Aufbau der Quantenmechanik', *ibid.*, 245–272; 'Thermodynamik Quantenmechanischer Gesamtheiten', *ibid.*, 273–291; 'Beweis der Ergodensatzes und des H-Theorems in der neuen Mechanik', *Zeit. Phys. 57* (1929), 30–70; 'Allgemeine Eigenwertetheorie Hermetischer Funktionsoperatoren', *Math. Ann. 102* (1929), 49–131.
[26] von Neumann, 'Mathematische Begründung'; E. Fischer, 'Sur la convergence en moyenne', *Comptes Rendus 144* (1907), 1022–1024; F. Riesz, 'Sur les systèmes orthogonaux de fonctions', *ibid.*, 615–619.
[27] von Neumann, 'Warscheinlichkeitstheoretischer Aufbau'.
[28] For a relatively comprehensible presentation of the theory underlying von Neumann's statement of the probability interpretation, see B. L. van der Waerden, *Group Theory and Quantum Mechanics* (New York, 1974), 12–16.

CHAPTER 9. THE UNCERTAINTY PRINCIPLE AND THE COPENHAGEN INTERPRETATION

[1] Heisenberg to Pauli, 28 October 1926, *PB*, 349–352 (Item 144).
[2] *Ibid.*
[3] *Ibid.*
[4] See W. Pauli, 'Über Gasentartung und Paramagnetismus', *Zeit. Phys. 41* (1927), 81–102, submitted on 6 December 1926, but not very helpful in the present context. On Heisenberg's letter to him of 4 November 1926, *PB*, 352–353 (Item 145), Pauli wrote the words "Ferromagnetismus geht nicht!"
[5] Heisenberg to Pauli, 15 November 1926, *PB*, 354–356 (Item 146).
[6] The precise status of the time variable in quantum mechanics, as compared with the other kinematic and mechanical quantities, was and continued to be somewhat problematic. See for example Jammer, *Philosophy*, 140–141.
[7] W. Heisenberg, 'Schwankungserscheinungen und Quanten-mechanik', *Zeit. Phys. 40* (1926), 501–506.
[8] Heisenberg to Pauli, 23 November 1926, *PB*, 357–360 (Item 148).
[9] See the discussion of Bohr's ideas below, this chapter.
[10] Interview with Heisenberg, *SHQP*; Heisenberg, *Physics and Beyond*, 77.
[11] P. Jordan, 'Kausalität und Statistisch', *Naturwissenschaften 15* (1927), 105–107, and 'Philosophical foundations of quantum theory', *Nature 119* (1927), 566–568.
[12] Heisenberg to Pauli, 5 February 1927, *PB*, 373–376 (Item 153).
[13] Heisenberg to Pauli, 23 February 1927, *PB*, 376–382 (Item 154).
[14] W. Heisenberg, 'Über den anschaulichen Inhalt der quantentheoretischer Kinematik und Mechanik', *Zeit. Phys. 43* (1927), 172–198.
[15] *Ibid.*; Bohr's criticisms are given in the postscript on 197–198. See Jammer, *Philosophy*, Chapter 3, for an extensive discussion of the measurement thought experiment.
[16] Heisenberg, 'Anschaulichen Inhalt'; Heisenberg to Pauli, 23 February 1927, *PB*, 376–382 (Item 154).

[17] Heisenberg, 'Anschaulichen Inhalt', 197.
[18] Heisenberg to Pauli, 23 February 1927, *PB*, 376–382 (Item 154).
[19] Heisenberg, 'Anschaulichen Inhalt', 173.
[20] Heisenberg to Pauli, 23 February 1927, *PB*, 376–382 (Item 154).
[21] M. Born, 'Zur Quantenmechanik der Stossvogänge', *Zeit. Phys. 37* (1926), 863–867.
[22] Jordan, 'Kausalität und Statistisch'.
[23] M. Born 'Quantenmechanik und Statistik', *Naturwissenschaften 15* (1927), 238–242.
[24] H. A. Senftleben, 'Zur Grundlagen der Quantentheorie', *Zeit. Phys. 22* (1923), 127–156, esp. 131.
[25] Heisenberg, 'Anschaulichen Inhalt', 197.
[26] Heisenberg to Pauli, 23 February 1927, *PB*, 376–382 (Item 154).
[27] Heisenberg to Pauli, 9 March 1927, *PB*, 383–385 (Item 156).
[28] Heisenberg, 'Anschaulichen Inhalt', 172–173.
[29] *Ibid.*, 172.
[30] Heisenberg, *Physics and Beyond*, 63.
[31] Heisenberg to Pauli, 9 March 1927, *PB*, 383–385 (Item 156).
[32] See for example N. Bohr, 'On the application of the quantum theory to atomic structure. Part I. The fundamental postulates', *Supplement to Proc. Camb. Phil. Soc.* (1924), 1 and 35.
[33] *Ibid.*, 35.
[34] N. Bohr, H. A. Kramers and J. C. Slater, 'The quantum theory of radiation', *Phil. Mag. 47* (1924), 795–802. See Chapter 5 above.
[35] A. Einstein, 'Quantentheorie des einatomigen idealen Gases', *S.–B. Preuss. Akad. Wiss.* (1924), 261–267, and *ibid.* (1925), 3–14. C. Ramsauer, 'Über das Wirkungsquerschnitt der Gasmolekule gegenüber langsamen Elektronen', *Ann. der Phys. 64* (1922), 513–540; *ibid. 66* (1922), 546–558; *ibid. 72* (1923), 345–352. C. Davisson, 'The scattering of electrons by a positive nucleus of limited field', *Phys. Rev. 21* (1923), 637–649; C. Davisson and C. H. Kunsman, 'The scattering of low speed electrons by platinum and magnesium', *ibid. 22* (1923), 242–258. Ramsauer's results had long been recognised as presenting a problem, while the implications of the Davisson-Kunsman experiments were realised during the Spring of 1925 during discussions on de Broglie's matter-wave theory between Born, Franck, and the research student Elsasser (Born, *My Life*, 231). They were published in W. Elsasser, 'Bemerkung zur Quantenmechanik freier Elektronen', *Naturwissenschaften 13* (1925), 711.
[36] See Note 48 to Chapter 5 above and, for physicists' expectations, Born to Bohr, 15 January 1925 and 24 April 1925, *BSC*.
[37] N. Bohr, speech to the Royal Danish Academy, reported in *Nature 116* (1925), 262.
[38] Bohr to Heisenberg, 18 April 1925, *BSC*: "Besonders durch Gesprache mit Pauli angeregt, quäle ich mich in diesen Tagen nach besten Kräften, mich in die Mystik der Natur einzuleben und versuche, mich auf alle Eventualitäten vorzubereiten, ja sogar auf die Annahme einer Kopplung der Quantenprozesse in entferten Atomen. Die Kosten dieser Annahme sind allerdings so gross, dass sie nicht in der gewöhnlichen, raumzeitlichen Beschreibung ermessen lassen."
[39] Bohr to Geiger, 21 April 1925, *BSC*, in Stuewer, *Compton Effect*, 301.
[40] See Note 38 above.
[41] Bohr to Franck, 21 April 1925, *BSC*: "Ich hatte schon lange die Absicht, Ihnen

wieder zu schreiben, denn der Zweifel an der Richtigkeit meiner Überlegnungen über die Stosserscheinungen, dem die Nachschrift in meinem letzten Brief Ausdruck gab, hat sich seitdem immer verstärkt. Es sind besonders die Ramsauerschen Ergebnisse der Durchdringung langsamer Elektronen durch Atome, die sich anscheinend dem angenommen Gesichtspunkte nicht einfügen. In der Tat dürften diese Ergebnisse unserer gewohnlichen raumzeitlichen Naturbeschreibung Schwierigkeiten ähnlicher Art darbieten wie eine Koppelung der Zustandsänderung entfernter Atome durch Strahlung. Dann ist aber kein Grund mehr, an einer solchen Koppelung und an den Erhaltungssatzen überhaupt zu zweifeln. Dies ist nur eine grosse Befriedigung denn, wie Sie hervorheben, wird ja dann so vieles bei den Stössen so ungemein viel einfacher. Auch waren die thermodynamischen Betrachtungen von Einstein ja sehr beunruhigend. Ich habe schon diesen Morgen an Fowler geschrieben, dass ich eine englische Arbiet über die Bremsung der α-strahlen zurückziehen und werde Ähnliches an Scheel betreffend die ihm zugesandte Arbeit schreiben. Ausserdem habe ich eben jetzt von Geiger gehört, dass seine Versuche für die Koppelung entschieden haben, und es ist wohl nichts anderes zu tun als unseren Revolutionsversuch möglichst schmerzlos in Vergessenheit zu bringen. Unsere Ziele werden wir aber doch nicht so leicht vergessen können und in den letzten Tagen habe ich mit allerlei wilden Spekulationen geqüalt, um eine adäquate Grundlage der Beschreibung der Strahlungsphänomene zu finden. Darüber habe ich viel mit Pauli diskutiert, der jetzt hier ist, und dem seit langem unser "Kopenhagener Pretsch" unsympatisch war." The earlier letter referred to appears to have been lost.

42 Bohr to Born, 1 May 1925, *BSC*: "Ganz abgesehen von der Frage der Richtigkeit derartiger Einwände gegen Ihr Theorie, möchte ich gern betonen, dass ich der Ansicht bin, dass die Annahme einer Koppelung zwischen den Zustandänderungen in entfernten Atomen durch Strahlung einer einfache Beschreibungsmöglichkeit des physikalischen Geschehens mittels anschaulicher Bilder ausschliesst. Mit meinen Äusserungen in den Brief an Franck über die Koppelung war nur gemeint, dass ich den Verdacht bekommen hatte, dass schon für die Stosserscheinungen solchen Bildern ein noch geringeren Anwendbarkeit zukommt als gewöhnlich angenommen. Dies ist ja zunächst eine rein negativ Aussage, aber ich fühle, besonders wenn die Koppelung wirklich eine Tatsache sein sollte, dass man dann in noch hoheren Grade wie bisher seine Zuflucht zu symbolischen Analogien nehmen muss. Eben in letzter Zeit habe ich mir den Kopfe zerbrochen in solche Analogien mich hineinzuträumen."

43 N. Bohr, 'Über die Wirkung von Atomen bei Stössen', *Zeit. Phys. 34* (1925), 142–157, esp. 154.

44 Interview with W. Heisenberg, *SHQP*.

45 See Chapter 4 above.

46 See Chapter 2 above, and Heisenberg to Pauli, 5 February 1927, *PB*, 373–376 (Item 153), where Heisenberg himself makes a similar statement.

47 The main argument appears to have been between Ehrenfest, supporting the five-dimensional theory, and Pauli and Heisenberg, opposed to it. In the Spring of 1927 Bohr sided with Klein, who continued to work with the wave mechanics, against Heisenberg; but his position the previous Autumn is unclear. See Ehrenfest to Pauli, 25 January 1927, and Heisenberg to Pauli, 5 February 1927, 4 April 1927, 16 May 1927, and 31 May 1927, *PB*, 371–376, 390–397 (Items 152, 153, 161, 163 and 164).

48 Interview with W. Heisenberg, *SHQP*; Heisenberg, *Physics and Beyond*, 76, and in Rosental, *Niels Bohr*, 104.

[49] Heisenberg, 'Anschaulichen Inhalt'.
[50] Pauli to Bohr, 6 August 1927, *PB*, 402–406 (Item 168). The context was the quantum electrodynamics of Jordan.
[51] Jordan recalled (Jammer, *Philosophy*, 68) that only Pauli's skillful diplomacy avoided a serious conflict between Bohr and Heisenberg; but this trouble seems to have been on a personal level, arising out of a misunderstanding, quite unintentional, between Heisenberg and Klein, corrected by Pauli at Heisenberg's request: Heisenberg to Pauli, 16 May 1927 and 31 May 1927, *PB*, 394–397 (Items 163, 164).
[52] Heisenberg, 'Anschaulichen Inhalt', 197–198.
[53] Heisenberg to Pauli, 14 March 1927, *PB*, 387–388 (Item 158).
[54] Heisenberg to Pauli, 4 April 1927, *PB*, 390–393 (Item 161).
[55] Heisenberg to Pauli, 16 May 1927, *PB*, 394–396 (Item 163).
[56] Heisenberg, 'Anschaulichen Inhalt'; the title speaks for itself.
[57] For further discussion of Heisenberg, Bohr and the meaning of *Anschaulichkeit* see Miller, 'Beyond Anschaulichkeit' and 'Visualisation lost and regained'.
[58] Heisenberg to Pauli, 31 May 1927, *PB*, 396–397 (Item 164).
[59] Ehrenfest to Pauli, 24 January 1927, and Heisenberg to Pauli, 5 February 1927, *PB*, 371–376 (Items 152, 153). This debate was also related to that on the interpretation of spin, which had occupied both Pauli and Heisenberg the previous Spring.
[60] See Note 51 above.
[61] Heisenberg to Pauli, 31 May 1927, *PB*, 396–397 (Item 164).
[62] N. Bohr, 'The quantum postulate and the recent development of atomic theory', *Atti del Congresso Internazionale dei Fisici*, Vol. 2 (Bologna, 1928), 568–588; see also *Nature 121* (1928), 580–590 and *Naturwissenschaften 16* (1928), 245–257.
[63] *Ibid.*, (*Nature*), 580.
[64] *Ibid.*, 580.
[65] *Ibid.*, 580.
[66] *Ibid.*, 581.
[67] *Ibid.*, 582–584.
[68] Pauli to Bohr, 17 October 1927, *PB*, 411–413 (Item 173).
[69] For example, P. Jordan, 'Die Quantenmechanik und die Grundprobleme der Biologie und Psychologie', *Naturwissenschaften 20* (1932), 815–821.
[70] Pauli to Bohr, 6 August 1927, *PB*, 402–406 (Item 168).
[71] Bohr to Pauli, 13 August 1927, *PB*, 406–407 (Item 169). M. Born and W. Heisenberg, 'La mécanique des quanta', in Institut International de Physique Solvay, *Électrons et Photons* (Paris, 1928), 143–181.
[72] P. Jordan and W. Pauli, 'Zur Quantenelektrodynamik ladungsfreier Felder', *Zeit. Phys. 47* (1928), 151–173.
[73] Born and Heisenberg, 'Mécanique des quanta'; N. Bohr, 'Le postulat des quanta et le nouveau développement de l'atomistique', in Institut International de Physique Solvay, *Électrons et Photons*, 215–247; Dirac, in the discussion on Bohr's paper, *ibid.*, 261–263.
[74] de Broglie was advocating his theory of double solution, itself as radical a departure from classical physics as was the Copenhagen interpretation. Schrödinger was pursuing his semi-classical interpretation with the abandonment of the classical energy concept. Einstein was already talking in terms of a statistical or ensemble interpretation as being all that the theory could unambiguously support (from a completely opposite viewpoint this actually came very close to Dirac's position). Lorentz had criticisms to make of

everybody, but knew not what to do himself. The various positions are discussed and referenced in Jammer, *Philosophy*.

[75] This mood was apparent from the proceedings and was also recalled by Heisenberg: interview with W. Heisenberg, *SHQP*.

[76] Jordan's early probings in this direction have been discussed above. The subject then dominated his and Dirac's work, and Pauli's correspondence, from February 1927. See P. A. M. Dirac, 'The quantum theory of the emission and absorption of radiation', *Proc. Roy. Soc. A114* (1927), 243–265, and 'The quantum theory of dispersion', *ibid.*, 710–728; Jordan and Pauli, 'Quantenelektrodynamik'; *PB*, 385ff.

[77] W. Heisenberg and W. Pauli, 'Zur Quantendynamik der Wellenfelder', *Zeit. Phys. 56* (1929), 1–61; *59* (1930), 168–190.

[78] See Jammer, *Philosophy*, Chapter 6.

[79] *Ibid.*, 76.

[80] Born and Heisenberg, 'Mécanique des quanta'.

CHAPTER 10. CONCLUDING REMARKS

[1] See especially the work of Hanle, and also that of Wessels, cited in the bibliography.

[2] See especially Cassidy, 'Werner Heisenberg' and 'Heisenberg's first model'; see also MacKinnon, 'Heisenberg'.

[3] See Kragh, 'Methodology and philosophy of science'.

[4] See Kragh, 'Niels Bohr's second atomic theory', and the introductions to the volumes of the *Collected Works* of Bohr.

[5] See Serwer, 'Unmechanischer Zwang'.

[6] See for example, Klein, 'First phase', and Raman and Forman, 'Why Schrödinger'.

[7] Forman, 'Weimar culture'.

[8] Brush, 'Chimerical cat'.

[9] Confusingly, although the language of *Anschaulichkeit* may itself be traced to Kant, the way in which the word was used changed significantly: see Miller, 'Beyond Anschaulichkeit'.

[10] The clearest indication of this is in the overwhelmingly hostile reception afforded to the purest Kantian of the quantum mechanics period, Eddington.

SELECT BIBLIOGRAPHY

A. UNPUBLISHED PRIMARY SOURCES

The great majority of the known extant source material for the history of quantum mechanics is to be found in the microfilm archive, *Sources for the History of Quantum Physics* (*SHQP*), copies of which are located at the University of California at Berkeley, the american Philosophical Society at Philadelphia, the American Institute of Physics in New York, the Niels Bohr Institute in Copenhagen and (shortly) the Science Museum in London. The archive contains the *Bohr Scientific Correspondence* (*BSC*) and the *Bohr Manuscript Collection* (*Bohr MSS*), also the *Pauli Correspondence*. It is frequently augmented, but the original catalogue remains invaluable, including not only items in the archive but also those in other collections, together with biographical details of the physicists: Kuhn, T. S., *et al.*, (eds.): 1967, *Sources for History of Quantum Physics* (Philadelphia). Interviews with many of the quantum physicists are also to be found in these archives.

Apart from the *SHQP* set of archives, the only other archive to be used in this study is the *Weyl collection* at the ETH Zürich (*ETHZ*).

B. PUBLISHED CORRESPONDENCE AND PAPERS

The most important sources for this study have been the published correspondence and papers of Pauli and Bohr:

Pauli, W.: 1979, *Wissenschaftlicher Briefwechsel mit Bohr, Einstein, Heisenberg u.a., Band 1:1919–1929* (Berlin, Heidelberg, New York).

Bohr, N.: 1975, *Collected Works*, Vol. 3 (Amsterdam).

The works of Bohr (Volumes 1, 2 and 4 are also relevant) contain a selection of his correspondence as well as previously unpublished papers. Other useful collections of correspondence include:

Przibram, K. (ed.): 1967, *Letters on Wave Mechanics* (New York).

Born, M. and A. Einstein: 1971, *The Born-Einstein Letters* (New York).

Einstein, A. and M. Besso: 1972, *Correspondance 1903–1955* (Paris).

Einstein, A. and A. Sommerfeld: 1968, *Briefwechsel* (Basel).

C. REPRINT COLLECTIONS AND TRANSLATIONS

As well as the reprints and translations of individual papers there are also several thematic collections. Among the most useful of these are:

de Broglie, L. and L. Brillouin: 1928, *Selected Papers on Wave Mechanics* (London).
de Broglie, L.: 1953: *La Physique Quantique Restera-t-elle Indeterministique*? (Paris).
ter Haar, B.: 1967, *The Old Quantum Theory* (Oxford). [But note that the translations here are not very reliable.]
Bohr, N.: 1924, *Theory of Spectra and Atomic Constitution* (Cambridge).
Bohr, N.: 1934, *Atomic Theory and the Description of Nature* (Cambridge).
Hermann, A. (ed.): 1962–1964, *Dokumente der Naturwissenschaften*, Bd. 1–4 (Stuttgart).
Lorentz H. A. *et al.*: 1923, *The Principle of Relativity* (London, 1923; New York, 1952).
Ludwig, G.: 1967, *Wave Mechanics* (Oxford).
Schrödinger, E.: 1928, *Collected Papers on Wave Mechanics* (London).
van der Waerden, B. L.: 1967, *Sources of Quantum Mechanics* (Amsterdam).

Collected editions of the work of individual physicists include:

Bohr, N.: 1974–, *Collected Works* (Amsterdam).
Born, M.: 1963, *Ausgewahlte Abhandlungen* (Gottingen).
Ehrenfest, P.: 1959, *Collected Scientific Papers* (Amsterdam).
Fermi, E.: 1962, *Collected Papers* (Chicago).
Kramers, H. A.: 1956, *Collected Scientific Papers* (Amsterdam).
Pauli, W.: 1964, *Collected Scientific Papers* (New York).
Von Neumann, J.: 1961, *Collected Works* (Oxford).
Weyl, H.: 1968, *Gesammelte Abhandlungen* (Berlin).

At the time of writing a collection of Heisenberg's papers is also in preparation.

D. SECONDARY WORKS

The standard and invaluable general history of quantum theory is Jammer's *Conceptual Development of Quantum Mechanics*. This contains a few mistakes, notably in the treatment of the causality issue (see Hendry, 'Weimar Culture', 161–162) and in the transmission of Born's recollections (for which see instead the originals, published since), but it is for the most part a reliable and comprehensive guide to the published primary material. More comprehensive but still to be assessed is the massive multi-volume work by Mehra and Rechenberg. Compared with these works none of the other general histories, only a few of which are listed below, add anything significant.

There is a shortage of good biographical treatments of the quantum physicists, but the situation is gradually improving. The *Collected Works* of Bohr and *Briefwechsel* of Pauli (see Section B above) both include excellent and extensive introductory and editorial material and will when complete constitute good scientific biographies. Otherwise, the best studies of Bohr are those by Rosenfeld, Honner and Stolzenburg, to be supplemented by the volume edited by Rozental. A major biography of Pauli is currently in preparation, but meanwhile see Richter's book, the introduction to Pauli's *Papers* (Section C above) and the article by Fierz. For Heisenberg and Born, the most thorough treatments are their respective autobiographical works, those of Heisenberg being particularly interesting and that of Born both accurate and informative. There is also a popular but rather good biography of Heisenberg by Hermann. Of the many biographies of Einstein the most valuable are those by Pais, Frank, Seelig and Kuznetsov. Of the full length biographies available of other physicists, those by Reid (Hilbert), Scott (Schrödinger), Douglas (Eddington), Klein (Ehrenfest) and Segré (Fermi) deserve special mention, though Scott in particular has been surpassed by more recent work. For Sommerfeld, see especially the article by Born.

In general the thematic treatments of aspects of the history of quantum mechanics, many of them dealing largely with the work of a single physicist, are rather better than the specifically biographical works. The intellectual and social context of quantum mechanics has as yet received relatively little attention, but Holton's *Thematic Origins of Scientific Thought*, Miller's papers on *Anschaulichkeit*, and Forman's work on the causality issue (but see also the corresponding paper by Hendry) are essential reading. The papers of Brush are also interesting in this respect. Much has been written on the old quantum theory of Planck and Einstein, and selected references are included below. Of particular interest as background to the history of quantum mechanics are the works of Kuhn, Klein and McCormmach, together with Hendry's paper on the wave-particle duality and Stuewer's book on the Compton effect. Relativity theory has also received a lot of attention, though most of this has been devoted to the special theory, for which see especially the papers of Holton and Miller. Interesting contributions to the history of general relativity theory include the books of Pauli, North and Mehra, as well as the excellent papers by Earman and Glymour.

The history of Bohr's atomic theory leading up to the creation of matrix mechanics is particularly well documented in a set of papers fitting into a natural sequence, by Heilbron and Kuhn, Heilbron, Forman ('Landé'), Cassidy, Forman ('Doublet riddle'), Serwer and Hendry. The present work is particularly indebted to this set of papers. The best treatment of matrix mechanics itself remains that by van der Waerden in his *Sources*, though this should be supplemented by the recollections of Heisenberg and Born and by the papers of Serwer, Hendry and MacKinnon. On wave mechanics the work of de Broglie has been treated by many writers, but never very well, while that of Schrödinger and its origins are covered in an excellent group of papers by Klein ('Einstein and the wave-particle duality'), Raman and Forman, Hanle and Kragh. For the evolution of quantum mechanics the best works remain the two books by Jammer and the article by van der Waerden.

Bernkopf, M.: 1967, 'A history of infinite matrices', *Arch.Hist.Exact Sci.* 4, 308–358.

Bloch, F.: 1976, 'Heisenberg and the early days of quantum mechanics', *Phys. Today* (December), 23–27.

Bohr, N.: 1949, 'Discussions with Einstein', in Schilpp, P. A.: 1949, below.

Bohr, N.: 1961, 'Reminiscences of the founder of nuclear science and of some developments based on his work (Rutherford memorial lecture)', *Proc.Phys.Soc.* 78, 1083–1115.

Bopp, F. (ed.): 1966, *Werner Heisenberg und die Physik unserer Zeit* (Braunschweig).

Born, M.: 1943, *Experiment and Theory in Physics* (Oxford).

Born, M.: 1949, *Natural Philosophy of Cause and Chance* (Oxford).

Born, M.: 1956, *Physics in My Generation* (Oxford).

Born, M.: 1963, 'Arnold Sommerfeld, 1868–1951', in Born, M.: 1963, *Ausgewählte Abhandlungen* (Section C above).

Born, M.: 1978, *My Life. Recollections of a Nobel Laureate* (London).

Bromberg, J.: 1976, 'The concept of particle creation before and after quantum mechanics', *Hist.Stud.Phys. Sci.* 7, 161–183.

S. Brush,: 1976, 'Irreversibility and indeterminism: Fourier to Heisenberg', *J.Hist.Ideas* 37, 603–630.

Brush, S.: 1980, 'The chimerical cat: philosophy of quantum mechanics in historical perspective', *Soc.Stud.Sci.* **10**, 393–447.

Cassidy, D.: 1976, 'Werner Heisenberg and the crisis in quantum theory, 1920–1925', Ph.D. dissertation, Purdue University.

Cassidy, D.: 1979, 'Heisenberg's first core model of the atom: the formation of a professional style', *Hist.Stud.Phys. Sci.* **10**, 123–186.

Clark, R. W.: 1971, *Einstein: The Life and Times* (New York).

Cropper, W. H.: 1970, *The Quantum Physicists* (Oxford).

Dirac, P. A. M.: 1963, 'The evolution of the physicist's picture of nature', *Sci.American* **208**, 45–53.

Dirac, P. A. M.: 1971, *The Development of Quantum Theory* (New York).

Dirac, P. A. M.: 1977, 'Recollections of an exciting era', in Weiner, C.: 1977, below.

Earman J. and C. Glymour: 1978a, 'Einstein and Hilbert: two months in the history of general relativity', *Arch.Hist.Exact Sci.* 17, 291–308.

Earman J. and C. Glymour: 1978b, 'Lost in the tensors: Einstein's struggles with covariance principles, 1912–1916', *Stud.Hist.Phil. Sci.* 9, 251–278.

Elkana, Y. (ed.): 1974, *The Interaction between Science and Philosophy* (Atlantic Highlands, New Jersey).

Elsasser, W. M.: 1978, *Memoirs of a Physicist in the Atomic Age* (New York).

Enz, C. P.: 1973, 'W. Pauli's scientific work', in Mehra, J.: 1973, below.

Fierz, M.: 1974, 'Wolfgang Pauli', *Dict.Sci.Biog.* **10**, 422–425.

Fierz M. and V. S. Weisskopf (eds.): 1960, *Theoretical Physics in the Twentieth Century* (New York).

Forman, P.: 1967, 'The environment and practice of atomic physics in Weimar Germany', Ph.D. dissertation, University of California at Berkeley.

Forman, P.: 1968, 'The doublet riddle and atomic physics circa 1924', *Isis* **59**, 156–174.

Forman, P.: 1970, 'Alfred Landé and the anomalous Zeeman Effect, 1919–1921', *Hist.Stud.Phys. Sci.* **2**, 153–261.

Forman, P.: 1971, 'Weimar culture, causality, and quantum theory, 1918–1927', *Hist. Stud.Phys.Sci.* 3, 1–116; republished in a shortened and in some ways improved version in Chant, C. and J. Fauvel (eds.): 1980, *Darwin to Einstein: Historical Studies on Science and Belief* (Harlow, Essex).

Forman, P.: 1978, 'The reception of an acausal quantum mechanics in Germany and Britain', in Mauskopf, S. H. (ed.): 1980, *The Reception of Unconventional Science* (Boulder, Colorado).

Frank, P.: 1947, *Einstein, His Life and Times* (New York).

Gerber, J.: 1969, 'Geschichte der Wellenmechanik', *Arch.Hist.Exact Sci.* 5, 349–416.

Goldberg, S.: 1969, 'The Lorentz theory of the electron and Einstein's theory of relativity', *Am.J.Phys.* 37, 982–994.

Goudsmit, S. A.: 1961, 'Pauli and nuclear spin', *Phys.Today* (June), 18–21.

Goudsmit, S. A.: 1976, 'It might as well as be spin', *Phys.Today* (June), 40–43.

Hanle, P. A.: 1977a, 'The coming of age of Erwin Schrödinger: his quantum statistics of ideal gases', *Arch.Hist.Exact Sci.* 17, 165–192.

Hanle, P. A.: 1977b, 'Schrödinger's reaction to de Broglie's thesis', *Isis* **68**, 606–609.

Hanle, P. A.: 1979a, 'The Schrödinger-Einstein correspondence and the sources of wave mechanics', *Am.J.Phys.* 47, 644–648.

Hanle, P. A.: 1979b, 'Indeterminacy before Heisenberg: the case of Franz Exner and Erwin Schrödinger', *Hist.Stud.Phys. Sci.* **10**, 225–270.

Hanson, N. R.: 1961, 'Are wave mechanics and matrix mechanics equivalent theories?', in Feigl, H. and G. Maxwell (eds.): 1961, *Current Issues in the Philosophy of Science* (New York).

Hanson, N. R.: 1963, *The Concept of the Position* (Cambridge).

Heelan, P. A.: 1975, 'Heisenberg and radical theoretic change', *Zeit. allgem. Wissenschaftstheorie* 6, 113–136.

Heilbron, J. L.: 1964, 'A history of atomic structure from the discovery of the electron to the beginning of quantum mechanics', Ph.D. dissertation, University of California at Berkeley.

Heilbron, J. L.: 1967, 'The Kossel-Sommerfeld theory and the ring atom', *Isis* 58, 451–482.

Heilbron, J. L. and T. S. Kuhn: 1969, 'The genesis of Bohr's atom', *Hist.Stud.Phys. Sci.* 1, 211–290.

Heisenberg, W.: 1960, 'Errinerungen an die Zeit der Entwicklung der Quantenmechanik', in Fierz, M. and V. S. Weisskopf: 1960, above.

Heisenberg, W.: 1967, 'Quantum theory and its interpretation', in Rozental, S.: 1967, below.

Heisenberg, W.: 1971, *Physics and Beyond: Encounters and Conversations* (New York).

Heisenberg, W.: 1973, 'Development of concepts in the history of quantum theory', in Mehra, J.: 1973, below.

Heisenberg, W.: 1974, 'Wolfgang Pauli's philosophical outlook', in his 1974, *Across the Frontiers* (New York).

Hendry, J.: 1980a, 'Weimar culture and quantum causality', *Hist.Sci.* 18, 155–180.

Hendry, J.: 1980b, 'The development of the wave-particle duality of light and quantum theory, 1900–1920', *Ann.Sci.* 37(1980), 59–79.

Hendry, J.: 1981, 'Bohr-Kramers-Slater: a virtual theory of virtual oscillators and its role in the history of quantum mechanics', *Centaurus* 25, 189–221.

Hermann, A.: 1962, 'Max Born', in *Dokumente der Naturwissenschaften,* Bd. 1 (Section C above).

Hermann, A.: 1975, 'Erwin Schrödinger', *Dict.Sci.Biog.* **12**, 217–223.

Hermann, A.: 1976, *Werner Heisenberg, 1901–1976* (Bonn).

Hesse, M.: 1961, *Forces and Fields* (London).

Hoffmann, B. and H. Dukas: 1972, *Albert Einstein, Creator and Rebel* (New York).

Holton, G.: 1973, *The Thematic Origins of Scientific Thought* (Cambridge, Mass.).

Holton, G.: 1978, *The Scientific Imagination: Case Studies* (Cambridge, Mass.).

Honner, J.: 1982, 'The transcendental philosophy of Niels Bohr', *Stud.Hist.Phil. Sci.* **13**, 1–30.

Hörz, H.: 1966, *Werner Heisenberg und die Philosophie* (Berlin).

Hund, F.: 1974, *The History of Quantum Theory* (New York).

Jammer, M.: 1966, *The Conceptual Development of Quantum Mechanics* (New York).

Jammer, M.: 1974, *The Philosophy of Quantum Mechanics* (New York).

Klein, M. J.: 1959, 'Ehrenfest's contribution to the development of quantum statistics', *Kon. Ned. Akad. v. Wetenschappen* **B62**, 41–62.

Klein, M. J.: 1964, 'Einstein and the wave-particle duality', *Natural Philosopher* 3, 1–49.
Klein, M. J.: 1970a, *Paul Ehrenfest: The Making of a Theoretical Physicist* (Amsterdam).
Klein, M. J.: 1970b, 'The first phase of the Bohr-Einstein dialogue', *Hist.Stud.Phys. Sci.* 2, 1–39.
Klein, M. J.: 1977, 'The beginnings of quantum theory', in Weiner, C.: 1977, below.
Konro, H.: 1978, 'The historical roots of Born's probabilistic interpretation', *Jap.Stud. Hist.Sci.* 17, 129–145.
Kragh, H.: 1979a, 'On the history of early wave mechanics', IMFUFA tekst 23, Roskilde University Centre.
Kragh, H.: 1979b, 'Methodology and philosophy of science in Paul Dirac's physics', IMFUFA tekst 27, Rosklide University Centre.
Kragh, H.: 1979c, 'Niels Bohr's second atomic theory', *Hist.Stud.Phys.Sci.* 10, 123–186.
Kragh, H.: 1982, 'Erwin Schrödinger and the wave equation: the crucial phase', *Centaurus* 26, 154–197.
Kramers, H. A. and H. Holst: 1923, *The Atom and the Bohr Theory of its Structure* (Copenhagen).
Kronig, R.: 1960, 'The turning point', in Fierz, M. and V. S. Weisskopf: 1960, above.
Kubli, F.: 1970, 'Louis de Broglie und die Entdeckung der Materiewellen', *Arch.Hist. Exact Sci.* 7, 26–78.
Kuznetsov, B.: 1965, *Einstein* (Moscow).
McCormmach, R.: 1970, 'Einstein, Lorentz and the electron theory', *Hist.Stud.Phys.Sci.* 2, 41–87.
MacKinnon, E.: 1976, 'De Broglie's thesis: a critical retrospective', *Am.J.Phys.* 44, 1047–1055.
MacKinnon, E.: 1977, 'Heisenberg, models and the rise of matrix mechanics', *Hist.Stud. Phys.Sci.* 8, 137–188.
MacKinnon, E.: 1982, *Scientific Explanation and Atomic Physics* (Chicago).
Margenau, H.: 1944, 'The exclusion principle and its philosophical importance', *Phil. Sci.* 11, 187–208.
Mehra, J.: 1972, 'The golden age of theoretical physics', in Salam, A. and E. Wigner: 1972, below.
Mehra, J.: 1974, *Einstein, Hilbert, and the Theory of Gravitation* (Dordrecht).
Mehra, J. (ed.): 1973, *The Physicist's Conception of Nature* (Dordrecht).
Mehra, J. and H. Rechenberg: 1982–, *The Historical Development of Quantum Theory* (New York): Volume 1 (in two parts), *The Quantum Theory of Planck, Einstein, Bohr and Sommerfeld*; Volume 2, *The Discovery of Quantum Mechanics*; Volume 3, *The Formulation of Matrix Mechanics and Its Modifications*; Volume 4 (in two parts), *The Fundamental Equations of Quantum Mechanics*, and *The Reception of the New Quantum Mechanics*. (Further volumes in preparation.)
von Meyenn, K.: 1980, 1981, 'Pauli's Weg zum Ausschiessungsprinzip', *Phys. Bl.* 36 (1980), 293–298, 37(1981), 13–20.
Meyer-Abich, K.: 1965, *Korrespondenz, Individualität und Komplementarität* (Wiesbaden).
Miller, A. I.: 1978, 'Visualisation lost and regained', in Wechsler, J. (ed.): 1978, *On Aesthetics in Science* (Cambridge, Mass.).
Miller, A. I.: 1982, 'Beyond Anschaulichkeit', in *Laszo Tisza Festschrift* (Cambridge, Mass.).

Moore, R.: 1966, *Niels Bohr* (Copenhagen).
Mott, N. and R. Peierls: 1977, 'Heisenberg', *Biog.Mem.Fell.Roy.Soc.* **23**, 213–252.
Newman, M. H. A.: 1957, 'Weyl', *Biog.Mem.Fell.Roy.Soc.* **3**, 305–328.
Nisio, S.: 1973, 'The formation of the Sommerfeld quantum theory of 1916', *Jap.Stud. Hist.Sci.* **12**, 39–78.
Nobel Foundation: 1964, *Nobel Laureates in Physics, 1942–1962* (Amsterdam).
Pais, A.: 1979, 'Einstein and the quantum theory', *Rev.Mod.Phys.* **51**, 869–914.
Pais, A.: 1982, '*Subtle is the Lord . . .' The Science and Life of Albert Einstein* (Oxford).
Peierls, R.: 1960, 'Pauli', *Biog.Mem.Fell.Roy.Soc.* **5**, 175–192.
Peterson, A.: 1963, 'The philosophy of Niels Bohr', *Bull.Atom.Sci.* **19**, No. 7 (September), 8–14.
Pyenson, L.: 1974, 'The Göttingen reception of relativity', Ph.D. dissertation, The Johns Hopkins University.
Radder, H.: 1982, 'Between Bohr's atomic theory and Heisenberg's matrix mechanics. A study of the role of the Dutch physicist H. A. Kramers', *Janus* **69**, 223–252.
Raman, V. V.: 1972, 'Relativity in the early twenties', *Indian J.Hist.Sci.* **7**, 119–145.
Raman, V. V. and P. Forman: 1969, 'Why was it Schrödinger who developed de Broglie's ideas?', *Hist.Stud.Phys.Sci.* **1**, 291–314.
Reid, C.: 1970, *Hilbert* (London).
Richter, S.: 1979, *Wolfgang Pauli* (Aurau).
Robertson, P.: 1979, *The Early Years. The Niels Bohr Institute, 1921–1930* (Copenhagen).
Rosenfeld, L.: 1936, 'La première phase de l'evolution de la théorie des quanta', *Osiris* **2**, 149–196.
Rosenfeld, L.: 1945, *Niels Bohr – An Essay* (Amsterdam).
Rosenfeld, L.: 1970, 'Niels Bohr', *Dict.Sci.Biog.* **2**, 239–254.
Rozental, S. (ed.): 1967, *Niels Bohr* (Amsterdam).
Salam, A. and E. Wigner (eds.): 1972, *Aspects of Quantum Theory* (Cambridge).
Schilpp, P. A. (ed.): 1949, *Albert Einstein: Philosopher-Scientist* (Evanston, Illinois).
Scott, W. T.: 1967, *Erwin Schrödinger: An Introduction to his Writings* (Amherst, Mass.).
Seelig, C.: 1956, *Albert Einstein. A Documentary Biography* (London).
Segré, E.: 1970, *Enrico Fermi, Physicist* (Chicago).
Segré, E.: 1980: *From X-rays to Quarks* (San Francisco).
Serwer, D.: 1977, 'Unmechanischer Zwang: Pauli, Heisenberg, and the rejection of the mechanical atom, 1923–1925', *Hist.Stud.Phys.Sci.* **8**, 189–256.
Small, H. G.: 1971, 'The helium atom in the old quantum theory', Ph.D. dissertation, University of Wisconsin at Madison.
Sopka, K. R.: 1980, *Quantum Physics in America, 1920–1935* (New York).
Stolzenburg, K.: 1977, 'Die Entwicklung des Bohrschen Komplementaritätsgedankens in den Jahren 1924 bis 1929', Ph.D. dissertation, University of Stuttgart.
Stuewer, R.: 1975, *The Compton Effect – Turning Point in Physics* (New York).
Tonnelat, M. A.: 1965, *Les Théories Unitaires de l'Électromagnétism et de la Gravitation* (Paris).
Tonnelat, M. A.: 1973, 'L'influence de la relativité sur l'oevre de Louis de Broglie', in *Louis de Broglie. Sa Conception du Monde Physique* (Paris).
Uhlenbeck, G.: 1976, 'Personal reminiscences', *Phys.Today* (June), 43–48.

van der Waerden, B. L.: 1960, 'Exclusion principle and spin', in Fierz, M. and V. S. Weisskopf: 1960, above.
van der Waerden, B. L.: 1967, *Sources of Quantum Mechanics* (Amsterdam).
van der Waerden, B. L.: 1973, 'From matrix mechanics and wave mechanics to unified quantum mechanics', in Mehra, J.: 1973, above.
Weiner, C. (ed.): 1977, *History of Twentieth Century Physics* (New York).
Wessels, L.: 1979, 'Schrödinger's route to wave mechanics', *Stud.Hist.Phil.Sci.* 10, 311–340.
Weyl, H.: 1944, 'David Hilbert and his work', *Bull.Am.Math.Soc.* 50, 612–654.
Wiener, N.: 1956, *I am an Mathematician* (New York).

INDEX

STUDIES IN THE HISTORY
OF MODERN SCIENCE

Editors:

ROBERT S. COHEN (Boston University)
ERWIN N. HIEBERT (Harvard University)
EVERETT I. MENDELSOHN (Harvard University)

1. F. Gregory, *Scientific Materialism in Nineteenth Century Germany.* 1977, xxiii+279 pp.
2. E. Manier, *The Young Darwin and His Cultural Circle.* 1978, xii+242 pp.
3. G. Canguilhem, *On the Normal and the Pathological.* 1978, xxiv+ 208 pp.
4. H. C. Kennedy, *Peano.* 1980, xii+230 pp.
5. L. Badash, J. O. Hirschfelder, and H. P. Broida (eds.), *Reminiscences of Los Alamos, 1943–1945.* 1980, xxi+188 pp.
6. L. L. Bucciarelli and N. Dworsky, *Sophie Germain: An Essay in the History of the Theory of Elasticity.* 1980, xi+147 pp.
7. D. J. Struik, *The Land of Stevin and Huygens.* 1981, xiv+162 pp.
8. I. Langham, *The Building of British Social Anthropology.* 1981, xxxii+392 pp.
9. Z. Bechler (ed.), *Contemporary Newtonian Research.* 1982, vi+241 pp.
10. W. T. Sullivan, III, *Classics in Radio Astronomy.* 1982, xxiv+348 pp.
11. L. Anderson, *Charles Bonnet and the Order of the Known.* 1982, xvi+160 pp.
12. P. L. Farber, *The Emergence of Ornithology as a Scientific Discipline: 1760–1850.* 1982, xxi+191 pp.
13. T. Lenoir, *The Strategy of Life.* 1983, xii+314 pp.

According to the court, it was a situation he would eventually exploit in an exceptionally repugnant attempt at murder, the target of which would be neither a patient nor a fellow medical professional but rather someone much closer to home: his infant son.

Stewart, a resident of Columbia, Illinois, a small town twelve miles south of St. Louis, was dating a woman, Jennifer Jackson, who in 1990 unexpectedly became pregnant. After eschewing the abortion that Stewart recommended, the pair married and a few months later brought into the world their son Bryan.[17]

In the months that ensued, the infant's development was problem-free but Stewart and Jackson's marriage was not, and before long they found themselves on the brink of divorce. By chance, it was during this same period, the winter of 1992, that eleven-month-old Bryan was hospitalized for asthma at St. Joseph's Medical Center in Lake Saint Louis. And it was while preparations were being made to release the infant, who had largely recuperated, that Stewart made his move. He was purportedly convinced that divorce was imminent and he had no intention of paying child support for the next seventeen years.

It was on February 6, 1992, a cold Thursday morning, that the phlebotomist showed up at St. Joseph's to visit his son. A lab coat was draped over Stewart's arm, one that was described as unusual given the season.[18] Its presence was also inexplicable, since Stewart worked at Barnes Hospital, not St. Joseph's, meaning there was no reason for him to be carrying it in the first place. Prosecuting attorney Tim Braun would later contend that the garment concealed a syringe and a vial of HIV-infected blood. "The inference that we would have is that he had blood in there," says Braun, "and if somebody looked in the pocket of the coat, he would say, 'Oh, I had that left over from my other job' or 'I forgot to take that out.'"[19]

According to Jennifer Jackson, Stewart arrived at the hospital room and suggested she get a drink while he visited their son.[20] And it was when she returned from the cafeteria a few minutes later that she found Stewart sitting in a rocking chair with Bryan in his lap, the infant wailing uncontrollably. Stewart had injected the baby with HIV-tainted blood while his wife was out of the room, published reports contend.[21]

Two hours later, Bryan become intensely ill. The reason for his sudden turn for the worse perplexed his doctor, who described the infant's symptoms as similar to those observed when a person receives a transfusion of incompatible blood. Bryan had not undergone a transfusion, however, therefore the physician was unable to understand what she was witnessing.

As it turned out, Bryan would suffer an ever-worsening succession of medical problems in the ensuing months and years, his still-developing immune system being unable to resist the onslaught triggered by the virus. Then, when he was five years old, his immune system collapsed entirely and multiple life-threatening conditions emerged. What had brought about these problems remained a mystery, however, until Bryan was tested for HIV infection and, in short order, diagnosed with "full-blown AIDS" as it was referred to in 1996. When the boy presented at Children's Hospital in St. Louis where the diagnosis was made, his T-cell count was zero, his temperature was 105 degrees, his liver and spleen were severely damaged, and his digestive system had ceased to function. He was nearly deaf as well, a side effect of the ten daily medications he required, some of which he received through a tube inserted in his abdomen. At this juncture, doctors predicted he had three to five months to live. And while every effort was being made to ease his suffering and prolong his life, attention was also coming to bear on how he may have contracted the infection.

Among the possibilities that were ruled out were blood transfusions, since he had never received one. Sexual abuse was also eliminated by means of an interview and a physical examination. And needle-stick injuries were excluded, although there had been two intravenous drug users living in his home at one point. Both were tested and neither was found to harbor the virus, however. In short, no standard means of infection could be established. But there did remain another possibility, one that pointed directly to Brian Stewart, and it was anything but standard.

After purportedly infecting Bryan in early 1992, Stewart ended his relationship with Jackson in August of that same year. Jackson stated that during their breakup Stewart denied that Bryan was his son and said he would not be providing for him. And he told her why. "You won't need to look me up for child support anyway, because your son's not going to live that long," he said.[22] He then gave an estimate that was right on the mark, saying the child would not live past five years of age. And he told a family friend the same thing, a woman who would later testify against him in court.[23] As if this weren't enough, Jackson said her ex-husband threatened to have her, Jackson, eliminated as well, boasting that he could arrange it without being caught.[24]

As it happened, DNA testing confirmed that Stewart was indeed Bryan's biological father, and twice Jackson found it necessary to take legal action against him to obtain child support payments. Then, in 1998, a lengthy investigation wrapped up, with Stewart being charged with first-degree assault, a charge that would be upgraded to murder if the boy perished before the trial.

As to the rationale for the assault charge, it centered on the sequence of events that Stewart's visit to Bryan's hospital room appeared to set into play, along with his subsequent prediction that his son would not live past five years of age. Then, too, the phlebotomist had a powerful financial motive to commit the crime, together with the means to do it and the opportunity to pull it off.

As it came to pass, young Bryan did not die. Thanks to the advent of

HIV protease inhibitors in the mid–1990s, he received a drug cocktail that averted his death and handed him a quality of life better than that which he had ever known. For the first time, he was genuinely healthy. Still alive and well today, his viral load—the amount of HIV in his system—remains so low as to be undetectable on nucleic acid amplification tests. To further differentiate himself from his estranged father, he has since changed his name to Brryan with a double "r," has a girlfriend, and oversees the non-profit organization he founded, called "Hope Is Vital" (HIV). As well, he serves as a speaker for Upward Bound Ministries.[25] "I went from full blown [AIDS], on my death bed, given five months to live, and here I am 19 years old, living on 15 years of borrowed time," Bryan said in 2011.[26]

His father's life would not be as auspicious. Found guilty by a jury, the judge in the case, Ellsworth Cundiff, handed down the maximum sentence, life in prison. "I believe when God finally calls you," Cundiff told the phlebotomist, "you are going to burn in hell from here to eternity."[27]

Unfortunately, Stewart was not alone in using HIV-contaminated blood in an attempt to take the life of someone close to him. A handful of other episodes occurred in the 1990s, the most significant of which culminated in a landmark criminal case, the State of Louisiana vs. Robert J. Schmidt. In a story carried by newspapers around the world, prosecutors sought to introduce into evidence the findings of a phylogenetic, or evolutionary, analysis of HIV in an effort to establish a link between two individuals diagnosed with the intractable infection. If allowed into court, this innovative form of DNA analysis would set a precedent and thereafter constitute a formidable tool in cases centering on biologically-based crimes.

The Richard J. Schmidt Case

A forty-four-year-old gastroenterologist, Richard Schmidt, at the time of his alleged offense in 1994, was living with his wife and three children in Lafayette, Louisiana, a relaxed Southern city nestled in the heart of Creole and Cajun country. Unbeknownst to his spouse, he had a fourth child elsewhere in the city, a son named Jeffrey, the mother of whom was a thirty-one-year-old nurse, Janice Trahan. Since 1984, Schmidt and Trahan had been having an affair, with the pair also working together at the same hospital and with Schmidt occasionally serving as the nurse's personal physician.

According to Trahan's account, the two had previously decided to divorce their spouses and marry each other, but while she carried out her end of the agreement, Schmidt did not follow suit. Exasperated, the nurse announced that she was ending their affair, and to underscore her decision began dating another man. This was when criminality erupted in what was already a troubled relationship.

It was on August 4, 1994, a humid night in St. Louis, that Schmidt showed up at his estranged girlfriend's home, uninvited, unannounced, and unwanted. In a darkened bedroom, she lay sleeping, with their three-year-old Jeffrey beside her. Schmidt, who in the past had given Trahan vitamin B-12 injections to help manage her fatigue, told her he had come to give her another one. This time, however, she refused it. So he proceeded all the same. "Before she could do anything more, he jabs her in the left arm," says District Attorney Michael Harson.[28] Because the room was dark and the injection swift, Trahan was unable to see the contents of the syringe. At once, Schmidt left the house and the couple never had physical contact again. The affair was over, and this had been the physician's parting gift.

Court documents reveal that Trahan described the August injection as being qualitatively different from the previous ones. Not only did it hurt as he administered it, but her arm remained painful the next day to the extent that she asked a colleague to examine it. Something wasn't right, Trahan felt.[29]

Twelve days later, on August 14th, she began experiencing flu-like symptoms, a transient reaction some people experience shortly after exposure to HIV. It indicates that the body has detected the virus's presence and begun creating antibodies in an attempt to eradicate it. Trahan's symptoms included fever, malaise, and swollen lymph nodes.

In the months that followed, she experienced other bouts of illness and for this reason sought Schmidt's opinion. Even though she had broken up with him, the two still worked together and Schmidt was familiar with Trahan's medical history. And his response was to order lab tests. Shortly thereafter, Schmidt told Luis Mesa, a lymphoma specialist and a consultant in Trahan's case, that the blood work had included a test for HIV antibodies, the results of which revealed that she wasn't infected.[30] This second doctor subsequently made a note of these findings in Trahan's medical chart so there would be a permanent record of them. What Mesa did not know is that no such test had actually been performed. Schmidt had knowingly provided false information about the HIV status of his former lover and nurse who, on a daily basis, was in close contact with the hospital's patients.

Finally, in December, the still-ailing Trahan was evaluated by yet another doctor, and this one ordered an HIV test despite the fact that her chart showed she had recently received one. To Trahan's dismay, the findings confirmed that she did in fact carry the human immunodeficiency virus, a baffling development since she had not, to her knowledge, been exposed to the microorganism.

As things stood, the results of an HIV test she took in the spring of 1994 were negative, meaning the infection had to have occurred in the summer or autumn of that year. Taking this information into account as well as reviewing her symptom history and lab data, pulmonologist Ernest Wong, another of her physicians, was able to hazard a guess as to the precise date of transmission. "Dr. Wong estimated that Trahan had been infected with HIV in the first week of August 1994," states a court document.[31] Certainly it appeared that the flu-like symptoms she experienced approximately two weeks after the August 4th injection had been her body announcing its detection of the virus. Accordingly, Trahan began to suspect that the inordinately painful shot Schmidt have given her may have been responsible for her condition.

Contributing to her worries, a further workup revealed that she also was infected with the hepatitis C virus (HCV), which, combined with the HIV in her system, placed her at grave risk. And while it was possible she could have become infected with both pathogens at the hospital where she worked, it was unlikely considering that her duties did not place her at elevated risk. "Trahan was not in an 'exposure prone' job such as surgery, nor was she exposed to HIV patients," reads a document from the Louisiana Court of Appeal.[32]

Without delay, Trahan secured representation, with her legal team seeking a DNA analysis of the HIV in her system along with that of the sole AIDS patient in Schmidt's caseload, a man named Donald McClelland. But what Trahan's attorneys were seeking—a meaningful comparison of the viruses harbored by Trahan and McClelland—was something of a long shot due to the mutability, the inconstancy, of the human immunodeficiency virus itself.

Whereas DNA testing had already been used successfully in other legal cases to identify or exclude suspects, there was a problem when using it to compare HIV samples from different individuals. "HIV," states an *ABC News* report on the Schmidt case, "mutates immediately upon being transmitted to another host, and it continues mutating at a very rapid rate."[33] This means the virus in Trahan's bloodstream would not be identical to that of McClelland because several months had elapsed and the viruses had continued to change independently. For this reason, a procedure was needed that could, in effect, look back in time to determine if the distinctive strains carried by Trahan and McClelland had evolved from the same "family"; that is, a technique that could reveal if the two pathogens shared a common ancestry when compared to other strains of the virus. And this is where phylogenetics entered the picture, defined by the Oxford Dictionary as "the branch of biology that deals with phylogeny, especially with the deduction of the historical relationships between groups of organisms."[34]

Starting in September 1995, a research team led by Michael Metzker set about performing phylogenetic analyses of thirty samples of HIV, Metzker being an associate professor of molecular and human genetics at the Baylor College of Medicine in Houston. For the study, he and his team obtained HIV from Trahan and McClelland, along with twenty-eight additional samples drawn from HIV-infected individuals in the Lafayette area. In a blind study design to prevent bias, one in which the researchers were initially kept in the dark about which sample belonged to which participant, the team compared the genetic sequences of the thirty samples to determine if any of them appeared to be related. And the results were stunning. It seems the genetic sequences of two samples were extremely similar, the ones belonging to Trahan and McClelland.

Two years later, fresh blood samples drawn from Trahan and McClelland were sent to the University of Michigan laboratory of David Mindell, a molecular geneticist. The purpose was to double-check the Metzker analysis. And, indeed, Mindell's phylogenetic analysis confirmed the findings obtained by the Metzker team.

Here it is worth noting that the researchers stressed that these phylogenetic findings did not explicitly prove that Trahan had received McClelland's virus. A direct match would be necessary to establish such a connection, and, as noted earlier, such a correspondence was not possible. But the analysis of the gene sequences did reveal that they were located on precisely the same part of the phylogenetic tree, whereas if they were unrelated they would have been situated on different parts of the tree. "The close relationship between the victim and patient samples," writes the Metzker group in the *Proceedings of the National Academy of Sciences*, "[was] supported by both the genes that we examined, using all major methods of phylogenetic analysis (parsimony, minimum evolution, and likelihood), and a broad range of evolutionary models."[35]

The question now was whether the DNA findings would be admitted as evidence in a court of law. The defense team opposed it, among other arguments insisting that the comparison of Trahan and McClelland's viruses was unreliable since HIV is a changeable microorganism, and, additionally, that contamination may have occurred. In response, the prosecution's experts reiterated the strength of the microorganisms' degree of likeness—"as closely related," testified one scientist, "that two individuals could be"—while stressing that their phylogenetic procedure was highly reliable, that it was designed expressly to take into account the virus's changeability, and that the Trahan-McClelland analysis had been repeated by an independent molecular geneticist and contamination was not an issue.[36] Upon hearing these arguments of the prosecution and defense, the judge decided to allow the DNA analysis into the proceedings as circumstantial evidence, circumstantial because it did not, in itself, prove unequivocally the existence of a connection between Trahan and McClelland's viruses. The court then turned its attention to other evidence.

The additional material exhibited at the trial, which took place in the autumn of 1998, included a notebook Schmidt appears to have hidden in his office, a record of the blood samples extracted on the premises during the summer of 1994. All of the logbook's entries were accompanied by stickers indicating they had been forwarded to laboratories for testing, except for two: a blood sample drawn from Leslie Louviere, a hepatitis C patient, and one from Donald McClelland, the AIDS patient mentioned earlier, whom Schmidt purportedly phoned at home and asked to come to his office and provide a blood sample. The date of the Louviere blood draw was August 2nd, while that of McClelland was August 4th, the same date that Schmidt injected Trahan at her home. Prosecutors drew attention to a notation in Schmidt's logbook that instructed his staff to affix lavender stoppers to this particular set of samples.

Offered as evidence, too, were billing statements revealing that the two patients' insurance companies had not been charged for these August blood draws. Instead, the fees had been canceled by hand on the billing forms that Schmidt's office typically submitted to insurers.

In terms of Schmidt's defense, namely that Janice Trahan was a scorned lover who had leveled her allegations against him in order to wreck his life, it failed to sway the jurors. Similarly, they rejected the defense team's attempts to paint Trahan as a sexually vigorous woman who may have contracted HIV by having had sex with other men and not disclosing it to the court, just as they discounted the claim that the results of the phylogenetic analyses were unreliable.

In the end, the jury found the doctor guilty of attempted second-degree murder, and the judge, having the latitude to sentence him to anywhere from ten to fifty years in prison, chose fifty. A few weeks later, Schmidt surrendered his license to practice medicine in the state of Louisiana, while patients Donald McClelland and Leslie Louviere, whose medical conditions had become public knowledge owing to the legal proceedings, filed their own lawsuits against the gastroenterologist.

In the summer of 2015, Schmidt, who still maintains that he is innocent, sought parole, but the three-member panel denied it by a unanimous vote. If he exhibits satisfactory behavior, he will be eligible for early release in 2023, at which time he will be seventy-four years old.

Janice Trahan, for her part, survived and remarried. At Schmidt's 2015 parole hearing, she attended and spoke out against his discharge from prison.[37]

In the years since the case concluded, the crime and resultant legal proceedings have enjoyed a certain renown. Besides inspiring an episode of the American crime-drama series *Law & Order* and being reenacted in the documentary-style television show *Forensic Files*, it also proved to be a case for the law books. The fact is, Louisiana vs. Schmidt set a precedent in that it marked the first time that phylogenetic research, the basis of which is the concept of evolution, was allowed into evidence in a United States courtroom.

As such, it represented a turning point in the legal recognition of the scientific merit of evolutionary theory, as Metzker and his team pointed out in a journal article about their research. "It is ironic," they added, "that this case originated in Louisiana, which enacted the Balanced Treatment for Creation-Science and Evolution-Science Act in 1982."[38] (Five years later, this law would be overturned due to its unconstitutionality.)

In their journal account, Metzker and his colleagues also spelled out some of the ways in which the field of phylogenetics holds considerable promise in helping to protect the public. "[T]hese studies have broad applications for the identification of putative sources of existing and new pathogens that can cause food-borne infections," they write, "and hazardous agents that can be used for the purpose of biological warfare."[39] From a forensic science perspective, then, Louisiana vs. Schmidt was significant in that it validated, and imbued with respect, an important new tool for use by the American legal community in detecting and identifying biological agents employed in biocriminal and bioterrorist assaults.

As noted in Chapter Four, biological attacks can be executed using a microbe—a living organism such as those that cause AIDS, anthrax, or plague—or a toxin, which is the poisonous product of a living organism. In the latter category, a toxic agent of considerable concern to biocrime and bioterrorism experts is ricin, a fast-acting, lethal poison that, in some countries, is easy to obtain. Then, too, the means by which it may be weaponized is not exactly a closely held secret. "In the underground terrorist literature, there are simple cookbook solutions of how to do this," says Drew Campbell Richardson, supervisory special agent with the FBI.[40]

Accordingly, the final case to be revisited in this chapter, a case that will be covered at some length, focuses on ricin. In this respect, the crime is different from those that have been presented thus far, all of which employed microorganisms rather than their toxins. Also unlike most of the previous cases, the crime occurred in a non-medical setting despite the fact that both the perpetrator and the victim were physicians. And lastly, the episode, perhaps more than any of the preceding ones, illustrates the remarkable degree of psychopathology that may be present in the medical professional who seeks to commit a biologically-based crime. The case is that of Debora Green, and it has long been noted for its meticulously crafted, and excessive, cruelty.

The Debora Green Case

Throughout most of her brief medical career, Debora Green was a doctor whose abilities were respected by her peers and her patients. It was when she was twenty-eight years old, however, that her emotional stability came into question while she was practicing emergency medicine at Jewish Hospital in Cincinnati, Ohio. It seems that she complained of being bored with her training program and, at the same time, began displaying over-the-top outbursts in the emergency room. Among other triggers, insignificant medical problems, because they failed to interest or challenge her, led increasingly to tantrums. And such idiosyncrasies were not minor league. Green's self-centeredness coupled with her hair-trigger temper soon undercut her ability to perform on the job writes Ann Rule, author of *Bitter Harvest*, a book offering an in-depth look at the physician and a key source of material for the present discussion.[41]

As it stood, Green's fiery, erratic disposition was not an entirely new development. She had always had a "temper problem" but was able to control it so that it did not intrude into her career, according to her second husband, Mike Farrar, a man four years her junior and a resident in internal medicine at the University of Cincinnati College of Medicine when he met her. Farrar recalled that he first became aware of her overblown outbursts in the early years of their relationship.

"She was very volatile; she would fly off the handle and do things that were really embarrassing," he says.

> I remember one time she got into an argument in a Kroger parking lot with two people who took the space we were headed for. Debora got out of the car and just gave them hell. I was shocked. I asked her, "What are you *doing*?" ... (S)he walked with them all the way to the door.[42]

As could be expected, such emotional instability eventually created problems in the Jewish Hospital emergency room, where Green found herself increasingly at odds with her colleagues and patients. "Her bedside manner," writes Rule, "was more confrontational than comforting."[43] In the early 1980s, then, Green made the decision to leave the field of emergency medicine and seek a change of direction, and to this end embarked on a residency in internal medicine. Once again, though, her unapproachability together with her explosive outbursts compromised her effectiveness.

In due course, Green became further detached from those around her, as well as losing interest in arguably the greatest love of her life, the practice of medicine itself. No longer devoted to her training, she spent more and more time at home with the couple's two small children, Tim and Kate. It did not come as a shock, then, when the skilled physician with a purported IQ of 165 failed her hematology and oncology boards on two occasions. "She was the strangest doctor I've ever worked with in my life," said a colleague at this point.[44]

Mike Farrar, by comparison, was warm and enthusiastic, well-liked by his colleagues and patients, and successful in his career. Upon completing a three-year residency in internal medicine, he served an additional year as chief resident of the same program, then completed another three-year residency in cardiology. Subsequent to this, he was appointed to the faculty of the University of Cincinnati School of Medicine and named medical director of the facility's heart transplant unit.

On the home front, meanwhile, he and Green continued to struggle as a married couple. Whereas their problems began on their Tahitian honeymoon when Green, rather than have sex with him, decided to read a book, tensions mounted in the ensuing years and were compounded by the stresses inherent in their medical careers.[45] Partly for this reason, they decided to make a fresh start in Kansas, where Farrar, in the summer of 1986, was offered a plum position in a practice near the town in which he had spent his childhood.

Here in the heartland, Green joined a private practice, with her area of expertise being oncology despite her lack of certification in this specialty centering on the treatment of cancer. When the group practice decided not to retain her services at the end of her first year, she opened a practice of her own, but it failed to prosper and therefore she took a job with a peer review organization. When this likewise failed to pan out, she worked out of her home as an independent reviewer, scrutinizing medical records that were mailed to her, before finally throwing in the towel and ending her medical career altogether.

By this point, the couple's third child had been born, a daughter named Kelly, even though the marriage itself was now hollow. Contributing to the pair's difficulties, Farrar discovered that Green was using narcotics, which, he believed, she might be obtaining by writing the prescriptions and misrepresenting herself as a patient at local pharmacies.

In the months that followed, conditions at home steadily worsened in tandem with Green's mental state. "Sometimes she beat herself on the head with books, or beat on her thighs until she left bruises," writes Rule. "Or, worse, she behaved the same way in a public place while their children cringed and strangers stared at her."[46] According to Farrar, Green also tried relentlessly to turn their children against him, filling them with distortions designed to instill a hatred of him. So it was that the cardiologist, no longer able to cope with the situation, announced in 1994 that he was moving out of the house and planned to file for divorce. And Green's response was as frenzied as he had anticipated, replete with "screaming, a lot of profanity, hitting herself," says Farrar. "It was awful."[47] Little did he know that matters would get even worse.

Shortly after leasing an apartment, Farrar received an urgent phone call at work. It seems the family's home was on fire. A mysterious incident that eventually was dismissed as accidental, the cardiologist's colleagues were convinced that Green had torched the house. Whatever the case, the couple reconciled in the wake of the crisis and purchased a new residence in the same neighborhood.

It was with this jumble of troubles festering in the couple's background that Green and Farrar volunteered in 1995 to help chaperone a group of sixth-graders on a summer trip to Peru. Their son Tim was to be among the youths. As for the itinerary, the excursion would entail a week in the jungle and another week on the Amazon River, with Farrar providing medical care for the group.

Also on this trip was a woman, a former nurse named Margaret Hacker, who was in many respects in similar emotional circumstances as Farrar.[48] "She was a married woman," writes Rule, "and she was a confused and miserable married woman."[49] Margaret's husband, who did not accompany her, was a clinically depressed anesthesiologist who would soon commit suicide by over-sedating himself.

In the tropical forest and on the river, Farrar bonded with this woman who shared his personal anguish while helping him treat illnesses in the group. But although the two became close, they did not make love during the trip. This would happen only after they returned to Kansas in what would prove to be an intense, meaningful relationship. It was one that offered them an island of respite from the distress of their marriages—until, that is, Green discovered the affair, which occurred when she hid under the bed and eavesdropped on a phone conversation between them. She did not confront them about their relationship, however, preferring instead to address it in a far more twisted manner.

It was a few evenings later, after Farrar had returned home from work, that Green hatched her plot. Explaining that she and the children had already eaten dinner, she handed him a chicken salad sandwich that she had set aside specially for him. And even though their relationship was broken beyond repair at this point and marked by constant quarrels, Green, on this night, sat in the kitchen with her disaffected husband and talked to him while he ate his meal. She was also ready with a reply when he complained that his sandwich tasted bitter. "We all had them—and nobody else's tasted funny," she told him.[50]

A couple of hours later, Farrar was stung by a burning sensation in his stomach, followed by diarrhea and vomiting. Intermittently during the next few days, waves of pain and nausea swept over him, causing him to be unable to perform his job. He did not suspect Green of poisoning him, however, despite their ruptured relationship and the fact that she was a spiteful woman with a bachelor's degree in chemical engineering. Then, too, he was unaware she had found out about his extramarital affair. Given his recent travels, he believed that he must have contracted a tropical disease in Peru, since he had only been back from South America for a few weeks and there was still time for an ailment to manifest. Farrar therefore was relieved when his symptoms began to subside four days later, although the forty-year-old physician remained concerned about them, especially since an extensive medical workup had failed to identify their cause.

A few days later, the symptoms erupted again, but even more severely this time perhaps because he was now in a weakened state. As to their nature, they entailed "shaking rigors" and "torrential" vomiting—"twenty or thirty episodes over a few hours"—accompanied by a 104.4 fever and dehydration.[51] Worsening matters, as Farrar's condition spiraled downward, he developed an infection that spread throughout his bloodstream, a serious complication known as *sepsis*. "[H]is physician considered his condition life-threatening," reads a court document.[52] Fortunately, the treatment team, as before, managed to stabilize him in the intermediate care unit and a week later discharged him to the family home to recuperate. He had no way of knowing his ordeal was far from over.

The third time it happened was on August 25th shortly after he enjoyed a home-cooked spaghetti dinner. "He was home for a few hours, when, after eating, he became violently ill again," court papers state.[53] Readmitted to the hospital, Farrar received treatment until September 11th, during which the etiology of his condition still could not be determined. Even though all manner of tropical diseases were considered, most notably an atypical form of typhoid fever as well as "tropical sprue" or gluten-sensitivity enteropathy (GSE), intentional poisoning was not explored as the causative agent.[54] As so often happens in such cases, the members of his medical team, including those specializing in infectious diseases, assumed they were looking at a naturally-occurring illness, not an attempted homicide.

Discharged once again from the hospital, Farrar rejoined his family, with Green wasting no time serving him a plate of ham and cornbread. Even though he was beginning to intuit a connection between his wife's meals and the onset of gastrointestinal symptoms, he was still not sure if she was trying to murder him. He wondered instead if their stressful marital life might be triggering the features of his ailment. But what convinced him of the former hypothesis came about a few hours after dinner, when he again found himself in dire shape and back in the hospital, where he would remain for several days.

Looking thin and wasted after a thirty-pound weight loss—the formerly healthy doctor weighed only 125 pounds at this point—he returned yet again to the family home. Suspicious that his wife had contaminated his food and having renewed his decision to seek a divorce, Farrar nevertheless felt it best to stay in the house temporarily because he was worried about their children. His concern was that Green, who was steadily deteriorating, might endanger them, especially since she had taken to drinking heavily and spent a considerable amount of time inebriated. In fact, it was because he came home one day to find her unconscious that he decided to place her in the Menninger Clinic in Topeka. To this end, Farrar phoned the police on September 24th, who transported Green to a local emergency room for an initial assessment, which, to no one's surprise, did not go well.

Whereas the cops, according to the *Associated Press*, described Green as "drunk, profane, bizarre but cooperative" when they arrived at the couple's residence, her ensuing conduct in the emergency room was anything but compliant.[55] Pamela McCoy, the physician who evaluated her, recalls that Green quickly spun out of control, spitting at Farrar and calling him a "fuck hole."[56] Court documents add that Green also warned him about the divorce and custody battle she believed was on the horizon. "[Y]ou will get the children over our dead bodies," she told him.[57]

Certainly the prospect of death was already on Farrar's mind, but in the form of Green committing suicide. And it was during this ugly spectacle in the emergency room that he told McCoy about a disturbing discovery he had made a few days earlier when he looked through his wife's purse. Although he was searching for pills that she might use to kill herself, he instead found a collection of items that made no sense to him. Inside her bag were several packets of seeds from the castor bean plant, along with vials of potassium chloride and syringes. At the time, the connection between seeds and ricin, the highly toxic protein they contain, did not come to mind, although Farrar knew the beans were not for gardening since his wife had no interest in this pursuit. Regarding the potassium chloride and syringes, suicide once again did not appear to enter the picture, although Farrar realized that Green, if she had wished to kill him rather than herself, could have injected the chemical into his catheter while he was receiving intravenous fluids in the hospital. Yet she had not done so in the face of numerous opportunities. So it was, then, that the odd assortment of items stumped him. He did note, though, that a sales receipt for the castor-plant seeds dated back to early August when he first began experiencing gastrointestinal symptoms, and he shared this information with the emergency room staff on this fraught autumn night.[58]

Subsequently admitted to the Menninger Clinic, Green was diagnosed with bipolar depression and prescribed Prozac, Klonopin, and Tranxene. Four days later, however, she checked out of the facility and returned home to a husband who was now wise to her efforts to murder him. While she was away, it seems he had looked into castor-plant seeds and ricin poisoning and was sure she had tried to kill him. Accordingly, he moved out of the house after Green returned, leased an apartment nearby, and continued his relationship with Margaret. And then the unthinkable happened.

The date was October 23, 1995. Green and their three children were living in the family's three-story, eighteen-room estate in Prairie Village, a moneyed enclave situated in the Kansas City metropolitan area near Mission Hills and Country Club Plaza. Ten-year-old Kate, who had been training with the State Ballet of Missouri, was set to perform the role of Clara in the upcoming Nutcracker Ballet, while seven-year-old Kelly was to play an angel. Tim, for his part, was engrossed in sports.

On this fateful evening, Farrar took the children to a hockey match, after which he left them with their mother and drove back to his own apartment. But Green's alcohol abuse continued to gnaw at him, prompting him to phone her three hours later to discuss her drinking, which they both knew could interfere with her psychiatric medications. "[H]e told [her] she had better straighten up or he would call [the] authorities," reads a court document.[59] Farrar further revealed that he knew she had been poisoning him and had decided to seek custody of their three children.[60] "[H]e also told officers that he was very angry during the conversation and that it ended abruptly about midnight."[61]

Twenty-one minutes later, a Prairie Village police dispatcher received an anonymous phone call: the family's house was engulfed in flames. Tim, Kelly, and Kate were trapped inside, while Green stood outside, watching it burn. When Tim pleaded for help on the intercom, his mother told him to "stay in the house and let the professionals rescue you," rather than instruct him to climb out of his bedroom window.[62] As a result, the thirteen-year-old was consumed by the inferno, as was Kelly and their two beloved dogs. The deaths were due to a combination of heat and smoke inhalation.

Kate, on the other hand, took matters into her own hands and jumped from a bedroom window and onto the garage. Here, she looked down at her mother, who held out her arms and motioned for her daughter to leap into them. When Kate jumped, however, Green did not catch her, although the ten-year-old was not seriously injured in the fall.

In the aftermath of the blaze, Mike Farrar was so devastated by the deaths of his son and daughter that the authorities handed over Kate to his parents for temporary care. Even so, he was able, within twenty-four hours, to file for divorce and seek sole custody of his surviving daughter. He also refused Green's request to stay at his apartment until she could make arrangements for other lodgings.

Tellingly, Green, rather than appearing traumatized like her husband, was described as "talkative, even cheerful" in her interview with the police in the hours after the fire, an interview in which she already seemed comfortable speaking about Tim and Kelly in the past tense.[63] Soon, she would also be observed to chant phrases about their deaths, her behavior betraying a profound instability.[64] But while her reaction to the calamity was the polar opposite of that of her husband, the two did share an intense interest in the police inquiry into the conflagration that took their children.

The following month, the investigation wrapped up and yielded unmistakable evidence of criminal activity. Not only had accelerants been used at multiple locations in the house, but the escape routes the children would most likely use had been doused with especially large quantities of combustible material. And virtually all of the evidence, including a charred book found in the remains of Green's bedroom, a book recounting a family that was purposely burned to death, pointed to Green herself. In view of that, she was promptly charged with arson, along with two counts of first-degree murder and two counts of attempted first-degree murder. Bond was set at three million dollars.

In due course, Green confessed to the crime, although her legal team argued that she was not in control of herself on that terrible night and therefore should not be held accountable for her actions. What surprised observers, though, was one of the counts of attempted murder: it pertained not the fire, but to the doctor's efforts to poison her spouse.

When Farrar, in the moments after the tragedy, told investigators his wife had poisoned him on four separate occasions, they sent samples of his blood to the county crime laboratory, as well as to the FBI lab in Quantico, Virginia, and the Naval Research Laboratory in Washington, D.C. The purpose was to test for evidence of ricin poisoning; for instance, by testing for the presence of antibodies to the toxin, which would indicate that he had been in contact with it. And here it is important to note that exposure to ricin, unlike many other toxins, is nearly always deliberate. In fact, international treaties to which the United States is a signatory classify it as a chemical and biological weapon and prohibit its use.[65] Therefore, it was highly significant to the criminal case against Green when the analyses confirmed that her husband had indeed been poisoned with this protein of the castor oil plant.

Unfortunately for Farrar, while the lab tests were being performed, the sepsis with which he had previously been diagnosed recurred, meaning there was still bacteria lurking in his system. Weak and at times unable to work, he sought medical attention in mid–November, during which a battery of tests, including an echocardiogram, confirmed the presence of *Streptococcus viridans* in his bloodstream. Also diagnosed was endocarditis, an associated infection of the lining of the heart. On November 22nd, he underwent surgery at North Kansas City Hospital to repair a leaky mitral valve and install a catheter so he could receive, in his own apartment, intravenous antibiotics while he recuperated. His recovery would not progress as hoped, however.

In early December, the distraught cardiologist found himself suffering from a new set of symptoms, severe headaches being the most disruptive. And both he and his medical team were understandably concerned. To his doctors' dismay, a cerebral arteriogram revealed the presence of a brain abscess that most likely had been initiated by the heart surgery a few weeks earlier. Presumably, material loosened during the previous procedure had traveled to the right frontal lobe of his cortex. Without delay, he underwent a craniotomy and the abscess was drained. Furthermore, because it was a risky procedure that he might not survive, his lawyers videotaped his testimony prior to the operation. As it turned out, though, the surgery was successful, such that his taped deposition was discarded and he testified in person in the case against his wife.

To avoid the death penalty, Green agreed to a plea deal in which she would serve a minimum of forty years in prison with no possibility of parole. And, sure enough, she did go willingly to a Kansas penitentiary, although she later attempted to reclaim her innocence by blaming Tim, her deceased son, for having set the house afire, as well as Farrar and the woman with whom he was having an affair. Her efforts were to no avail, however. Also futile was her attempt in 2015 to have the plea deal vacated so that she might have a new sentencing hearing. Unsuccessful unless an appeal overturns the decision, the one-time physician will remain confined until at least 2035.[66]

Regarding Mike Farrar, he survived the multiple poisonings and their long-term effects, although did require additional hospitalizations in the months after the incidents and his two open-heart surgeries. In terms of his relationship with Margaret, the woman to whom he had become close while in Peru, it did not endure. On a more positive note, he was awarded full custody of his daughter Kate, and, moving beyond the malevolence that had nearly cost him his life, eventually returned to the practice of cardiology and to a more peaceable life. In 1997, Mike married an attorney.

The only remaining issue—or more precisely, question—is how Green extracted the ricin from the castor-plant seeds, since simply eating them does not result in poisoning. Unfortunately, she did not provide the details of her method in spite of having plead guilty to the crime. And yet, her silence on the matter notwithstanding, various possibilities have been considered, with the two likeliest techniques having been described by Seth Carus of the Center for Counterproliferation Research at the National Defense University. "Green must have prepared the beans in a way that exposed the ricin inside," says Carus, "or she extracted the ricin from the beans and then added the toxin to the food."[67] Whatever her method, it was undoubtedly an intricate and rather tricky one. And while Green may not give away her secrets, the ordeal itself has drawn further attention to the dangers posed by this plant toxin. Alongside this increased attention to ricin as a murder weapon, the Debora Green case, a horrific example of filicide and attempted mariticide, has also served to remind us of the way in which an individual who is well-versed in fields such as chemical engineering and internal medicine may misuse their knowledge to inflict pain, illness, and death upon others, a dreadful desecration of the Hippocratic Oath.

◆ 6 ◆

The Oregon Conspiracy

Salad Bars, Salmonella and the Orange People Sect

In Oregon, at the end of the Oregon Trail, sits Wasco County, which in the 1980s was home to 13,640 people.[1] Over half of this number was made up of new arrivals to the area, an unconventional assemblage whose members hailed from around the world and dubbed themselves "Rajneeshees."[2] Dressed in orange, red, and violet garments so as to mirror the colors of the sunrise, they were ensconced in a sprawling commune, where they lived in accordance with the teachings of Bhagwan Shree Rajneesh, an Indian guru who had also recently relocated to Wasco County. It would be in the summer of 1984 that his operation would abruptly bring even more people, four thousand of them, to this rustic region in which John Wayne westerns were once filmed, with their arrival contributing to the political, social, and legal conflicts that were already inundating this rural area.

For several weeks, members of the sect, which the local media depicted as a cult, had been traveling to cities from coast to coast in an effort to bring back to the Oregon collective as many homeless people as possible, preferably male veterans. The plan was to use these newcomers, all of whom were at least eighteen years old and therefore of voting age, to pack the local voter rolls so the sect could manipulate a crucial election. Of course, this meant the Rajneeshees would need to control the newcomers' actions, most of all in the voting booth, which appears to have been accomplished by requiring them to vote pro–Rajneesh if they hoped to remain at the settlement. The Rajneesh medical staff also medicated many of the new arrivals with Haldol and other pharmaceuticals.[3] "As the homeless people were bused into the commune," writes Hugh Milne, the guru's former bodyguard, "the Rajneesh medical unit ordered huge supplies of tranquillisers and mood-altering drugs."[4]

Before the sect's leadership would face accusations of voter fraud, however, it would first have to contend with a more immediate matter: where to house the thousands of street people it had brought onto the premises. And the solution, it turned out, was to place them in an oversized tent city situated on the commune grounds, as well as in the tiny, makeshift houses that also dotted the property. And this turn of events would spark yet another face-off between the Rajneeshees and the local population in a larger conflict that had been brewing for nearly three years.

Owing to the sudden influx of thousands of indigents to the local region—a hostile incursion in the eyes of Wasco County's original residents—three county commissioners were required to travel to the Rajneeshees' enclave to inspect its living quarters. The law required that the lodgings comply with official housing regulations. Unfortunately, the visit would mark the sect's first known venture into biological aggression.

On August 19th, the three men, Wasco County judge Bill Hulse, commissioner Ray Matthew, and commissioner Virgil Ellett, drove to the commune, which was located in a rural area twenty miles from the nearest town. Rambling and verdant, the Rajneeshees had cultivated the land, creating an oasis in the formerly overgrazed terrain. As the officials neared it, however, members of the sect stopped the men's vehicle and instructed them to proceed in a van provided by the commune, with a belligerent Rajneeshee telling the commissioners, "Snakes should sit in the back seats."[5] In this manner, the three men continued on to the site, inspected it for violations (of which there were many), then returned to their own car. Curiously, though, it now had a flat tire. And it was not the first time an official vehicle had encountered problems at the site. Only a few weeks earlier, Wasco County planner Dan Durow, whose position required that he also inspect the commune, found the road blocked by heavy equipment belonging to the Rajneeshees, who claimed it had somehow malfunctioned.

So it was that the three officials stood beside their car while a sect member replaced the tire, a brief interval during which a pair of colorfully-clad Rajneeshee women arrived on the scene flourishing a pitcher of water and three glasses. The one who had prepared the drinks called herself Ma Anand Puja—she was born Dianne Yvonne Onang in the Philippines—and was a nurse as well as the secretary-treasurer of the Rajneesh Medical Corporation. Each of visitors accepted the glass that was offered to him and drank the cool water on the scorching summer day.

As things stood, one of the officials, Virgil Ellett, was well-disposed toward the sect while the other two were critical of it. And corresponding to their views, Ellett would remain in fine health after drinking the water whereas his fellow commissioners, Matthew and Hulse, would have a very difference experience.

After leaving the commune, Matthew traveled a hundred miles away to a cabin at Camp Sherman to spend the evening. And it was here that he awoke during the night suffering from a ferocious illness. Unaware of the cause, the seventy-one-year-old man remained alone in the cabin for the next two days recuperating from the puzzling affliction.

In a neighboring town, meanwhile, an acute illness also roused Bill Hulse from sleep, with the commissioner vomiting so uncontrollably that his wife rushed him to the emergency room. Here, medical tests revealed the existence of a "highly toxic substance" in his kidneys, a substance that was later determined to contain *Salmonella* bacteria.[6] Hulse, for one, was not surprised. Given the chain of events, he was convinced that Puja had poisoned him, and he even made a public statement to this effect. As to the severity of his illness, it was life-threatening. "Hulse remained in the hospital four days," reports *The Oregonian*, "with doctors telling him he would have died without treatment."[7]

The Rajneeshees, for their part, denied poisoning the men and chided them for making what the sect insisted were false allegations. Much later, however, the Rajneeshees would fess up to the crime.

The following year, the woman who was behind the poisonings, Ma Anand Puja, headed up another plot, one in which she slipped onto the eighth floor of St. Vincent Hospital in Portland clutching a syringe loaded with a drug that would trigger cardiac arrest. Her plan: inject it into the intravenous tube of Wasco County commissioner James Comini, who had undergone ear surgery and was recovering in an isolation room.[8] The sect's membership was convinced that he and other officials were determined to expel it from the region. Luckily for Comini, Puja's plan to assassinate him failed, since he wasn't connected to an IV line when she entered his room. So it was that the ill-intentioned nurse rejoined the hit squad that awaited her in the hospital parking lot for the two hundred mile drive back to the commune. "The murder scheme was just one of many increasingly desperate attempts to save the guru's empire," writes investigative journalist Les Zaitz, who would eventually land on the sect's hit list, too.[9] It was an empire whose Oregon operation became the sect's international headquarters starting in July of 1981 with the arrival of the spiritual leader Bhagwan Shree Rajneesh, a purported mystic who offered an alluring philosophical approach that fused Eastern spiritualism and Western capitalism.

The Rise of Rajneesh

Rajneesh was born Chandra Mohan Jain on December 11, 1931, in the town of Kuchwada, situated in the heart of India. Upon completing high school, the son of a cloth merchant attended Hitkarani College two hundred miles away in the city of Jabalpur, but owing to conflicts with the faculty, soon transferred to the newly inaugurated D. N. Jain College in the same city. It was during this period that the young scholar later claimed to have found enlightenment. Graduating with a degree in philosophy, he proceeded to obtain a master's degree, then secured a teaching position at Raipur Sanskrit College in Raipur. Once again, though, tensions arose between Rajneesh (Jain) and the school's administration, tensions that were mainly ideological in nature, prompting the future guru to relinquish his teaching position and accept a professorship in philosophy at the University of Jabalpur.

While he was performing his academic duties at this time, Rajneesh also spearheaded a spiritual movement that would prove to be rather controversial. "In long monologues that drew eclectically on Zen Buddhism, Hindu traditions, progressive psychology, Gurdjieff, and Nietzsche, [Rajneesh] proclaimed a doctrine of self-fulfillment without and beyond constricting rules and narrow morality," writes Carl Abbott. "Enlightened capitalism and the pleasures of the world were to be embraced and combined with open-ended spiritualism, crafting an ideal that he called Zorba the Buddha."[10]

In terms of the inner growth process itself, the professor and nascent spiritual leader taught that satisfying one's physical needs is intricately associated with spiritual development and enlightenment. He further insisted that the need for sexual gratification was a singularly important one. "Among his teachings was the notion that sex was the first step toward achieving 'superconsciousness.'"[11] Rajneesh thus became a vocal proponent of "free love"—the immersion in, and celebration of, the human sexual experience with one or more partners outside of conventional social parameters. This stood in opposition to traditional marriage, according to the guru, which led to possessiveness.

Then again, free love, as taught by Rajneesh, was intended to apply to heterosexuals, the spiritual leader having little concern for his LGBT devotees. Despite his outward exhortations about sexual liberalism, Rajneesh's understanding and acceptance of same-sex relationships remained mired in the prejudices and ignorance of earlier centuries and therefore were excluded in his enthusiastic encouragement of natural, uninhibited sex. In fact, he eventually suggested that gay people should remove themselves from society. "There are deserts, there are islands uninhabited," he said. "Just give them to homosexuals."[12] As for contact between gay and straight people, Rajneesh taught that segregation should be absolute. "[T]here should be no communication between them."[13]

In certain respects, Rajneesh's model, notwithstanding its extreme homophobia, shared similarities with that of another pioneer of the era, American psychologist Abraham Maslow, who in 1943 proposed a hierarchical theory of motivation. Maslow believed that a person's physical requirements must be satisfied before the individual can fully advance to higher-level needs, such as those involving love, self-actualization, and spirituality. But whereas Maslow's theory won acceptance from his peers, or at least their respect, Rajneesh's ideas provoked disdain, not least because he encouraged his followers to seek out and rejoice in unrestrained sexual activity. Then, too, his views on prosperity were at odds with traditional Eastern thought. "He believed that wealth, which more traditional gurus shunned, was actually the precondition for spirituality," writes American psychiatrist James Gordon, who traveled to the ashram to scrutinize and study with the religious figure. "Only wealthy people living in a wealthy community had the freedom and leisure to transform themselves."[14] While Rajneesh's emphasis on the material world would alienate traditionalists, it would attract seekers who were eager to embark on a fresh path; a path that, in several respects, was in line with the counterculture of the 1960s and the emerging human potential movement.

Into the 1970s he continued to draw devotees, even as he dropped his birth name, Chandra Mohan Jain, and became Bhagwan Shree Rajneesh and made his home at an ashram in the ancient city of Poona. It was a religious community that scores of his followers had forged, men and women who lived, worked, meditated, and studied in accordance with the guru's teachings and referred to themselves as "sannyasins," or disciples. By all accounts, they were devoted to Rajneesh and embraced his teachings on an array of subjects, among them the significance of what was termed "dynamic meditation." Incorporating physical movement, such as dance, into the contemplative experience, dynamic meditation, which is still practiced by devotees today, is a five-phase process unique in its fusion of the physical, mental, and spiritual.

As could be predicted, Western media soon became besotted with the colorful religious leader, most notably his tantalizing prescription of free love. Accordingly, the press wasted no time saddling Rajneesh with the tabloid nickname "sex guru." It was a perspective that was furthered, if unintentionally, by those in the United States who were personally familiar with life at the ashram in Poona and who described it as a "continuous orgy."[15] Clearly, the spiritual leader's Western disciples remained devoted to him and to his distinctive philosophy upon their return from India. When James Gordon met with some of Rajneesh's sannyasins in San Francisco in the late 1970s, he found them quick to laud their spiritual mentor as well as rather haughty. "They seemed as a group self-consciously hip," Gordon writes, "and smug about the man they breathlessly called 'Bhagwan.'"[16]

By this point, the organization had set up meditation centers in several countries, among them the United States, in what was turning out to be a thriving global enterprise. But the principal ashram in Poona, which now had six thousand sannyasins, was also attracting its share of opposition from the public because of its controversial practices and escalating scandals. Among the numerous complaints against the sect was the physical violence that was said to be typical of its encounter groups. It was a notorious feature that author, professor, and literary and social critic Christopher Hitchens underscored when he visited the Poona ashram to film a BBC documentary about the spiritual hermitage.[17] The filmmaker appears to have been taken aback by the cruelty that he observed in the ashram's encounter groups, aggression that was often directed at the female participants. Hitchens writes,

> In a representative scene, a young woman is stripped naked and surrounded by men who bark at her, drawing attention to all her physical and psychic shortcomings, until she is abject with tears and apologies. At this point, she is hugged and embraced and comforted, and told that she now has "a family." ... It was not absolutely clear what she had to do in order to be given her clothes back, but I did hear some believable and ugly testimony on this point.[18]

The treatment of male participants was likewise brutal, Hitchens reports, even more than for the females. "In other sessions involving men, things were rough enough for bones to be broken and lives lost," he writes, although the case on which the latter allegation is based remains unconfirmed.[19] As could be expected, such practices shocked the public and fueled the rage of the local community, with the same hostility being on display in the United States as well. When *Time* magazine described Rajneesh's ashram as the "Esalen of the East," Richard Price, co-founder of the esteemed Esalen Institute in Big Sur, California, vehemently rejected the comparison, with "particular attention to the violence endemic in the Poona encounter groups."[20]

Other scandalous features that riled opponents included the sect's alleged entrapment of powerful figures in dubious sexual situations, as well as the guru's fleet of Rolls-Royces; this from a man who asked his followers to dispose of their material possessions upon joining his operation. As for the intensity of the critics' animosity, it was excessive at times, even to the point of causing the Rajneeshees to become suspicious of outsiders and hence more guarded, more isolated from the population at large.

Worse still, acts of violence, including bombings, began to occur, although it is unclear whether they were the work of opponents attacking the operation or the sect's own sannyasins committing the crimes and pinning the blame on outsiders. A pair of arson attacks directed against the ashram's bookstore and health clinic, for instance, was blamed on outside adversaries, although many believed the Rajneeshees had burned their own businesses in order to feign victimization while cashing in on the insurance.[21] "The fortuitous nature of the fires and the bombings made me see the 'steadily mounting anti-sannyasin violence' in a new light," says Hugh Milne, the aforementioned bodyguard for Rajneesh. "[I]t became clear that it was all being engineered so that it would look as though we were being persecuted."[22]

Then, too, cash was an issue for the ashram and a generous insurance pay-out would surely have been welcomed. Although the Rajneesh Foundation banked tens of millions of dollars that the operation earned from its various business ventures, it nevertheless owed the Indian government four million dollars in unpaid taxes.[23] This is because the government had revoked the Foundation's tax-exempt status once it discovered that the operation was bringing in a staggering amount of revenue yet without performing charitable services. By all accounts, it was proving to be a complicated and frustrating legal battle for the government, especially since the Rajneeshees had placed their own members in high-level positions within the nation's tax assessment and collection system. The result: the government was stymied by calculated internal obstacles in its attempt to compel the sect to pay what it owed.

Evidently, it was for such reasons that Rajneesh's chief lieutenant, personal secretary, and confidante Ma Anand Sheela (formerly Sheela Silverman), set out to relocate the operation to the United States, where it would serve as the global headquarters. Not only would this sidestep the immediate tax situation in India, but it would also provide the principal ashram with room to expand. The latter, it seems, had become a problem at the Poona site, which had outgrown its facilities and needed space to grow. And although a team of sannyasins had identified a sizable property elsewhere in India that they sought to purchase and develop, the government's land-use regulations prohibited the parcel from being used for the sect's purposes. By moving to the U.S., then, the Rajneesh operation would be able to acquire a larger property.

As to the cover story for the impending relocation, Rajneesh and Sheela claimed it was because the guru was ill and needed to undergo medical treatment in the United States. It was the same explanation they would offer to officials in the U.S. And taking the guru at his word, American immigration officials would grant Rajneesh a visa, a temporary one, which would permit him to enter the country to receive medical care. It appears, though, that he may have had no intention of returning to his homeland, since efforts to dismantle the six-acre ashram in Poona began shortly after he departed.

As the time arrived for Rajneesh to move to the United States, he took a vow of public silence, explaining that it marked the next phase of his spiritual evolution. That said, he would continue speaking to a handful of trusted insiders, Ma Anand Sheela foremost among them and the woman to whom he would entrust the day-to-day functioning of what had become an extensive, international enterprise. And indeed, the venture truly was far-reaching. As of 1981, the Rajneesh entity operated four hundred meditation centers worldwide for use by 200,000 disciples.[24]

In America

In the same way that Bhagwan Shree Rajneesh left millions of dollars of debt behind in India, so he also left behind ten thousand sannyasins who had forfeited their earthly possessions to live and study with him in Poona. If these devoted men and women wished to remain in his presence, it would be up to them to come up with the funds to trek halfway across the world.

Rajneesh, himself, set out on this journey in late May of 1981, accompanied by a small crew of hand-selected disciples. Arriving in the United States on June 1st, the group settled in a thirty-room, hilltop mansion, an imposing edifice built in the manner of a Rhineland castle and overlooking the city of Montclair, New Jersey. Although the sect applied for the majestic estate to be tax exempt, the ensuing public outcry caused it to withdraw the request. The group also set about publicizing the programs offered by its local meditation center, running five national advertisements focusing on love, laughter, and other pleasurable experiences. "On July 22," reports the *New York Times*, "the Montclair center ran an advertisement in Time magazine headlined 'Sex' and reading, in part: 'Never repress it! Search all the nooks and corners of your sexuality.'"[25]

Predictably, the tantalizing promotional campaign aroused curiosity in some quarters, but it also invited criticism, most of all from the citizens of Montclair. From parents to public officials, local opponents, like those in Poona, condemned what they regarded as a free-love cult, a sham religion, and a moral threat to the community. This is despite the fact that the sannyasins, thus far, had not behaved in a manner that violated the law. Rather, the situation seems to have been that of a fundamental clash of values, with the incoming devotees' beliefs and lifestyles differing markedly from those of the established, larger community and causing the original residents to feel threatened. In all likelihood, the townspeople would have not been so unnerved had they been privy to the plans that were unfolding within the sect itself.

At precisely this moment, Ma Anand Sheela and her second husband were discreetly purchasing a vast amount of ranchland, over 64,000 acres of it, near the Cascade Mountains in Oregon, where Bhagwan Shree Rajneesh would take up residence at the end of August. Known as the Big Muddy Ranch, the property spread across rural Wasco and adjoining Jefferson counties. "This expanse was to become both a fully-functional urban center and a spiritual mecca for followers from around the world," write Chloe Prasinos and Steven Jackson.[26] It is not known to what extent Sheela and her spouse, when buying the $6,000,000 property, inquired into Oregon's land-use regulations that applied to the ranchland, but they surely would have been struck by the similarities to those laws that had vexed the sect in Poona. Just as India's land-use policies prevented the Rajneesh operation from developing property that was not zoned for the ashram's purposes, so the regulations in Oregon would place firm limits on the organization's new digs in the Pacific Northwest.

Of all the American states in which to settle, the Rajneesh sect, for whatever reason, selected one having among the most restrictive land usage laws on the books. "Oregon's land use planning program has been cited as a pioneer in U.S. land use policy," write Hannah Gosnell and her colleagues, alluding to the state's 1973 Land Conservation and Development Act.[27] "The program was a response to rapid population growth in western Oregon during the 1950s and 1960s, which raised concerns in the state about the loss of forests and farm land to development."[28] Among the law's principal objectives: "the protection of forests and agricultural lands, and the protection and conservation of natural resources."[29] It was this same respect for the environment that had been on display in Wasco County when officials previously decided that the Big Muddy Ranch and surrounding territory—the property that Sheela and her husband were now acquiring for Rajneesh—would be protected from urbanization. To make sure it would not be used for such purposes, officials classified the land as "EFU," meaning "exclusive farm use" only.

Rajneeshpuram, Oregon

Shortly after the forty-nine-year-old guru, along with Sheela and a small team of sannyasins, moved onto the Oregon ranch, an ambitious effort was put forth to develop the site as rapidly and extensively as possible. The purported aim was to transform the existing overgrazed property into a natural paradise, a modern-day Eden boasting fertile farmland and a commune for those who worked it. What was not acknowledged was the sect's plan to construct a town as well, one with buildings and roads and schools, despite the "exclusive farm use" law. Equally misleading, the organization's leadership told local officials that the commune would be home to forty people, when in fact hundreds and eventually thousands of Rajneesh's devotees would take up residence at the site, all of them dressed in the sect's distinctive colors. These sannyasins would become known in the region as the "Orange People," a neutral descriptor that both locals and sannyasins considered inoffensive, although the nickname would later be replaced by such derogatory terms as the "Red Menace," "Red Vermin," and "Red Rats."[30]

In terms of the commune's demographics, the men and women who were devoted to Rajneesh's entrancing message of personal and spiritual liberation were a remarkable lot. "His feel-good philosophy attracted sannyasins who were overwhelmingly well-educated," writes Rachel Graham Cody, "affluent urbanites with every intention of remaining in the world—on their own terms."[31]

Cody's observations were bolstered by a University of Oregon survey conducted within two years of the commune's founding.[32] Most of the sannyasins, it found, were Caucasians in their early thirties who were married and had relocated to the Oregon ranch from large cities. The genders were more or less evenly divided. In terms of education, 64 percent were college graduates, with 36 percent holding masters' degrees or doctorates, most often in the social sciences, arts, and humanities. Yet many of the locals failed to be impressed by the Rajneeshees' urban roots and prolonged schooling. "It didn't mean they had any horse sense," said Margaret Hill, the former mayor of a nearby town and critic of the sect.[33] Still, the record would show that the sannyasins were a politically astute group and thus well-equipped to take on the local community when matters took an unpleasant turn. This would happen the following year when the commune's leadership put into motion a strategy for land development that was unpalatable to the established populations of the two counties.[34]

The structure and composition of commune's leadership, it should be noted, was unique. Although Bhagwan Shree Rajneesh had taken a vow of public silence, he nevertheless conversed daily with his right-hand woman, Sheela, who, in turn, managed the commune with the backing of a coterie of female sannyasins. Rajneesh believed that women, not men, should run it, although he himself would remain the head of the venture. Even so, Sheela quickly acquired more and more power, including the presidency of the Rajneesh Foundation International, and ultimately rose to the level of de facto chief of the entire international operation. "Sheela controlled the finances and directed the operations of almost all of the organizations created to administer the cult's activities," says Seth Carus of the Center for Counterproliferation Research.[35] Sheela was also fast to force her decisions on the sannyasins, her management style being classically authoritarian. "Because she was accountable only to the Bhagwan, members of the cult had to accept her decisions or risk being expelled," Carus adds.[36] This included those decisions about which many sannyasins had doubts; specifically, doubts that the guru had been involved in the decision-making process, since the outcomes were at odds with his teachings. Significantly, it would be Sheela at the helm when the sect eventually entered into the ominous realm of bioterrorism.

In its formative days, the Oregon operation appeared to function smoothly and efficiently, with amiable relations between the orange and red-clad sannyasins and the residents of Antelope and The Dalles, the latter being a nearby town and the Wasco County seat. Townspeople regarded the Rajneeshees, as the sannyasins now dubbed themselves, as a benign curiosity, not as a threat as was the case in India and New Jersey. But the locals' perception shifted in response to the Rajneeshees campaign of expansion and land-use conversion.

A few months after its founding, the commune began building houses and other structures on the rural land. As to the homes, EFU zoning laws permitted lodging for only ninety people on the ranch, so the sect found itself at odds with local authorities when it sought to dramatically increase its housing for the disciples who were arriving from around the world. Then, too, zoning laws required that the Rajneeshees conduct any business that was not farm-related at a location off the ranch, the nearest spot being the incorporated town of Antelope eighteen miles away. For the Rajneeshees, this proved to be an annoyance, having to drive to Antelope each time they wished to carry out business.

In an attempt to dispense with the troubling regulations, the Rajneeshees petitioned Wasco County for permission to hold an election, one that would decide whether the Big Muddy Ranch—that is, the commune—could incorporate. If so, it could alter the zoning regulations. As it happened, the county had no choice but to consent to the election, since the law required it if at least a hundred Oregon citizens petitioned for it. And so it was that the Rajneeshees handily prevailed in balloting that was held in May of 1982, and at once set about fulfilling their agenda.

After renaming the property "Rajneeshpuram"—the sannyasins whimsically called the commune "Rancho Rajneesh"—Sheela and her cadre of helpers turned to more commercial matters. "As a municipality," writes Carl Abbott, "the community now controlled land-use decisions and was no longer constrained as rural land, so long as it adopted a comprehensive plan that met Oregon's state planning goals."[37] Because the City of Rajneesh now had the authority to approve building permits, it sanctioned a flurry of them, over three hundred for business enterprises and residences. "That authority was the key to changing cow pastures into card rooms," reports *The Oregonian*.[38] By late 1982, Rajneeshpuram was home to a jewelry shop, numerous pizza parlors, cafés, and restaurants, a shopping mall with clothing stores in which the sect's characteristic orange and red garments were sold, and organizations such as the Rajneesh Foundation International and the Rajneesh Institute for Therapy.[39] During the brief lifespan of Rajneeshpuram, the commune would create "almost a million square feet of buildings, including a large meeting hall, and numerous residences, almost all of which [were] below building codes," reports the Center for Land Use Interpretation.[40] "[The Rajneeshees] hoped to blend spirituality and materialism while building an intentional community that could also serve as a destination resort and luxurious pilgrimage center for sannyasins from all over the world," writes sociologist Marion Goldman.[41]

During this period, the Rajneeshees' presence also increased substantially in neighboring Antelope, previously a sleepy hamlet comprised of forty-seven people, among them retired ranchers seeking solitude in a settlement long known for it.[42] As could be expected, the sudden influx of Orange People caused the townspeople to feel as if the sect was, in effect, annexing their home. And flare-ups were inevitable. As the townspeople increasingly shunned the Rajneeshees, the latter, in turn, arrived in clusters in Antelope each day, harassing public servants by making excessive demands for needless services while also surveilling and photographing the residents as they went about their daily business. In addition, sect members bought up much of the town's real estate. And while the residents of Antelope, feeling overrun, made a desperate attempt in 1982 to disincorporate the town so as to discourage the Rajneeshees from using it for their business affairs, the sannyasins, whose numbers were now greater in Antelope, retaliated by ensuring that the disincorporation attempt failed at the voting booth.

Later that year, the Rajneeshees took even bolder action. When a second election was held, one for membership on the Antelope city council, the Rajneeshees, now comprising the majority of the town's citizens, won all but one of the seats. Exercising its power, the Rajneesh-controlled council wasted no time raising taxes, a tactic that would burden the elderly retirees subsisting on fixed incomes. The council also changed the town's name from Antelope to Rajneesh, as well as changing Main Street to Maylana Bhagwan Street and the city dump to Adolf Hitler Landfill and Recycling Center.[43] Longstanding residents complained that the identity of the laid-back western town was slipping away—or, more precisely, being hijacked—but their grievance fell upon deaf ears, with the sect, unapologetic, rebuffing the original residents' desire to preserve their way of life. In her memoir, Sheela blamed the victims—that is, the residents of Antelope—for the sect's aggressive annexation of their town, which the Rajneeshees now planned to transform into the entrance to their commune several miles away. "[T]hey forced us to take over the city," Sheela writes.[44] And other Rajneesh leaders insisted, if condescendingly, that the sect knew what was in the best interests of the original residents. "In a funny way I don't have a lot of sympathy for someone who wants to live in a stagnant place," remarked the newly elected mayor, Ma Prem Karuna.[45] Regarded by locals as an exercise in arrogance, the takeover of the town led to further hostility toward the sect. "Antelope, better dead than red," proclaimed a popular anti–Rajneesh bumper-sticker.[46] "[B]ag the Bhagwan," read another.[47]

For the Rajneeshees, there were other positive developments during this period in addition to assimilating the town of Antelope. The sect opened a discotheque, restaurant, and bar in Portland, the objective being to raise capital for the commune. Even more ambitious, it purchased the Martha Washington Hotel in the same city, a stately edifice constructed in 1923 and designed in Georgian Colonial Revival style. Re-christened the Hotel Rajneesh, it catered to visitors traveling to the city of Rajneeshpuram.

Meanwhile, in Rajneeshpuram itself, construction flourished and included more houses, businesses, and an airstrip, with such ambitious growth proving to be a minor boon to the region's economy. In large measure, this was because the Rajneeshees purchased their equipment and building materials locally. All parties appear to have benefited, then, even if a sizable portion of the original land was no longer used for farming.

As for the land that did remain intact, it enjoyed a stunning rejuvenation. Not only did overgrazed pastures become lush once again, but new species of plant life imported by the sannyasins thrived in their new home. It was a transformation that was both swift and sweeping, and it was due largely to the innovative farming methods the Rajneeshees introduced to the region. Not surprisingly, the sect made a considerable amount of money from agriculture and other dealings during its early years in the Pacific Northwest. Unlike Bhagwan Shree Rajneesh, however, the preponderance of sannyasins did not personally share in the profits.

As had been the case in India, the guru's circumstances were strikingly different from those of his disciples. Whereas the latter lived humbly off the land, Rajneesh immersed himself in earthly possessions, acquiring, for instance, a cache of jewelry valued at over one million dollars.[48] Even more conspicuously, a Rajneesh trust held ninety-nine Rolls-Royces for him, the collection being the largest of its kind in the United States. Each day, Rajneesh would drive one of the luxury automobiles through the commune's streets, greeting his followers with the Namaste hand gesture.[49]

It is also worth noting that the Rajneesh operation in Oregon raked in a substantial amount of money from donations, money that came largely from the sect's hundreds of meditation centers around the world. The headquarters in Rajneeshpuram solicited these donations with claims that it needed financial help to protect itself from religious persecution. Among other allegations, the Oregon group insisted that the local community was impugning its reputation and, more ominously, threatening to harm the sannyasins themselves. And although many observers in the region looked upon these claims as a ploy to make money, it is true that the sect was indeed receiving threats. And not just threats. In July of 1983, a domestic terrorist bombed the Hotel Rajneesh in Portland, a man described at his trial as a member of a "militant, fundamentalist Muslim organization" who kept a bomb-making workroom in his Los Angeles home.[50] But the local population felt physically threatened, too. The Rajneeshees had assembled a security force of their own by this point, with sannyasins trained at the state's police academies and armed with Israeli-made Uzis and Galil assault rifles.[51] The previously sedate region was devolving into a powder keg.

Further ratcheting up the tension was a lawsuit filed at the end of 1983 by Oregon Attorney General David Frohnmayer. Taking the position that the incorporation of Rajneeshpuram, a religious city, breached that cornerstone of American democracy—the separation of church and state—Frohnmayer argued that the city was ineligible to exercise governmental powers or share in the state's revenues. Even more significant, he insisted that Rajneeshpuram, being in violation of federal and state constitutional requirements, disincorporate.

Descent into Bioterrorism

With conditions between the sect and the local community marked by mistrust and animosity bordering on rage, the stage was set for the sinister acts that would transpire in the summer and autumn of 1984. The catalyst: a Wasco County election scheduled for November, one in which three seats for county commissioner would be up for grabs. It was an election of the utmost importance, at least in the Rajneeshees' view, since two of the officials currently occupying the seats had been critical of the sect and were hindering its efforts to continue expanding in the region. Unseating this pair of commissioners was therefore of great consequence to the Rajneeshees' long-term growth project. And to ensure the desired result, Sheela and her accomplices set out to install at least two of their own sannyasins on the commission, thereby handing majority control to the sect. It was a turn of events the Rajneeshees would conspire to bring about through the one-two punch of voter fraud and bioterrorism.

The first scheme, voter fraud, entailed the launch of the sect's Share-A-Home campaign, which it touted as a humanitarian program geared toward helping the nation's homeless by transporting them to Rajneeshpuram to live, work, and grow spiritually. Busing in 4,300 street people, mainly veterans, the sect signed them up to vote as soon as they arrived, while instructing them to cast their ballots for the Rajneeshee candidates or face expulsion from the commune.[52] The illicit maneuver would quickly fail, however. Recognizing the bald attempt at voter fraud, Oregon's Secretary of State temporarily suspended the registration process and, pointing to Oregon's twenty-day residency requirement for voters, assembled a team of legal experts to interview each of the indigents in order to establish their eligibility. Not surprisingly, most of the new arrivals did not meet the state's criteria. The upshot: the Rajneeshees promptly canceled the Share-A-Home program and expelled the majority of the homeless it had bussed to Oregon.

The next stage of the plan was even more disturbing. Whereas the stab at voter fraud was designed to help the pro–Rajneesh candidates by packing the rolls with sannyasin voters, the bioterrorism scheme was intended to undermine the non–Rajneeshee candidates by incapacitating those citizens who might cast their ballots for them.

Ma Anand Puja, the thirty-eight-year-old nurse who had earlier contaminated the drinking water of two county commissioners—the episode recounted at the outset of this chapter—is the person who prepared the pathogens that would disable the local citizenry. Fascinated with poisons and death and feared by the other sannyasins, she was suspected by some of having been involved in the 1980 death of Sheela's first husband, a man whose last moments were reportedly videotaped.[53] It was further rumored that Puja, whom the Rajneeshees called Dr. Mengele behind her back, had infected at least one sannyasin with HIV in order to observe the results.[54] "There was something about Puja that sent shivers of revulsion up and down my spine," recalls Satya Bharti Franklin, a former sannyasin, who added that the nurse was the "alleged perpetrator of sadistic medical practices."[55]

To sabotage the upcoming Wasco County election, Sheela, Puja, and a team of twelve lower-ranking sannyasins explored the idea of flying a small aircraft laden with bombs into the county courthouse with the pilot parachuting to safety.[56] Less theatrical but more pragmatic, they weighed the benefits of using biological agents to lay low the local population. Among the pathogenic microorganisms they considered was the bacterium that causes typhoid fever and the virus that causes hepatitis. They also discussed the protozoan *Giardia lamblia*.

As to the means of delivery, potential methods included depositing dead beavers into the local water supply, since beavers harbor microbes that are harmful to the human population. It was an approach that was ruled out, however, when the team learned that workers at the regional water plant had installed protective grids over the tanks. Other methods of adulterating the water supply were considered as well, one of which appears to have been attempted but ultimately failed. At the end of the day, it was the idea of contaminating the local food supply that prevailed.

Regarding the pathogen, the decision was made to deploy *Salmonella* enterica Typhimurium, a bacterium the Rajneeshees already had in stock. They had ordered it several months earlier from a commercial supplier, VWR Scientific in Seattle, in a standard transaction that did not raise eyebrows because it was purchased by the Rajneesh Medical Corporation.[57] The underlying reason for the acquisition: Puja wished to experiment with the pathogen at the Rajneeshees' biological research unit, which, in reality, was a secret germ warfare facility. Situated in a gorge far removed from the sannyasins' lodgings, it consisted of two dozen buildings where, in the summer of 1984, she instructed a lab tech to culture the *Salmonella* in substantial quantities. Although the technician initially objected owing to the dangerousness of the microbe, he carried out the directive in due course. By all accounts, Puja could be very persuasive. Subsequent to this, Sheela, Puja, and their collaborators set about testing the pathogen's effects in a nearby town, the microorganism having been suspended in a liquid medium to make it dispersible.

One strike took place at an Albertson's Market, where Puja entered the grocery store with an eyedropper loaded with *Salmonella* and sprinkled it onto a bin of lettuce.[58] Another sannyasin, meanwhile, entered the Wasco County courthouse and deposited the same substance on doorknobs and urinal handles throughout the building. Still another moistened her hand with the liquid, then shook hands with an elderly gentlemen at a political gathering. And then there was the Rajneeshee who was dispatched to four schools and nursing homes in the area, where she was to slip the adulterated substance into the food supplies. Although the sannyasin did in fact visit the assigned locations, however, she later claimed that she could not bring herself to poison the pupils and elderly nursing-home residents.[59] Then again, she found herself "under close observation" by the staffs of the facilities, which is unsurprising in light of the sect's dubious reputation.[60] Unwilling or unable to complete the crime, the sannyasin wound up pouring the substance onto the street before returning to the commune. "When I got back I lied," she testified before a grand jury. "I said to them that I had done it."[61]

These early experiments, with the exception of the one that was purportedly aborted, were valuable in that they furnished Sheela, Puja, and their accomplices with hands-on experience with *Salmonella* poisoning. Such familiarity would prove advantageous when deploying the bacterium in the upcoming biological offensive.

Another advantage of *Salmonella* is that it would be difficult to detect, and it would most likely incapacitate, but not kill, those being targeted. And this was in line with the sect's objective of alarming and sickening the locals. As noted in Chapter Four, *Salmonella* poisoning is characterized by fever, headache, abdominal cramps, diarrhea, and bloody stools within hours or days of infection, and, while dreadful, usually resolves within a week. Among those in whom the infection may be more dangerous are young children, older adults, and ailing individuals with weakened immune systems. In some cases, hospitalization may become necessary, with *Salmonella* poisoning occasionally proving to be fatal if the infection spreads to the bloodstream or other parts of the body and remains untreated. While death resulting from the bacterial infection is infrequent, the illness itself would be sufficient to deter most people from leaving their homes until it had run its course, thereby making it an ideal agent for the Rajneeshees' purposes.

The poisonings were slated to take place in September. Even though the election was still several weeks away, Sheela and Puja decided to implement a mass contamination at this juncture to ensure that it would work. To this end, the pair, explaining that they "wished to do an experiment," instructed a half-dozen trusted sannyasins to help spread *Salmonella* throughout The Dalles, according to a witness.[62]

This time, restaurants would be the main focus, with at least ten of them being targeted in The Dalles area, from taco shops and pizza parlors to large, full-service restaurants. All of the eateries were popular establishments, and were patronized not only by the townspeople but Interstate motorists as well.

According to a retrospective account prepared by Oregon Congressman Jim Weaver and preserved in the Congressional Record, the Rajneeshees' bioterror plot entailed two waves of mass poisonings. The first was a small, limited attack initiated on September 10th, with a second being a much larger one implemented on September 23rd. The latter was a Sunday, and therefore a busy day for The Dalles' eateries.[63]

To execute their plot, the Rajneeshees, in both attacks, dispensed with their standard orange and red garb and donned street clothes in order to appear less conspicuous, then dined in the targeted restaurants. During their meals, they discreetly slipped *Salmonella* into food products, even into the coffee creamers, with special attention to the offerings of salad bars.[64] "Using a plastic bag filled with a brown liquid they nicknamed 'salsa,' they poured the salmonella filled slurry directly into salad dressings, splashed it on produce, put it in water, and generally got it everywhere they could," writes journalist Dylan Thuras.[65] Unfortunately for the residents of The Dalles, the attacks were successful. "Within hours, emergency rooms were flooded with sick patients," writes Scott Keyes.[66]

Typical of those arriving at the hospital were the Turners, a married couple and owners of a furniture store who, along with their two-year-old daughter, were poisoned at Sunday brunch.[67] Also representative was the Carlton family, the father being a state trooper who, along with his wife and three-year-old son, ended up in the emergency room after eating the bacteria-laden food.[68] Perhaps most disturbing was a couple, a husband and his pregnant wife, whose baby would be born displaying the effects of *Salmonella* poisoning. And so it went, with a staggering 751 people suffering from gastroenteritis as a result of the attacks, forty-five of whom required hospitalization.[69] Also worth noting, the total number of illnesses, while steep, does not include the Interstate motorists who were not residents of The Dalles but had simply stopped briefly in the town to enjoy a meal before continuing on to their destinations. "The actual number of victims was probably much higher because many out-of-state travelers may have been infected as well," writes Carus.[70] Fortunately, no one is known to have died, but the fact that over seven percent of the population of The Dalles was stricken made the ordeal the largest *Salmonella* outbreak in American history and the largest bioterrorism attack in modern times.

Regarding the response of public health workers, while some of the residents of The Dalles were certain it was the work of the Rajneeshees, health officials remained open to other possibilities and set about looking into the matter without delay. On September 17, in the wake of the first wave of infections, a member of the Wasco-Sherman Public Health Department began reviewing the incoming reports of *Salmonella* poisoning in The Dalles and, the following week, notified the state health department of the situation. The concern was that the number of infections was mounting, with a significant percentage of the cases necessitating hospitalization. Twenty-four hours later, the state health department reached out to the Centers for Disease Control in Atlanta and requested the federal agency's help in pinning down the cause of the outbreak. And within another forty-eight hours, a pair of CDC medical epidemiologists were on the scene. For the next six weeks, a team drawn from these county, state, and federal agencies worked together to track down the source of the infections.[71] "With an outbreak this large," reads a CDC account of the event, "investigators were initially optimistic that they would be able to find a common pattern or thread that could explain the occurrence of illness in so many people."[72]

Investigative procedures included analyzing the stool samples of those diagnosed with gastroenteritis coupled with one-on-one interviews with the victims about their recent experiences. For the purposes of comparison, the team also interviewed hundreds of people in The Dalles area who had not been ill. The plan was to sift through these personal accounts until a mutual factor was identified; a specific item, practice, or location common to all of those who had become sick but not to those who had remained healthy.

The ensuing laboratory analyses confirmed that *Salmonella* enterica Typhimurium had indeed been the causative agent in the avalanche of illnesses. The results of the interviews were another matter, however. Although the team queried the victims about their food preparation practices as well as the food products they had eaten prior to becoming ill, a common denominator did not emerge. "[T]he investigators could not identify a single food item or contamination of a single food item that could have accounted for the *Salmonella* Typhimurium gastroenteritis outbreak," reads the CDC account.[73] One pattern stood out, though: a disproportionate number of victims had consumed a variety of foods at the salad bars in certain restaurants.

Curiously, when laboratory analyses were conducted on samples of the contaminated foods in these bars, none of the ingredients used in preparing them was found to contain *Salmonella*. An unexpected finding, it could only mean that the bacterium was deposited on the food after it had been prepared and placed on the salad bars.

Inspections of the different types of salad bars implicated in the poisonings also revealed striking incongruities. For instance, one restaurant,

The Portage Inn, had a salad bar in its main dining room that was open to the public, and it operated a second salad bar in its private dining room, which was open only to banquet guests.[74] In both the public and private dining rooms, the same food was served, including the contents of the bars, and for this reason it was expected that the patrons of both salad bars would have been infected. Yet the investigative team discovered that it was only the patrons of the public bar who had contracted *Salmonella*; those who had dished up their food from the private one remained healthy. At least at The Portage Inn, then, this suggested that whatever or whoever had contaminated the food after it had been placed on the buffet-style bars only had admittance to the public one, such as a customer. A restaurant employee would have had access to both the public and private bars and therefore could have contaminated the food in the two of them.

Then there was the nature of the food itself. *Salmonella* was found in foods that were not naturally hospitable to the bacterium, such as lettuce salad, while being absent in many of those in which one would expect to find it—eggs, milk, fish, and chicken. And while several restaurants under investigation were discovered to have tainted lettuce, it had been grown by, and purchased from, different farms and vendors. Conversely, a name-brand product might be contaminated at one restaurant, but an identical product might be pathogen-free at another eatery. To many observers of the investigation, it was obvious that the massive *Salmonella* outbreak, in failing to correspond to any known pattern of food contamination, was anything but a naturally-occurring phenomenon.

In the end, however, the public health team pointed its finger at the restaurants' staffs, claiming that unhygienic practices had fouled the food at the ten establishments. A highly improbable conclusion, it meant that multiple workers at multiple eateries had engaged in unsanitary practices on precisely the same days and with the same pathogen at play. Not only that, the workers had, for the most part, contaminated those foods in which *Salmonella* is seldom found, and only after the food had already been prepared and made available to the public. Of course, it was an explanation that beggared belief. As to why the team did not consider the possibility of a Rajneesh-engineered bioterrorism plot, particularly when witnesses had reported observing Rajneeshee diners behaving oddly in the affected restaurants, the reason is unclear. Some have suggested it was because the evidence of a bioterror attack, while compelling, was circumstantial, while others emphasized the non-forensic nature of the public health team.[75] In any event, most people remained perplexed by the team's pronouncement and could not fathom how the extensive investigation could culminate in such a conclusion.

For the next several months, food handlers would continue being blamed for the historic *Salmonella* outbreak. But helping to absolve them would be Congressmen Jim Weaver, convinced, as he was, that the restaurant workers were not behind the poisonings. "I received daily printouts from the CDC investigation that made it only too clear that it was virtually impossible for the food handlers to be the source," says Weaver.[76]

His frustration mounting, the congressman, on February 28, 1985, stood before the U.S. House of Representatives and accused the Rajneeshees of having orchestrated a sweeping bioterror attack on the population of The Dalles. After summarizing the case and the public health team's findings, Weaver wrapped up his speech with a formal request. "I conclude my story by calling for an intensive police investigation of the salmonella outbreak in The Dalles," he said.[77]

Despite the fact that the lawmaker had presented a mountain of persuasive circumstantial evidence to support his argument, Weaver's allegations and his call for an formal inquiry were rebuffed. The Rajneeshees, as could be predicted, feigned indignation and vehemently denied his accusations, crying religious persecution. It was the sect's knee-jerk response to criticism. Yet public health officials were antagonistic as well. "Quotes from the health authorities stating I was wrong were printed in the Oregon press, condemning my speech," Weaver wrote in the *New York Times*.[78]

For the time being, then, there would be no investigation into the possibility that the religious sect had perpetrated a bioterror attack on an American city. The final report that was released in 1985, like the preliminary one, persisted in laying blame at the feet of The Dalles' restaurant workers. "The report infuriated locals, who were convinced that the cult was responsible for the outbreak, as well as law-enforcement officials, who now lacked the 'probable cause' needed to open a criminal investigation of the group," writes Judith Miller and her colleagues.[79]

At the commune, meanwhile, Bhagwan Shree Rajneesh maintained his public silence, except when compulsory. He broke it, for instance, when the Immigration and Naturalization Service ordered him to present himself to its local office and answer its agents' questions. Among other misdeeds, the INS suspected the guru of officiating at hundreds of sannyasin marriages in order to enable the foreign partners in the relationships (the couples were not actually romantically involved) to be granted U.S. citizenship.

Also during this period, 1984–1985, the guru continued driving his beloved Rolls-Royces around the grounds of the commune each day, as well as using drugs much of the time. "[H]e was taking large doses of Valium and inhaling nitrous oxide, sometimes twice daily," writes James Gordon, the aforementioned psychiatrist who had become deeply involved with the sect.[80]

Beyond the seemingly contented guru, however, life within the city of Rajneeshpuram had come to be characterized by internal strife and power struggles, with scores of sannyasins harboring immense hostility toward

Sheela. Widely disdained, she was accused of usurping the leader's power for her own ends.

Outside of the gates of the commune during this same period, the sect, having given up trying to rig the county commissioner election, was still battling with local officials over land-use regulations and other legal and political matters. Then, in the latter part of 1985, the dam gates broke and the sect's downward spiral into bioterrorism at The Dalles was thrust into the bright glare of the national media, together with a host of other illegal acts.

Paradise Lost

It was on September 16, 1985, in Rajneeshpuram, that Bhagwan Shree Rajneesh dispensed with his vow of silence and announced in an impromptu press conference that Sheela and nineteen of her cronies, Puja among them, had fled the country. The "gang of fascists," as he called them, were en route to Europe.[81] Explaining that he had just been informed that this renegade group had committed all manner of crimes, he claimed as his sources those loyal sannyasins who had long shunned Sheela and her accomplices and had stayed behind after the apostates bolted.

During the press conference and a follow-up presser the next day, Rajneesh claimed that Sheela and her minions had mismanaged the commune's finances, stolen large sums of money, committed arson and bombings, and tried to poison his dentist, his personal physician, and a district attorney. The gang had targeted him too, he declared, seeking to "kill or incapacitate" him using substances concocted in "a secret tunnel behind Sheela's house."[82] Other allegations: the group had tried to contaminate the local water supply, and had also sought to perfect a mass-casualty biological agent that would be undetectable by standard methods and would eliminate its targets slowly, over the course of time. Lastly, Rajneesh stated that he suspected the gang of having engineered the 1984 *Salmonella* outbreak in The Dalles. Denouncing Sheela and her minions for having created "a Stalinist regime," the religious leader asked officials to visit the site and investigate his allegations.

Reaction to Rajneesh's press conference ranged from shock, disbelief, and in some cases relief among the sannyasins, to the suspicion that the guru had known all along about the gang's illicit deeds and was simply being disingenuous. Now that the "fascist gang" had fled the country, it was suspected, he was desperately seeking to distance himself from it by feigning ignorance and indignation. A high-ranking sannyasin who had participated in the *Salmonella* project, for instance, a man who was later placed in the federal witness protection program after testifying for the prosecution, reported that Sheela, prior to implementing the bioterrorism plot, played a tape-recording of the guru responding to her plan to poison the people of The Dalles. "If it [is] necessary to do things to preserve [my] vision, then do it," Rajneesh said on the tape, according to the witness who listened to it.[83] Sheela interpreted the guru's response on the barely audible recording, a tape that may or may not have been altered, to mean that he approved of the proposed bio-attack.[84]

As to the investigation, it commenced on October 2nd, with fifty investigators arriving at the commune and establishing a base of operations for what would be an across-the-board inquiry. Among the participating organizations were the U.S. Customs Service, the Federal Bureau of Investigation, the Oregon State Police, and the U.S. Immigration and Naturalization Service, with the National Guard being present as well.[85] The multi-agency team was assembled by David Frohnmayer, the Oregon Attorney General who, as noted earlier, was pursuing a case against the city of Rajneeshpuram on the grounds that its incorporation violated the separation of church and state. (Investigators would soon unearth the sect's hit list, and on it they would find David Frohnmayer's name, along with that of U.S. Attorney Charles Turner and nine others in the fields of politics and journalism.[86])

Scouring the massive Rajneeshee compound, investigators discovered not only the secret tunnel under Sheela's house, the one the guru had described in his press conference, but also the buildings that comprised Puja's germ warfare factory. Here, investigators collected potential evidence of the type that would likely be present in most any laboratory, such as masks and gowns, syringes, and lab equipment, but they also tagged a more telling object: a freeze-dryer Puja had purportedly purchased so she could culture the AIDS virus. Fortunately, it was an attempt that failed. Also in the search, the team found suspicious manuals, including *How to Kill: Volumes 1–4*, *The Anarchist Cookbook*, *The Perfect Crime and How to Commit It*, *Deadly Substances*, and *The Handbook of Poisons*.[87] Most incriminatingly, though, the team discovered a handful of invoices documenting the numerous pathogens Puja had ordered and which bioterrorism experts associate with biowarfare, among them the microorganisms that cause typhoid fever, tularemia, shigellosis, gonorrhea, and certain types of respiratory and urinary tract infections.[88] Dated September 25, 1984, the pathogens had arrived at Puja's laboratory in the midst of the *Salmonella* outbreak in The Dalles. Equally damning, the strain of *Salmonella* enterica Typhimurium that Puja had ordered from the medical supply company matched that which was recovered from the victims' bodies. Officials therefore concluded that the Rajneeshees had in fact planned and executed the biological attacks in The Dalles, and quite possibly carried out poisonings in other Oregon locations as well, among them Portland and the capital city of Salem.

And there was more to be discovered beyond the evidence of bioterrorism. "Investigators uncovered what remains the largest, most sophisticated illegal electronic eavesdropping system in American history," writes Miller and her co-authors. "[A] Rajneesh security team had bugged entire floors of their [on-site] hotel, many of their disciples' homes, the ranch's public pay phones, the Zorba the Buddha café, and even the Bhagwan's bedroom."[89] And there was evidence of many other criminal acts as well, quite serious ones.

Indictments would swiftly follow on October 23rd, and would charge Bhagwan Shree Rajneesh with violations of federal immigration laws, while also charging others in the sect, Sheela and Puja foremost among them. And although Rajneesh would be hustled aboard a Lear Jet and be flown eastward shortly before the indictments were handed down—presumably, the plan was to sprint him out of the country—police were on the tarmac when he touched down in North Carolina. At the airport in Charlotte, they arrested and charged him with lying to the Immigration and Naturalization Service about his intent to remain in the country temporarily for medical treatment, and for performing four hundred counterfeit marriages so his followers could gain American citizenship. Besides the religious leader and his disciples traveling with him on the jet, authorities also seized over twenty suitcases stuffed with cash in various currencies, designer eyewear, watches, jewels, and a pistol.

As for Sheela, the mastermind of the sect's most audacious deeds, the authorities tracked her down in southern Germany, where she had recently appeared on a television broadcast. Arresting her, they charged the sect's unsanctioned leader with, among other crimes, arson, immigration fraud, and attempted murder, and extradited her to the United States. The latter allegation pertained to the two county commissioners whose water had been spiked while they were inspecting the commune and to the mass poisoning of the residents of The Dalles, which was classified as "product tampering." It was the first major use of the recently-enacted "Tylenol laws," created after an unknown perpetrator laced the popular painkiller with potassium cyanide.

Regarding Puja and the eighteen other sannyasins who also fled to Germany, officials traced them to a luxury hotel and arrested them, too, with the charges against Puja being the most serious. Then there were those disciples who had remained in the United States, but who had nevertheless taken part in the sect's wrongdoing and who would face criminal prosecution as well. All told, thirty-four sannyasins in the U.S. and abroad were charged with crimes ranging from burglary, racketeering, drug trafficking, and arson, to electronic surveillance, conspiring to commit product tampering, immigration violations, and attempted murder.[90] Most would not serve time in prison but would instead cooperate with the prosecution in turning over evidence.

Legal Consequences

The Rajneeshees' days in court would be crushing for the sannyasins who had devoted their lives to the commune and to their leader. And the collapse of the American incarnation of the Rajneesh operation would be swift and absolute.

On November 14, 1985, Bhagwan Shree Rajneesh entered an Alford plea to two felony charges, meaning that he formally pleaded guilty even though he maintained that he was actually innocent. Paying four hundred thousand dollars in fines, he received a ten-year suspended sentence with the understanding that he would leave the country within five days and not to return for at least five years. Even then, he could return only with the express consent of the American government. Vowing never to step foot in the United States again—the moneyed maharishi now called the U.S. a "wretched country"—the guru returned to India, where he struggled to reestablish his base.[91]

Back in his homeland, Indian officials monitored the discredited spiritual leader, as well as keeping an eye on his disciples. Three weeks after his American and European staff arrived to join him, officials voided their visas, with the guru, in protest, seeking to relocate to another country. After being refused admission by numerous nations, he returned to India, where he would remain.

"Lonely, bankrupt, banished from the United States, humiliated by the immigration authorities of a dozen other countries, the sensual sage is now chasing anonymity in the suburbs of Bombay," journalist Sundeep Waslekar wrote in 1987.[92] Yet Waslekar's assessment, while accurate, was incomplete. Changing his name to Osho, a Japanese title of respect and the word for a highly-evolved Buddhist monk, Rajneesh continued promulgating his doctrine of enlightenment from his site in India until his death in 1990. Regarding the cause of death, the fifty-eight-year-old's demise was the result of a heart attack, according to official records, although it was speculated that it was actually due to AIDS, murder, or other medical conditions or acts.

On November 29, 1985, three of the restaurants that the Rajneeshees targeted with *Salmonella* in The Dalles sued for damages, with their claims climbing well into the millions of dollars. One week later, the State of Oregon filed additional charges against numerous Rajneesh-owned corporations and organizations, declaring that they had engaged in illegal activities. The federal law used to charge the Rajneesh operations was the Racketeer Influenced and Corrupt Organizations Act (RICO). Under the agreement that the various parties signed the following year, the Rajneeshees would pay five million dollars to the state of Oregon, to selected residents of The Dalles, and to the restaurants the sect had been singled out for the bio-attacks.

Meanwhile, on December 10, 1985, the church-state lawsuit against the City of Rajneeshpuram, the one that Attorney General Frohnmayer filed in 1983, went before the court. The decision: the incorporation of Rajneeshpuram had violated the separation of church and state clause, therefore the city was to be disincorporated. As it happened, the decision was overturned in

1986, but by this time it had little impact because the commune had been abandoned.

Also that same year—July 22, 1986, to be exact—Ma Anand Sheela pleaded guilty to federal and state charges, the former consisting of wiretapping, conspiring to tamper with consumer products, and immigration fraud, and the latter involving arson. It seems she had set fire to the Wasco County planning office at one point.[93] Puja also pleaded guilty to wiretapping and conspiring to tamper with consumer products. Each of the women received a prison sentence of four and a half years, with the stipulation, in Sheela's case, that she exit the United States after serving her sentence. And indeed, she moved to Switzerland after her release from prison, where she would be known as Sheela Birnstiel and would buy and operate nursing homes.

As for the town of Antelope, it regained its original name in due course, even as a wealthy Montana rancher purchased the nearby Rancho Rajneesh and converted it to a seasonal camp for Christian youth. Known today as the Washington Family Ranch, it plays host to an estimated one thousand children and adolescents each summer.

Unfortunately, those businesses that the Rajneeshees targeted would not be as lucky. "Most of the contaminated restaurants never recovered from the poisonings," writes Miller and her colleagues. "Dave's Hometown Pizza was the only restaurant to survive long after the attacks at its original location and under its original ownership."[94] In the main, this was because the public psyche had been indelibly stamped with an association between the ten restaurants and the *Salmonella* bacterium. An irrational but predictable fear of contamination, the people of The Dalles and Interstate travelers thereafter avoided even the remote possibility of infection at these eateries despite of the fact that the establishments had been decontaminated and certified as safe. In this way, the Rajneeshees' attack continued harming the city of The Dalles long after the assault itself.

◆ 7 ◆

Tokyo Under Siege

The Aum Shinrikyo Biochemical Attacks

In the shadow of Mount Fuji, Japan's majestic and sacred volcano, once stood the village of Kamikuishiki, a pastoral community renowned for its rolling meadows and breathtaking view of the nation's tallest peak. Long known for its tranquil ambience, the settlement exuded an easygoing attitude that residents found appealing and visitors sought throughout the year. But its serenity would be shattered forever in the spring of 1995. Early one morning, police officers in gas masks arrived in droves, stormed into a religious complex situated at the edge of the hamlet, and launched a week-long search of the premises. Their revelations would shock not only the villagers, but the entire East Asian nation and the global counter-terrorism community as well.

Front and center was the sect's three-story "religious sanctuary." Inside, police noted that its ventilation system was a remarkably sophisticated one, the type normally found in industrial operations and a discovery signaling that the lackluster, concrete edifice was something other than a spiritual refuge. Continuing their search, authorities came upon a cache of explosives and, even more alarming, a series of laboratories stocked with equipment suggesting that the sect had been experimenting with chemical warfare. Confirming this suspicion were the chemical compounds recovered at the scene, among them sodium cyanide, phosphorus trichloride, sodium fluoride, acetonitrile, and isopropyl alcohol, the latter four agents being the components of the lethal nerve gas, sarin. The amounts, moreover, were staggering, a supply so enormous as to require hundreds of large metal drums to store them and a fleet of forklifts to transport them. "Police have found tons of chemicals that newspapers here estimated could make enough sarin to kill millions of people," the *New York Times* reported while the investigation was still underway.[1]

A further search revealed that the sect had been exploring other methods of mass destruction as well. In two more buildings that contained laboratories, authorities came upon a six-foot-tall apparatus known as a germ incubator, and nearby they discovered a storage unit in which 160 casks of peptone were stockpiled. A protein derivative used to grow bacterial cultures, the presence of the latter substance in such immense quantities meant the organization had almost certainly been experimenting with biowarfare. Supporting this notion, a substantial amount of *Clostridium botulinum*, the bacterium that causes botulism, was retrieved from the labs, along with vials containing the Ebola virus and the bacteria that cause anthrax infection and Q fever in humans.

Over the next three years, more disturbing truths would come to light about the sect's forays into bioterrorism. For one, the organization had constructed an eight-story, windowless building near Tokyo devoted exclusively to weaponizing microbes, and it was in the process of erecting a second one intended for advanced bioweaponry near Mount Fuji. For another, the sect, in the early to mid–1990s, had secretly carried out numerous small-scale bioterror attacks across Japan, with pathogenic bacteria being deployed in all of them.

As to the reason for the 1995 raid at the sect's Mount Fuji headquarters, the one exposing its vast WMD operation, it was in response to a monstrous chemical attack in downtown Tokyo. The first large-scale terrorist offensive of its kind in the modern age, it was the product of a cadre of scientists who were members of the Aum Shinrikyo religious sect, a controversial group often described as a doomsday cult. The attack was meant to spark profound social unrest in Japan, with the ultimate aim of setting off a global apocalypse and ushering in a new spiritual era.

The Making of a Sect

From the beginning, the Aum Shinrikyo sect promised to provide a path to enlightenment. Pairing the Buddhist mantra "Om," or Aum, with the phrase "Supreme Truth," or Shinrikyo, the name embodied the group's lofty mission according to its founder Chizuo Matsumoto, who would later become known as Shoko Asahara. A man whose life experiences were at once unusual and painful, they would, in due course, contribute to the anguish of countless others.

Born in Yashushiro, Japan, on March 2, 1955, Asahara (Matsumoto) was diagnosed with infantile glaucoma shortly after delivery, a congenital condition rendering him sightless in his left eye and with only thirty percent vision in his right one. His parents, unfortunately, were impoverished and thus unable to provide for his special needs—they lived in a dirt-floor hovel with their three other children—and for this reason they sent him to live at a government-run boarding school for the blind when he was six years old. Here, Asahara would remain for the next fourteen years. It would not be a pleasant stretch for him, nor for those who had to live and attend classes with him.

His former teachers and fellow students recall that Asahara appeared to feel rejected by his parents' decision to send him away from home, as well as ashamed of his family's destitution and embittered by his visual impairment.[2] "Asahara," says classmate Ikuo Hayashi, "often talked about his sense of loss, his feelings of betrayal by his parents, and his profound feelings of loneliness and fear at being alone in a strange place."[3] These feelings of dejection and indignation would persist throughout Asahara's lifetime and color his actions over and again.

During his school years, the future guru was also viewed as short-tempered and belligerent, and was quick to bully the other students. Most often, he targeted those who were easy marks, such as younger pupils who, unlike him, were totally blind and therefore at the greatest disadvantage. "He would force them to pick up his noodles and cakes," says a former classmate, with others reporting that Asahara intimidated his classmates into giving him their money and belongings.[4] Yet even as he grew older, Asahara didn't change in this respect except to become more antagonistic and arrogant, alarming his teachers and peers with threats of violence—he once warned that he would torch the school—and declaring matter-of-factly that he planned to become the prime minister of Japan. "A dormitory roommate described living with him as 'hell,'" write David Kaplan and Andrew Marshall.[5]

Shortly after graduating from the residential school at the age of twenty, Asahara was arrested for assault when he became combative during an argument.[6] A year later, he sought entry into college, but failing the admissions exams at Tokyo University and Kumamoto University, he embarked instead upon an informal study of traditional Chinese medicine. During this same period, he also married and used his in-laws' money to set up an acupuncture practice. Although it proved to be a lucrative business, it turned out to be a risky one because it involved the prescription of herbal potions and Asahara was not a registered pharmacist. Even more damning, some of his treatments appeared to be fraudulent. "One tonic, called Almighty Medicine, was simply tangerine peel in alcohol solution," write Kaplan and Marshall. "[Asahara] charged up to $7,000 for a three-month course of treatment."[7] It was products like Almighty Medicine that caused a Japanese court in 1982 to convict Asahara of practicing pharmacy without a license as well as selling unregulated medications. The judge jailed him, fined him nearly two thousand dollars, and thereby brought to an end his brief excursion into herbal medicine.[8] But Asahara remained unbowed. "Humility was not one of his more striking attributes, for his self-image was suffused with a sense of grandeur and destiny," writes Ian Reader, a professor in the Religious Studies Department at Lancaster University.[9]

For a fresh start, Asahara, determined as ever, turned his attention to the study of astrology and traditional and emergent religions, among them Taoism and New Age systems, and through such explorations became inspired to join the Agon-shu sect. A twelve-year-old Japanese "new religion," most observers were critical of Agon-shu and dismissed it as a cult. Its philosophy was based on a unique interpretation of Buddhism, one that stressed the importance of relinquishing one's ties to society, which was said to possess negative karma that could be transmitted to the individual. It was a view Asahara would come to embrace and henceforth promote. But while he respected the teachings of the Agon-shu sect and would eventually borrow some of its concepts, he exited the group three years later and established a small yoga school in Tokyo, one that he registered as Aum, Inc.

At first, the operation was predominantly secular, with Asahara combining yoga with psychic capacities and claiming to possess the ability to levitate. To prove his talent in defying gravity, he circulated photos, albeit unconvincing ones, purportedly showing him hovering in the lotus position. In due course, he moved beyond such theatrics, although he did remain attracted to mysticism, including the prophecies of the fourteenth-century French physician and astrologer Nostradamus, whose forecasts Asahara would later fuse with his own predictions.

Ian Reader has proposed a reason for Asahara's enduring affinity for mysticism and the supernatural, and it stems from the powerlessness the visually-impaired guru endured as an impoverished, displaced youth. More to the point, Asahara's early circumstances and experiences may have instilled in him the need to transform himself into a commanding presence, even a transcendent one; a need suggested by his childhood fantasy of becoming Japan's prime minister, and later the trailblazer of a new spiritual era of humanity. Supernatural forces, he appears to have believed, could help bring about such a transformation. "This interest in esoteric practices, mystery and psychic powers implied a rejection of and an alternative view to the overarching paradigms of the secularised, scientific, rationalist worldview," says Reader. "It was, in effect, suggesting that the human being, rather than being bound to the world and at the mercy of rationalised events and physical limitations, contained the potential for transcendence on the spiritual and physical levels."[10]

During subsequent treks to the Himalayas and northern Japan which Asahara claims to have made, he persisted in his quest for enlightenment.

And it was during one of these journeys that a Hindu wise man ostensibly told to him that an apocalypse was imminent, one that only gentle mountain hermits ("shinsen") would survive. As the story goes, Asahara, now in possession of this prophetic knowledge, returned to Tokyo convinced that he alone could save the world from annihilation. To this end—and with the wholehearted support of three dozen of his yoga students—he formed an organization and in 1986 bestowed upon it the name Aum Shinsen-no-Kai. The title translates as the "Aum Mountain Hermits' Society."[11] Its short-term objective was to prevent a global cataclysm, while its long-term goal was to usher in a new spiritual age with Asahara himself guiding the masses.

Within a year of establishing the organization, Asahara set about refashioning it to render it more religious in nature, and he placed special importance on asceticism. Offering a customized version of Buddhism infused with elements of Hinduism and Christianity—he declared himself to be the incarnation of Christ—Asahara tightened the sect's organizational structure, separated its members from their loved ones and from society at large, installed within the sect severe controls, and, not coincidentally, acquired a lucrative livelihood in the process. In the fullness of time, his personal fortune, along with that of his organizations, would exceed one billion dollars. "Monks and nuns had to sever all ties to their families and friends and give everything they owned to Aum Shinsen-no-Kai to demonstrate their lack of attachment to the mundane world and their commitment to the community," write Richard Danzig and his associates.[12]

Continuing to evolve, Asahara, in 1987, changed the sect's name to Aum Shinrikyo and officially changed his birth name, Chizuo Matsumoto, to Shoko Asahara. By this point, the organization's orientation had become wholly religious, with the guru reiterating his prediction that human civilization was about to be supplanted by a more spiritual one. New and enlightened, it would initially be populated exclusively by his disciples, whom he maintained were superhuman.

Asahara, at this juncture, also redoubled his demands for self-denial among the sect's membership; this, despite the fact that he himself continued engaging in sexual relations both with his wife and with a number of female disciples, eventually fathering six children with his spouse and an unknown number with the sect members.[13] He sought to extend the reach of Aum Shinrikyo too, establishing a publishing program and, as its inaugural project, producing the journal *Mahayana*. The result of such efforts: a dramatic spike in membership—the number of participants soared into the hundreds and soon, the thousands—coupled with the formation of Aum branches throughout Japan and beyond.

In terms of the sect's membership, it attracted young adults who were dissatisfied with the status quo and in search of a deeper, richer meaning in life. As well, it enticed a considerable number of older members, and for the same reasons, among their ranks being scientists and other professionals. And as the membership soared, so did the sect's wealth. "Devotees flocked to the Mount Fuji center, where the cult charged over $2,000 for week-long meditation seminars," write Kaplan and Marshall.[14] It was only one of many profitable marketing ventures.

As it happened, within a few months of these seemingly positive developments, matters began to take a more troubling turn. For one thing, Aum Shinrikyo's application for recognition as a religious organization, a legal status that would entitle it to tax breaks and other benefits, was turned down by the Japanese government. Although it would later be approved, this initial rejection enraged Asahara, prompting him to cry religious discrimination. And there was another unsettling development: people began dying.

"In September 1988 the first unnatural death occurred among Aum members when Majima Terayuki died due to excessive ascetic practice," writes Martin Repp, editor of the journal *Japanese Religions*. "In order to avoid police investigation, Aum's leaders ordered his body to be burned and then disposed of, which is illegal."[15] In response to the disciple's death, concerned relatives of Aum members began demanding that the organization allow their loved ones to resume contact with their families, while disillusioned ex-members sought an official investigation into the sect's questionable activities.

Matters became even more alarming as the organization added murder to its repertoire. First, in early 1989, an Aum Shinrikyo hit team strangled one of the group's members, Shuji Taguchi, who was preparing to the reveal wrongdoing by the organization. Asahara ordered the assassination and hand-picked the disciples who could carry it out. Then, later that same year, a more notorious trio of killings took place as an indirect result of a series of exposés by the Japanese press.[16] The reports, which were scathing, presented the sect as a counterfeit operation that was raking in enormous sums of money and headed by a man who whose credentials were less than pristine. "Asahara was portrayed as an exploitative and manipulative leader with an overbearing ego," says Reader, a representation that fueled Asahara's paranoia and convinced him that he and his organization were the targets of the government and the media.[17] As things stood, he already believed the sect was a target of the British royal family and Jewish interests. The result is that the guru became even more convinced that he must assume power in Japan before public sentiment turned against him and the entire Aum enterprise.

As to the murders, they would transpire after the illuminating newspaper accounts were published, a time when the loved ones of Aum members joined forces with a band of ex-members to form an organization called the Aum Shinrikyo Victims' Society. For legal representation, the assemblage turned to Sakamoto Tsutsumi, a respected attorney who set about scrutinizing the background and current status of the sect and, in so doing, came to be perceived by Asahara as a grave threat.

Regarding Tsutsumi's discoveries, one of the first concerned a fundamental myth of the Aum Shinrikyo organization, namely that Asahara's DNA had been analyzed and was found to be uniquely non-human. Tsutsumi determined that the assertion was false, that Asahara's genetic code had never been examined. Moreover, when geneticists at Kyoto University did analyze the guru's DNA at the attorney's request, they found it to be ordinary in all respects. Additional evidence that Tsutsumi gathered, which apparently was voluminous, likewise shed a harsh light on Asahara's spurious claims and the sect's disturbing, even abusive, practices. But most unsettlingly to the guru was the fact that the attorney, who could not be bought off or otherwise "persuaded," planned to place the incriminating materials before the public. Of course, Shoko Asahara knew this would bring about his own downfall and the collapse of his organization.

So it was that Asahara issued an order to eliminate the troublesome Tsutsumi, with one of the men he selected for the job being Tomomitsu Niimi, the same high-level member who had killed Shuji Taguchi earlier that year. And indeed, Niimi and his collaborators proceeded to break into the lawyer's home, where they found Sakamoto Tsutsumi and his wife in a bedroom and strangled the two of them. As she was fighting off her killers, Mrs. Tsutsumi begged that the couple's fourteen-month-old son not be harmed. "Please save the child, at least," she said, according to a *New York Times* report.[18] Unfortunately, the infant began to cry, so the assassins suffocated it with a blanket. A trained assassin, Niimi would eventually injure or murder even more innocents as a member of the terrorist team that would conduct the infamous sarin gas attack on the Tokyo subway system.

Convinced that he was being blocked from his destiny of escorting humanity into a new age, Asahara decided to form his own political faction. With this aim in mind, he and twenty-five members of Aum Shinrikyo announced their candidacies in 1990 in Japan's upcoming parliamentary election. By this point, the sect claimed several thousand members and enjoyed a moderate degree of influence in certain segments of Japanese society, therefore the guru fully expected his party to be victorious. But he miscalculated badly. The sect's numbers were tiny compared to the hundreds of thousands of citizens who would be voting in Asahara's district, and a sizable swath of the population viewed the sect and its campaign as bizarre, even comical. And sure enough, the election delivered a decisive blow to the sect and its unrealistic aspirations. "Every Aum candidate lost heavily, and none more so than Asahara, who received only 1,783 votes out of the half million cast in his constituency."[19] Humiliated by the sweeping public repudiation, Asahara, an unstable man still haunted by his formative experiences of rejection and disability, now became emboldened to a pathological degree. He vowed to seize power by whatever means necessary.

Bioterror Attacks

Bioterrorism was Asahara's method of choice. Lethal pathogens were obtainable in Japan at the time, and, as noted earlier, their use as weapons is potentially more catastrophic than chemical agents, since certain microorganisms, unlike chemicals, are capable of reproducing and spreading far beyond the original victim pool. So it was that the guru, after his group's spectacular defeat in the parliamentary election, set out to assess the effectiveness of an in-house bioterrorism program, and he began by ordering his scientist-members to carry out small, covert attacks on selected components of Japanese society and on American military interests in Japan. At the time of their occurrence—the experimental assaults transpired between 1990 and 1995—they were not recognized as such; their true nature only came to light a few years later. In part, their exposure was the result of a *New York Times* investigation, one that made use of court testimony, interviews with Japanese and U.S. authorities, and confessions by the sect's members. The ensuing account of the group's bioterrorist activities is based in large measure on these findings, which the newspaper published in May of 1998.[20]

Asahara set into motion his bioterrorism program in 1990, with his first step being to appoint a chief of operations to oversee it. This person was Seiichi Endo, a thirty-year-old man who had joined the sect three year earlier and previously undergone graduate training in virology at Kyoto University. Endo and Asahara, in turn, assembled a cadre of young scientists from within the ranks of Aum Shinrikyo and offered them the opportunity to participate in the aberrant project. Certainly the undertaking would be a challenge, the group's mission being at once complex and consequential, and Endo himself being held accountable for its outcome. As to the objective, it was onerous but clear-cut. "He was to find a few lethal germs, feed them special foods, grow them to astronomical numbers and turn the resulting brew into a widely dispersible material, preferably a fine mist or powder that could easily penetrate human lungs," says science writer William Broad.[21]

1990 Botulism Attacks (Tokyo). Predictably, Endo and his collaborators were initially drawn to the deadliest toxin known to affect the human population, the one produced by *Clostridium botulinum*, which they decided to culture and deploy. The mission would take place a few weeks after the sect's trouncing in the election, when a four-person Aum Shinrikyo team traveled to Hokkaido, Japan's northernmost island and Endo's childhood home. Here, in a section of "rough country," they hoped to find botulinum samples. And indeed, on this landmass teeming with forests and lakes and volcanoes, they managed to amass an ample supply of the bacterium. The group then returned to the sect's headquarters and set to work growing the microorganism. A month later, their bioweapon was ready—or so they believed.

In April 1990, Asahara sent three customized vans from the sect's compound near Mount Fuji to Tokyo, sixty miles away. Inside the vehicles were large storage tanks loaded with aerosolized botulinum toxin, which the bioterrorists sprayed indiscriminately as the vans glided through the city streets. They next traveled to a pair of American targets, again to discharge the toxin: Yokohama North Dock, which is an important U.S. Navy base, and the Yokosuka Naval Base, a neighboring installation and home of the Seventh Fleet. Wrapping up their offensive, the group released the agent at the New Tokyo International Airport, known today as the Narita International Airport, a sprawling operation that holds the distinction of being the largest commercial air facility in Japan.

Back at the Mount Fuji compound, Endo and his team awaited news of the disaster they believed that they had put into motion. A week passed, however, with no reports of mass illness or death in the areas thought to have been contaminated by the poison. Presumably, the strain that the team had collected in northern Japan, although it did generate a toxin, did not produce one that would noticeably harm humans. Furthermore, because the number of botulinum strains number well into the hundreds, and since the vast majority of them do not make toxins that possess the lethality sought by Endo, the wayward scientist saw before him an unending series of time-consuming, trial-and-error attempts to bring about the wholesale suffering Asahara had ordered. Therefore, to cut his losses, he switched to another pathogen.

1993 Anthrax Attacks (Multiple locations). The findings from the *New York Times* investigation suggest that Seiichi Endo instructed a sect member, one who had held a license to practice medicine and had contacts inside the medical research community, to discreetly acquire samples of *Bacillus anthracis*.[22] Shortly thereafter, this person did manage to obtain the sought-after bacterium. As to the source, the pathogen was supplied by an individual employed by the University of Tsukuba, a renowned research institution situated forty miles northeast of Tokyo.[23] Officials at the university contend that they were unaware of such a transaction.[24]

With the anthrax bacterium in hand, Endo and his team now shifted their work to the organization's aforementioned eight-story, concrete edifice in Tokyo, where their objective was to grow immense amounts of the microbe and prepare it for dissemination. And for the next three years, the team members toiled in their laboratories until they succeeded in multiplying the anthrax into a staggering quantity, enough to eradicate the populations of entire nations. But first, Asahara was determined to attack Japan itself, so

Endo and his associates prepared a domestic offensive with a target date of June 1993.

It was on the 29th of that month, a Tuesday, that the Aum bioterror team began discharging aerosolized anthrax from the roof of its Tokyo edifice. Because anthrax is odorless, Endo and his crew did not expect the surrounding community—that is, the victim pool—to be aware of what was happening to it. But the experiment did not unfold as planned. Soon after the sect released the pathogen, calls from concerned citizens reporting a foul odor began pouring in to the police station in the city's Kameido district. Evidently, it originated from the sulfur and ammonium compounds that had been used when culturing the bacterium.

Next day, as more anthrax was released, forty-one additional complaints of stench were lodged, with still another 118 reports the day after that. Stumped, the police asked Aum Shinrikyo representatives for permission to inspect the sect's building, but the request was rejected. When asked about the persistent odor, an Aum disciple offered the off-the-cuff explanation that fellow members had been boiling beans. Because the unpleasant smell disappeared following the police visit, however, authorities considered the case closed, at no point suspecting the disciples of experimenting with biological weapons.

In terms of harm, no serious illnesses or fatalities were registered, although health workers did receive numerous reports of nausea, vomiting, and loss of appetite during the intermittent four-day offensive, along with accounts of sick birds and house pets. Hearteningly, a retrospective case-detection survey carried out six years after the attack, one that analyzed community health patterns in the region targeted by the assault, likewise found no upticks in anthrax-related conditions during the period in question.[25] And not only was no one seriously afflicted; there is a possibility, albeit a slim one, that the sect may have inadvertently helped protect the intended victims from infection. This is because, unbeknownst to the bioterror team, the anthrax deployed in the offensive was a form used to vaccinate cattle. "While they may have been trying to kill people," says microbiologist Paul Keim, "the most they could have done is actually immunize people from the disease."[26]

As to reason for the attack's failure, experts believe that at least five issues prevented the damage the sect was seeking to generate. "The use of an attenuated *B. anthracis* strain, low spore concentrations, ineffective dispersal, a clogged spray device, and inactivation of the spores by sunlight are all likely contributing factors to the lack of human cases," write Hiroshi Takahashi and his colleagues.[27]

Still determined to wreak biological, social, and political havoc, Asahara had his chauffeur, Shigeo Sugimoto, transfer more of the same anthrax to one of the sect's vans and head to the heart of Tokyo toward the end of July, where the driver sprayed the pathogen near the nation's legislature. Days later, when no reports of illness or death among Japan's lawmakers were forthcoming, Sugimoto tried again, on this occasion routing the van near the Imperial Palace, the emperor's principal residence. As before, though, no ill effects occurred, with part of the problem, as noted earlier, being that it was a vaccine strain of the pathogen.

1994 Sarin Attack (Matsumoto). Owing to the repeated failures of the team to induce mass illness by biological means, Endo and his collaborators turned to chemical weapons. They were aware that chemical agents, while they might not produce as many casualties as a biological attack, would be easier to devise and successfully deploy. And this did turn out to be the case.

The bioterror squad set out on a dry run on a summer night in June 1994, the target being a three-story apartment building in the picturesque mountain town of Matsumoto. Among others, the building's tenants included a handful of local officials who recently had enraged Asahara by not conceding to his wishes regarding an Aum project proposed for the town. And not only were the trio of legal authorities in the guru's crosshairs, but the townspeople as well, since they too had opposed the scheme. So it was that the terrorists parked a specially-equipped truck near the apartment building together with a companion vehicle that served as a mobile gas chamber. Around midnight, as the building's tenants slept, industrial-sized fans blew sarin gas from the vehicles toward the building. "The group wanted to see how many it could kill," reports the *Japan Times*, "and the citizens of Matsumoto, who had already angered Aum founder Shoko Asahara by vigorously opposing his plan to set up an office and factory at the land in southern Matsumoto, were the guinea pigs."[28]

At once, the building's residents, along with those in the surrounding neighborhood, began experiencing symptoms that ran the gamut from minor aggravations to blindness and seizures. As to the victims, their ages ranged from three years to nearly ninety, with 208 people seeking outpatient treatment and 54 requiring hospitalization.[29] In all, fourteen died, some of them in their apartments within minutes of the attack, while others expired in the hospital in the ensuing days. One perished many years later from her injuries.

The agent, sarin, is a nerve gas the Nazis developed during World War II. Within seconds of exposure, the nervous system short-circuits and an array of symptoms erupt. "The nose runs, the eyes cry, the mouth drools and vomits, and bowels and bladder evacuate themselves," says James Hamblin, physician and senior editor at *The Atlantic*.[30] The victim usually has no idea as to the cause of the symptoms, pure sarin gas being colorless, tasteless, and odorless. Once exposed, the victim may die within as few as ten minutes.

At the time of the attack, the authorities suspected it might be the work of North Korea, Japan's adversary. Within seventy-two hours of the incident, however, the police were unofficially pinning the blame on a professional photographer who lived nearby; a man who, like his neighbors, required hospitalization for exposure to nerve gas. Certainly the reason for the allegation was a stretch: in his home, police found chemicals used to develop photographs. Under considerable pressure to solve the crime, however, officials were quick to leak the news that they may have collared their man. It would not be until a more expansive sarin gas attack in Tokyo that the Matsumoto assault would be recognized as terrorism perpetrated by the Aum Shinrikyo organization, forcing officials to apologize to the much-maligned photographer.

1995 Botulism Attack (Tokyo). Despite the fact that the chemical attack had been successful, Asahara was still eager to create a biological weapon, and for this reason he ordered Endo and his team to once again orchestrate an attack making using of the botulinum toxin. Accordingly, the team, in short order, selected a particularly vulnerable target: the Tokyo subway system. It would be a circumstance in which scores of unsuspecting commuters would serve as a captive audience for the group's malignant experiment.

The plan itself was ingenious. A clever Aum disciple invented a prototypical briefcase inside of which was an apparatus designed to convert a liquid medium containing the pathogen into vapor for dispersal. On the sides of the case were vents for the toxin's release. "The bacillus was held in solution in vinyl tubes, which were mounted on a small ceramic diaphragm," write Kaplan and Marshall. "Powered by dry batteries, this device turned the solution into steam, which was then blown from the briefcase by a small electric fan."[31] Three such briefcases were to be used in the actual attack, with the apparatuses contained within them being triggered by the vibrations of passing commuters or subways.

In terms of the toxin, it was more than capable of sickening or killing those in its path. "It would take eighteen to thirty-six hours to do its slow and deadly work, giving morning commuters a case of botulism from which they would never recover," write Kaplan and Marshall. "Thousands would probably succumb to bouts of diarrhea and vomiting, swelling eyelids and creeping paralysis, until their heart or lungs ceased to work."[32]

The assault took place on Tuesday, March 15th, at Kasumigaseki Station, a heavily-trafficked spot during the morning rush hour. Arriving at the depot, sect members carefully positioned the briefcases near three ticket booths and walked away, knowing that subway workers, or perhaps commuters, would pick up the cases. Regardless of who did it, the outcome would be the same: jostling them would set off their internal mechanisms and release the toxin, which would waft through the station, stairwells, subway cars, and tunnels, all the while infecting those in its orbit. Yet when the moment arrived, this did not happen. Instead, two of the briefcases failed to work, and the one that did function discharged a harmless vapor, an exasperating turn of events for the bioterror team. As to the reason for the absence of illnesses and fatalities, it was determined that a sect member who had been involved in the mission's preparations, overcome with guilt about the horror that was about to descend upon the people of Tokyo, decided to replace the botulinum toxin with a more benign agent. "The fate of the disobedient cultist is unknown," write Kaplan and Marshall.[33]

The Tokyo Subway Gas Attack

Asahara's determination to attack the capital city was not diminished, however. With a preoccupation bordering on obsession, he remained consumed with conducting a large-scale assault on the Tokyo subway system and adamant that it take place at once. For him, the mission had assumed a mystical dimension that extended far beyond the realm of rational thought, with his immersion in the prospective calamity causing the guru to place unreasonable demands on the bioterror team. One mandate was that they create, essentially overnight, enough sarin gas to kill thousands of commuters and release it on the capital city's subways as soon as it was ready. To Asahara, time was of the essence. But manufacturing and weaponizing a large quantity of pure sarin would, of course, be virtually impossible within such a narrow time frame, so the team sought to placate the increasingly volatile guru by concocting, over the course of a few days, a less potent, or attenuated, form of the nerve agent. Since Asahara was highly unpredictable and emotionally explosive by this point, the team was loathe to disappoint him or otherwise incur his wrath.

"[P]rior to the Tokyo sarin attack, Asahara's connection to reality was so muddled that we must suspect him to have been, at least some of the time, psychotic," writes psychiatrist Robert Jay Lifton, winner of the National Book Award for his study of Hiroshima survivors.[34] Besides insisting that the bioterror team produce sarin on a wholly unrealistic schedule, Asahara was also convinced that he and his sect were about to be assailed by the authorities. Driving this fear, he supposedly received a tip-off on March 18th from a contact in the Army, a warning that the military might soon raid the sect's Mount Fuji compound. For the erratic Asahara, his paranoia mounting, such a possibility now became an urgent matter of self-defense. He considered it imperative that he annihilate the aggressor, meaning governmental and military authority and even civilization itself, if he was to protect himself and preserve the Aum Shinrikyo operation. "Destroying the world became the only means of staving off a sense of death and extinction," writes Lifton.[35]

In addition, Asahara was counting on the Tokyo mass murder to spark social upheaval and distract the police and the military, his hope being that the latter would scrap its presumed plan to inspect Aum Shinrikyo headquarters. Then, too, he was planning on the terror attack to help lay the groundwork for a scheme that was already in the works and planned for the coming November: a final, grand offensive in Tokyo in which Aum Shinrikyo would disseminate "sufficiently large amounts of sarin to initiate World War III."[36] The irrational guru was banking on Japan blaming the chemical assault on its number-one ally, the United States. "The great release planned for November 1995 was to be the actual harbinger of Armageddon," says Lifton.[37] As noted earlier, Asahara was convinced that he and his followers would emerge from the global catastrophe unscathed and supremely positioned.

So it was that five high-ranking Aum members were instructed to report to a room at the Mount Fuji compound at three o'clock in the morning on March 20, 1995. These disciples, all of them men, would be the ones to conduct the gas attack in Tokyo a few hour later, and they had had been summoned to this early Monday gathering, a top-secret training session, to practice releasing the poison when the fateful moment arrived. Among other key facts, the team learned that the nerve agent would be liquid sarin, and that each man would carry between thirty and forty-four fluid ounces of it in small, plastic bags swathed in newspaper.

Next, the men rehearsed releasing the ghastly agent. "Using umbrellas sharpened with a file," writes Haruki Murakami, "they pierced plastic bags filled with water rather than sarin."[38] In the actual attack, the liquid sarin would seep from the punctured bags and begin vaporizing, emitting an invisible gas that would quickly sicken or kill everyone in its path. A simple but effective procedure, the disciples later said they had taken pleasure in practicing it.

At the conclusion of the session, self-protection materials were distributed, antidotes foremost among them. "[The] physician [Ikuo] Hayashi handed out hypodermic needles filled with atropine sulphate to the team, instructing them to inject it at the first sign of sarin poisoning," says Murakami.[39] If administered shortly after exposure, atropine sulphate is often effective in preventing death or permanent disability.[40]

The Team

Regarding the bioterror team, Ikuo Hayashi, its oldest and most highly educated member, was a graduate of the Keio University School of Medicine in Tokyo. After receiving his medical degree, he underwent advanced training at Mount Sinai Hospital in Detroit, Michigan, then returned to Keio to serve as a heart and artery specialist at a leading hospital.[41] Subsequent to this, he was appointed chief of the Department of Circulatory Medicine at the

National Sanatorium Hospital in Tokaimura, Japan.[42] It was in the late 1980s, while he was serving in the latter capacity, that the middle-aged physician began to reflect on his life and work and found them lacking. "Somewhere along the line Hayashi seems to have had profound doubts about his career as a doctor and, while searching for answers beyond orthodox science, he became seduced by the charismatic teachings of Shoko Asahara and suddenly converted to Aum," says Murakami. "In 1990 he resigned from his job and left with his family for a religious life."[43]

Once he had settled into the sect, Hayashi was tapped to lead the organization's Ministry of Science and Technology, as well as being named the Minister of Healing. His work in Aum Shinrikyo extended far beyond healing, however. Within a short period of time, Hayashi's duties came to include not only overseeing a nine-bed medical unit in Tokyo, the loftily named Astral Hospital Institute, but also administrating electric shocks to disloyal disciples or injecting them with a homemade truth serum, the purpose being to extract information from them. And the doctor did more besides treat and torture Aum disciples. "[H]is skills were made use of in every form of its medicalized criminality: in abductions and incarcerations, in plastic surgery for disguise, and in drug production and use," writes Lifton.[44] Clearly, Ikuo Hayashi was an impeccable choice for the impending terrorist mission.

During the attack, Hayashi would be assigned to Tokyo metro's Chiyoda Line, where his task would be to discharge the poison. He would have an accomplice whose main task would be to serve as the getaway driver. This person was Tomomitsu Niimi, the previously mentioned assassin who strangled a sect member as well as an attorney and his family.

Elsewhere on the city's subway system, the Hibiya Line to be precise, two more Aum disciples were set to release the poisonous agent at different points along the route. Their names were Toru Toyoda and Yasuo Hayashi (no relation to Ikuo Hayashi), each of whom also had a driver assigned to him for the escape.

At the time of the terrorist mission, Toru Toyoda was a twenty-seven-year-old physicist who had graduated with honors from Tokyo University, completed a master's degree, and was preparing to pursue a doctorate when he decided instead to join Aum Shinrikyo. Not surprisingly given his background, he rose swiftly in the sect's Ministry of Science and Technology, specifically in its Chemical Brigade, where he became known for his dedication to its projects. He also became known for his resolve. "He's the type that never rests once he has set his mind on something—he likes to see things through to the end," writes Murakami. "Or perhaps he is more the type of person willing to martyr himself for a principle."[45] Stoic, with a demeanor suggesting a touch of arrogance, Toyoda would be shocked when informed about the sect's plan to launch the chemical attack in Tokyo and especially about his own role in it, but he would participate all the same, in part by employing emotional anesthesia to make it through the murderous undertaking.[46]

Ten years older than Toyoda was the other terrorist on the Hibiya Line, Yasuo Hayashi, whose academic background was in the nascent field of artificial intelligence. After graduation, Hayashi traveled extensively in India, where he developed an interest in religion before returning to his homeland and joining Aum Shinrikyo. A senior figure in the Ministry of Science and Technology, younger disciples described him as kind and paternal, a listening ear and a guiding hand. In sharp contrast to this depiction, Shoko Asahara, perhaps because of his escalating paranoia, suspected Hayashi of being an infiltrator, and for this reason increased the amount of sarin the scientist would transport in the subway attack.

The two remaining terrorists, Ken-ichi Hirose and Mosata Yokoyama, were assigned to the city's Marunouchi Line and provided with getaway drivers. The thirty-year-old Hirose, the recipient of a baccalaureate degree in applied physics from Waseda University, graduated at the top of his class. Upon finishing his post-graduate coursework, he joined the Aum sect and spent his days at the Ministry of Science and Technology. When told about the planned gas attack, he reported feeling "instinctual resistance," yet he acquiesced largely due to personal doubts.[47] "I shuddered to think of all the victims this would sacrifice," Hirose said. "On the other hand, I knew I couldn't be well-versed enough in the teachings to be thinking like that."[48]

The last disciple-terrorist, Mosata Yokoyama, was described by his fellow sect members as quiet and reserved, and, like Ken-ichi Hirose, appeared to have certain insecurities. Thirty-one years of age at the time of the attack, Yokoyama graduated from Tokai University with a degree in applied physics, worked for an electronics company for a few years, and thereafter took his Aum Shinrikyo vows. Nondescript in appearance and deportment, he was the undersecretary of the Ministry of Science and Technology.

The Multi-Point Assault

As a mass destruction event, the gas-attack scheme was shrewd. Shortly after 7:30 a.m., each of the five perpetrators was to enter the Tokyo subway system at a different station. As mentioned earlier, one man would board a train on the Chiyoda Line, while two others would board separate trains on the Marunouchi Line. The remaining pair would do likewise on the Hibiya Line. All five subways were inbound and would intersect at Kasumigaseki Station in the heart of Tokyo, near government offices and adjacent to the city's police department. By releasing their poisons simultaneously, the perpetrators would ensure that all five trains carried the noxious vapors to this crucial destination, perhaps the most congested spot in city, and from here it was expected to spread throughout the entire underground network.

Unfortunately for the commuters of Tokyo on this bright Monday morning, the terrorists' strategy proceeded according to plan. Sporting a surgical mask similar to those sometimes worn by Japanese commuters, each of the terrorists sat in a subway car holding a bag filled with liquid poison and wrapped in newspaper. One of the perpetrators, the physician Ikuo Hayashi, recalls his thoughts: "If I unleash the sarin here and now, the woman opposite me is dead for sure."[49] Like his accomplices on the other four trains, he was acutely aware of the impending horror, but he proceeded all the same.

A few minutes before 8:00 a.m., each man placed his sarin-filled bag, still concealed in a newspaper, on the floor of the subway car. On all five cars at precisely the same moment, the men thrust the tips of their umbrellas into the plastic bags and released the liquid, then arose and exited at the next stop. As the cars continued on their way to Kasumigaseki Station, the liquid began oozing from the bags and emitting fumes.

In Ikuo Hayashi's case, he found it difficult to pierce the bag, and although he finally succeeded, only a tiny amount of poison seeped out. It was sufficient nonetheless. As a result of the doctor's actions, 231 people were injured and another two died.

For the hapless passengers on the other four trains, the consequences were more widespread and intense, since greater quantities of sarin were discharged. Not only that, the poison's grim effects were soon being felt far beyond the cars in which it had originally been released. "The fumes were spread at each stop, either by emanating from the tainted cars themselves or through contact with liquid contaminating peoples' clothing and shoes," writes Kenneth Pletcher.[50] For the morning commuters, life became hell.

Kenji Ohashi, a middle-aged car salesman, had no idea what was happening. Although he noticed an unpleasant odor when he boarded a subway car on the Marunouchi Line—impurities in the sarin caused the smell—he took a seat all the same, and still being a bit sleepy at this early hour, closed his eyes. When he opened them again as his destination was being announced, Ohashi discovered a changed world. He could not have known that he had just been exposed to a potentially lethal poison and that his pupils were dilated.

"The lights on the platform were faint," he recalls. "My throat was parched and I was coughing; a really bad, chesty cough."[51] Pulling himself to his feet and exiting the car, he found his symptoms multiplying.

> My nose was running and my legs were shaky. It was hard to breathe.... I didn't have the foggiest idea what had hit me. Only everything had gone dark before my eyes. My lungs were wheezing like I was running a marathon and the whole lower half of

> my body was cold and trembling.... All through the morning my body was like ice. Even with an electric blanket, I was shivering. My blood pressure was up to 180.[52]

Ohashi spent the next two weeks in the hospital, during which he endured muscle cramps, near-blindness, and labored breathing. Even more distressing, he suffered headaches so excruciating as to render painkillers futile. And his misery continued long after he was discharged. For several months, the headaches and lethargy persisted and sharply reduced his ability to perform his job or even go to work. In due course, his managerial duties were handed over to another salesperson, thereby impacting Ohashi's career prospects. "I almost think I'd be better off dead," he told an interviewer a few years after the incident.[53]

It was the same story on the Hibiya Line when another noxious train came to a stop. "The doors opened, and passengers surged and tumbled from the train, gasping for breath," write Kaplan and Marshall. "Five collapsed on the platform, foaming at the mouth."[54] Within minutes, the disaster was evident at street level as well.

> Above ground it was pandemonium.... The pavements and soon the roads were blanketed with casualties lying where they had fallen, or clutching tissues to staunch blood flowing from their noses and mouths.... The commuters made little noise, since the nerve gas had crippled their lungs and stolen their voices.[55]

As the Aum team had hoped, a similar scenario was playing out on the Chiyoda Line. Aya Kazaguchi, a twenty-three-year-old woman in charge of a line of clothing at a garment company, found herself gasping for air. "It's like, there's this tight pressure in my chest, and as much as I try to inhale, no breath comes in."[56] Kazaguchi also realized that she was not alone. "[T]he people hanging on to the handstraps started coughing," she says, adding that a chorus of coughs soon echoed throughout the train.[57] From there, matters spiraled downward, as was the case on all of the contaminated cars as well as in the terminals.

It would be nearly forty-five minutes before the Japanese news media would pick up the story and, among other efforts to help, offer instructions to those in the vicinity of the attack. "[F]lee in the direction you see fewest bodies," was one such recommendation.[58] Still unsure of the cause, officials at first thought a gas line had ruptured, but a military physician who assessed one of the first casualties announced, at 10:30 a.m., that sarin was the culprit. Of course, this pointed to terrorism, since an agent of this type, widely regarded as a weapon of mass destruction, would not otherwise be found in the Tokyo subway system. And the sarin certainly did produce mass destruction.

Symptoms and crisis management. Thirteen people died and more than 5,500 were injured in the Tokyo offensive, with scores of victims being transported to nearby Saint Luke's International Hospital, a training and general facility as well as a designated field hospital in the event of a disaster.[59] Two years later, a team of doctors who were among those treating the injured at Saint Luke's published an article in the *Southern Medical Journal*, one in which they recapped the symptoms presented by the afflicted and highlighted the difficulties that arose in the medical management of the disaster.[60]

On the day of the attack, Saint Luke's treated 641 sarin victims, five of whom were in critical condition upon arrival. Diagnosed with respiratory arrest, two of them died. The symptoms of another 106 were judged to be moderately severe, and for this reason they were hospitalized overnight, the majority for observation. Nearly all of these patients, 105 out of 106, were diagnosed with miosis, or constriction of the pupils. "Other ophthalmologic symptoms were ocular pain, blurred vision, and visual darkness," write physician Sadayoshi Ohbu and his colleagues. "Dyspnea, nausea, vomiting, muscle weakness, coughing, agitation, and fasciculation were relatively common."[61] Also among those in the moderately-severe category were four pregnant women, one of whom would opt for an abortion owing to the potential effects of fetal sarin exposure. As for those whose symptoms were considered mild, they too suffered primarily from eye problems and received treatment through Saint Luke's outpatient services. Ohbu notes that the main factor in the severity of all of the patients' symptoms was the amount of the poison they had inhaled, with related factors including their proximity to the sarin package and the amount of time they had remained in the noxious environment.[62]

Unfortunately, some of those taken to Saint Luke's unknowingly carried the contaminant on their clothing, thereby transferring it to hospital personnel. "[N]early 50% of the medical staff working in this hospital site complained of some degree of Sarin gas exposure due to secondary contamination from patients," writes Kenichiro Taneda, another Saint Luke physician. "Approximately 10% of pre-hospital EMTs also experienced acute symptoms of secondary contamination due to vaporized Sarin, probably from the victims' clothes."[63]

And then were the psychological effects of the attack. Of a sample 610 sarin cases treated at Saint Luke's, nearly sixty percent reported symptoms of posttraumatic stress disorder a month after the assault. Dishearteningly, this number remained more or less the same at the six-month mark. In terms of specific symptoms, the Ohbu team writes that "[f]ear was seen in 32% of the victims, some of whom still cannot use the subway."[64] Headaches and malaise were the most common and enduring psychological effects, with other problems including insomnia (29 percent), flashbacks (16 percent), irritability (16 percent), depression (16 percent), and nightmares (10 percent).[65]

Turning to the victims' medical treatment, several issues came to light in the course of the Tokyo episode. Because the authorities were not immediately aware that sarin was the cause of the unprecedented number of people falling ill in the city's subways, caregivers who were first on the scene—paramedics, police, firemen—unwittingly placed themselves at risk. "No primary decontamination was performed on-site," writes Tomoyo Saito, a research fellow at Keio University, "and, more importantly, first responders and health care workers involved in the initial response were not wearing personal protective equipment."[66] At the receiving hospitals, medical staffs likewise placed themselves at risk during the early hours of the disaster through their contact with incoming sarin victims and contaminated first responders. Then, when it was determined that sarin was the source of the poisonings, further problems became apparent. First and foremost, the city's doctors found themselves at a disadvantage, very few of them having had experience treating patients suffering from exposure to this distinctive chemical weapon.

Contributing to the madness was the issue of "medical surge capacity" and the fact that Tokyo's healthcare resources were quickly overwhelmed. Despite the existence of a municipal emergency plan designed to provide structure and guidance during an urban disaster, the city's medical facilities were hard-pressed to attend to the sheer volume of victims—there were literally thousands seeking treatment on the same morning—meaning limited numbers of hospital beds as well as physicians to diagnose and treat the afflicted. The aforementioned Saint Luke's International Hospital was one of many that found itself filled beyond capacity. "Exhausted physicians who had sarin victims added to their existing caseloads treated patients in hallways," writes Robyn Pangi. "All hospital facilities, including chapels and halls, were used to treat sarin victims."[67] Routine surgeries were canceled and outpatient clinics shuttered if they were not required for victim management.[68] And there was still another challenge: the sarin antidote was in short supply. To be sure, Tokyo's private and public health services were stretched far beyond their limits as they strove to furnish timely medical care to those in need of it.

Aftermath: Assassination Attempts and Arrests

While these events were taking place in more than eighty hospitals and clinics across Tokyo and beyond, the five Aum Shinrikyo terrorists, elsewhere in the capital city, hurried to a designated safe house for refuge. Subsequent to this, they and their drivers returned to the sect's headquarters.

As for the police, they set to work assembling the pieces of the puzzle. Before long, they detected the connection between the Tokyo poisonings and the earlier gas attack on the apartment building in Matsumoto, home to the three judges who were adversaries of the Aum organization and who required hospitalization for sarin poisoning. The police also reopened three unsolved murders, namely, the strangling deaths of Sakamoto Tsutsumi and his wife and child, Tsutsumi being the attorney who was preparing a case against Aum Shinrikyo at the time of his death. And as the authorities' mounting suspicions reached the public's ears, the Aum organization reacted swiftly and visibly. On March 21st, the day after the nerve-gas offensive, the sect called a press conference during which it leapt into full denial mode replete with faux indignation, accusations of prejudice, and finger-pointing.

"I know what kind of substance sarin is," declared Yoshinogu Aoyama, an attorney speaking on behalf of the sect, "and only such parties as the American military could make and keep and use such a substance."[69] Many of those in attendance contend that Aoyama was not, in fact, blaming the nerve-gas attack on the United States, but rather suggesting that the Japanese government may have acquired the sarin from this Western ally and used it against its own people.[70]

Twenty-four hours later, police raided the Aum Shinrikyo offices in Tokyo and the sect's main compound near Mount Fuji. As Aum disciples decried what they portrayed as religious persecution by the government, law enforcement officials unearthed the sect's secret laboratories and retrieved biological and chemical warfare equipment along with the ingredients used to brew liquid sarin. Confiscated, too, were filmed accounts of the sect's experiments using the bacterium that causes Q fever, videotapes revealing that numerous disciples, evidently including the guru himself, had been accidentally exposed to the pathogen and sickened by it. Because the evidence being gathered appeared so incriminating, police arrested several disciples on the spot. Asahara was not one of them, though. The guru was in hiding.

As could be expected, the Japanese people were now up in arms, enraged that the religious group had carried out such an appalling act of violence against everyday commuters en route to work. And intensifying the anger and apprehension, Asahara's voice came on the radio at this juncture and proclaimed that the nation's citizens should accept their deaths without regret. It seems the guru had managed to purchase radio airtime in Russia, where his taped message was beamed back to his homeland in a broadcast clearly calculated to heighten the panic level of the population. Asahara's insinuation, of course, was that further attacks were imminent. And indeed, more offensives would be carried out, primarily small, individualized ones, although their aim would no longer be to trigger an apocalypse. "[T]he guru ordered that Aum resort to acts of terrorism in order to prevent his capture," writes Lifton.[71]

The first, an assassination attempt, was slated for March 30th, the objective being to murder the chief of the National Policy Agency, or so the authorities suspected.[72] The target, Takaji Kunimatsu, was a fifty-seven-year-old official who was leading the nation's principal inquiry into the controversial sect. By eliminating this top law enforcement figure, Asahara evidently believed both the manhunt and the investigation would come to a halt. In terms of the assassination attempt, it was to be a one-man job and was set to occur on a morning when Kunimatsu would be traveling without his bodyguards.

"According to TV interviews with Kunimatsu's secretary, who was helping him into his car at 8:25 a.m., a masked man standing behind a utility pole fired four shots, hitting Kunimatsu in the shoulder, the side and twice in the back," reports the *Chicago Tribune*.[73] The assailant, wearing a surgical mask, was not identified or captured. Although he was struck by four bullets, the stouthearted Kunimatsu survived and the investigation proceeded.

In the ensuing weeks, Aum members continued making bomb threats, the purpose being to frighten and intimidate those they regarded as enemies of their religion. In a small number of cases, disciples sent actual bombs, although none of the resultant injuries proved to be fatal.

In mid–April, Shoko Asahara, still in hiding, alarmed and angered the populace once again by declaring that his organization was about to deliver another catastrophic blow to Tokyo. A false alarm, it nevertheless managed to keep the citizenry on edge as well as costing the city financially, the entire metropolis being placed on alert as a precaution.

Even more ominous was an attempted chemical attack on May 5th. The Shinjuku subway station in Tokyo was to be ground-zero, a highly congested terminal at a critical railway hub. Here, an Aum terror squad placed four bombs containing sulfuric acid and sodium cyanide, enough to annihilate ten to twenty thousand people, beneath ventilation shafts in restrooms. Three of the bombs failed to detonate, however, and the remaining one caught fire, with observant subway workers extinguishing the flames before the poisonous mixture could be released. The terror unit, unfazed, set about preparing to conduct more gas attacks and bombings.

It would be on May 16th, for instance, that Aum disciples set out to murder Governor Yukio Aoshima, who was in the process of revoking the organization's status as a tax-exempt religious entity. Mailing a parcel containing an incendiary device to his office, the bomb mutilated the hands of the staff member who was unlucky enough to unwrap the package, although the governor himself escaped injury.

Yet it was also on this day that a pivotal event occurred. Police launched coordinated raids on one hundred thirty Aum Shinrikyo sites across Japan during which they arrested a large number of members, and, even more important, located and apprehended Shoko Asahara himself. "Police said they found Asahara sitting in silent meditation in a hidden room beneath the third floor of Truth Building No. 6 at the [Mount Fuji] compound," writes

Merrill Goozner.[74] Asahara's capture would constitute a turning point for the Aum Shinrikyo sect and its forty thousand members, who could now be found not only in Japan, but in Russia, the United States, and other nations as well.

Postscript

Shoko Asahara was charged with thirty-nine offenses, among them twenty-three counts of murder. Others included kidnapping, violating Japan's Arms Manufacturing Law, ordering and planning murders, and disposing of corpses.[75] His legal team entered pleas of not guilty on all of the charges, insisting that Asahara should not be held responsible for the deeds of his disciples. Although numerous Aum members, including its scientists, would testify against him, the guru himself would not speak at any point during his trial.

In the end, the court could not be swayed by the arguments submitted by Asahara's attorneys. At the conclusion of the guru's lengthy legal proceedings—they lasted from 1996 to 2004—he was found guilty and sentenced to death. According to court documents, he had "committed the crimes in the process of realizing his fantasy of expanding the cult through militarization and to reign as its king in the name of salvation."[76]

Interestingly, two years later Asahara's defense team sought to have his death sentence overturned. Claiming he had deteriorated during his incarceration—"he was often heard shouting obscenities and he masturbated during meetings with lawyers and psychiatrists"—they hoped their humanitarian plea would save his life.[77] To the lawyers' consternation, however, a psychiatrist reevaluated Asahara at the court's request and concluded that the guru was sane, the upshot being that the punishment was upheld.

Another 188 members of the sect were also charged during this period, their purported crimes extending from comparatively small offenses to homicide and experimentation with biological and chemical weapons. Of these defendants, 187 were convicted, with death sentences being handed down in a dozen cases. The latter included six of the men accused of hands-on involvement in the Tokyo subway attack. Seiichi Endo, head of the sect's biological and chemical warfare programs, would also join the ranks of the condemned. At this time, the men are awaiting execution even as the human rights organization Amnesty International seeks to have their sentences commuted.[78]

As for the Aum Shinrikyo sect, the government rescinded its official status as a religious organization in October of 1995, with the sect filing for bankruptcy a few months later. With Shoko Asahara out of the picture, his two pre-teen sons stepped in as co-leaders of the organization, although a former high-ranking disciple would, in due course, agree to take the reins, a man who had not been involved in the organization's biological or chemical warfare operations.

In 2000, the organization, still trying to purge its terrorist past, changed its name to Aleph, the first letter of the Hebrew alphabet and, to the remaining Aum Shinrikyo holdouts, a word denoting a new beginning. The sect also modified some of its teachings and apologized to the Japanese people for its predecessors' biological and chemical assaults. Yet the organization, still remarkably wealthy, has not lived up to all of its promises of redress. Whereas the court ordered the sect to pay nearly forty millions dollars to the victims of the Tokyo subway assault, its payouts have been partial—less than half the designated amount—prompting the Japanese government to step forward to offer monetary support to the victims. The casualties include those people are no longer able to work due to severely impaired vision or blindness.

Regarding the sect's home base in the village of Kamikuishiki near Mount Fuji, it was soon disassembled. Unfortunately for the peaceful, picturesque village itself, it too suffered a sad fate, being shunned by tourists who were loathe to visit it owing to Aum Shinrikyo's deeds. Then, too, several suicides occurred at the compound during the sect's collapse, so numerous as to bring further notoriety to Kamikuishiki. For these reasons, an imaginative effort was put forth in 1997 to reinvent and revitalize the settlement through the construction of a theme park based on Jonathan Swift's satirical novel, *Gulliver's Travels*. Featuring an enormous statue of the book's protagonist, Gulliver, staked to the ground by his hair—the figure was nearly 150 feet long—the exhibit included a miniature version of the hamlet Lilliput, with the hope being to attract Japanese families to the unusual spectacle and the associated gift shops. But alas, this would not be its fate. "The Fuji Gulliver's Kingdom theme park shut down," reports the *Japan Times*, "after a four-year attempt to dispel the site's negative image as a former Aum Shinrikyo base."[79] Evidently, parents were disinclined to take their children to an amusement area that had recently been the home of a biological and chemical warfare program, colossal statue aside. Today, Kamikuishiki no longer exists, having been rezoned and absorbed into two neighboring towns.

As to the fate of the Aum Shinrikyo sect, the organization still exists, albeit largely underground in various incarnations. In the Balkan nation of Montenegro, the government recently ejected nearly sixty foreign nationals for alleged involvement in the sect.[80] And in Russia, where the government has forbidden Aum Shinrikyo, it is estimated that the organization has up to thirty thousand practicing members. Many of these people have illegally sought donations for the organization, and some are suspected of having carried out threatening or aggressive acts. According to a 2016 report by the BBC, the Russian government is in the midst of an investigation into the sect and its strong-arm tactics. The allegation: "violence against citizens and injury to their health"[81] It is an ominous echo of the Aum Shinrikyo organization's early days in Japan.

Meanwhile, in Japan where Aum Shinrikyo remains legal—since World War II, the East Asian nation has been averse to interfering with organizations claiming to be religious—there are presently an estimated 1,500 to two thousand members of Aum's spin-off organization, Aleph, and a related group, Hikari no Wa. Since the nerve gas attacks, however, the Japanese government monitors these descendant entities so as to prevent future acts of violence against the people of Japan.

Fortunately, if there were to be another terrorist attack, the nation would be better prepared for it. In concert with a handful of governmental agencies, the Japanese medical community studied its response to the 1995 sarin attack, the result being an expanded and improved ability to contend with future biological and chemical offensives. Among other upgrades, 73 emergency rooms have been outfitted with chemical-substance analyzers, and medical settings across Japan now possess decontamination equipment and protective gear for their personnel.[82] Additionally, Japan has made it illegal to possess sarin gas, the Asian country being a signatory to the Chemical Weapons Convention which prohibits the formidable chemical weapon. By all accounts, the nation has learned from its dreadful experience with unmitigated violence, of which it has had more than its share.

On two mornings in 1945, the United States dropped atomic bombs on two civilian population centers, the cities of Hiroshima and Nagasaki, and in so doing vaporized, maimed, or sickened 130,000 men, women, and children in an unprecedented and unspeakable act of inhumanity. Fifty years later, on a morning in 1995, the Japanese people were again blindsided by unrestrained and unthinkable violence, this time a chemical attack carried out by domestic terrorists intoxicated with religious misconceptions. To be sure, nuclear, chemical, and biological weapons, whether wielded by state or non-state actors, continue to be among the greatest threats to human life, a dismaying reality to which the Japanese people can attest.

◆ 8 ◆

Lethal Letters

September 11th and the Anthrax Enigma

It was on the clear, bright morning of September 11, 2001, that the world watched in horror as a pair of hijacked airliners slammed into the Twin Towers in Manhattan, while a third bore into the west wall of the Pentagon near Washington, D.C., and a fourth plunged into a dewy Pennsylvania meadow. An elaborate terrorist offensive, it succeeded, in only seventy-seven minutes, in scarring indelibly the American psyche and forever changing the course of history.

During the remainder of that ill-fated September, the traumatized nation waited in anguish and in anger for an explanation; a justification, no matter how aberrant, as to why a band of terrorists had murdered thousands of its citizens and mutilated its iconic structures. In private government circles, meanwhile, another concern came to the fore. Counterterrorism officials worried that the other shoe was about to drop, their hypothesis being that the events of September 11th, monstrous as they were, represented only the opening salvo in what would become a series of assaults on the American people. And, sure enough, their fears were realized one week later when a biological offensive was stealthily set into motion against the distraught nation. Its first casualty: a British-born, sixty-three-year-old photo editor at the *Sun* tabloid, Robert Stevens, who lived in Lantana, Florida, a coastal community in Palm Beach County.

Anthrax in America

It was on September 18, 2001, that the sinister string of events commenced. A letter containing anthrax spores was dropped in the mail, one addressed to the offices of American Media, Inc., in Boca Raton, Florida, publisher of such tabloids as the *National Enquirer*, the *Globe*, the *Star*, and the *Sun*. The letter is not believed to have contained a message. Only the spores, invisible to the naked eye, are thought to have been present in the form of a tainted, powdery substance.

Days later, the missive passed through the American Media mailroom and, in due course, made its way into the hands of photo-editor Stevens, a fact investigators would deduce several days later when they detected traces of anthrax on the computer keyboard in his office. Since Stevens' job involved acquiring images and related materials by mail, he would have opened the envelope without a second thought. Likewise, the mailroom's employees would have handled it unreservedly, since it was part of the job. This appears to be how a seventy-three-year-old mail sorter and courier, Ernesto Blanco, along with a coworker, Stephanie Dailey, also came into contact with the envelope. Evidently, none of the three voiced any suspicions about the envelope or its contents at the time, at least not to law enforcement officials.

So it was that Stevens, his wife Maureen, and their twenty-one-year-old daughter traveled on Thursday, September 27th, from their Florida home to a resort in Lake Lure, North Carolina, where they enjoyed a brief family holiday. Little did they know it would be their last one, with Stevens becoming sick even before they returned home. "He started feeling ill on Sunday, September 30," writes David Wellman, "complaining of numbness and chills."[1] The photo editor did, however, manage to make it back to Lantana, a six hundred mile trip by car.

Prophetically, it was on this same Sunday that the *New York Times* published an article, "Some Experts Say U.S. Is Vulnerable to a Germ Attack."[2] Making the case that a biological assault was a realistic threat to the United States, the piece reasoned that such an offensive, unlike a conventional military strike, would not be immediately apparent. The pathogen would likely be inserted into the population in such a way as to mimic a natural occurrence and thereby avoid, or at least delay, attracting the attention of bioterrorism experts. "[P]eople will get sick and go to their hospitals," predicted one such expert, Asha George, "and the public health system will have to pick up on this."[3] Unfortunately, not all of the components of this system would be so efficient in the Stevens case, at least not initially.

Over the next forty-eight hours, the photo editor's condition plummeted, prompting his wife to rush him to the emergency room at nearby JFK Medical Center in Atlantis, Florida. At the time of his intake evaluation—it was 2:00 a.m. on October 2nd—he had a fever of 102.5, as well as confusion, swollen lymph nodes, and nausea.[4] Suspecting meningitis, doctors obtained samples of his cerebrospinal fluid and ordered a chest X-ray and CT scans. But even as these and other procedures were being performed, Stevens' symptoms compounded, and soon came to include seizures, unconsciousness, and the early stages of organ failure.

In terms of the diagnostic work-up, the findings pointed to inhalation anthrax, much to the staff's surprise. Also known as pulmonary anthrax, it is a rare form of the poisoning brought about by breathing in the bacterium's spores and, unlike other types of the infection, had not been diagnosed in the United States in a quarter of a century.[5] The prognosis, moreover, was grim, as revealed by government statistics. "Mortality has been essentially 100% despite appropriate treatment," reads a document from that period, one released by the Office of the Surgeon General.[6]

A repeat analysis of Stevens' cerebrospinal fluid performed at the Centers for Disease Control in Atlanta, Georgia, on Thursday, October 4th, confirmed the presence of *Bacillus anthracis*. At more or less the same time, Florida's state-operated laboratory in Jacksonville provided yet another verification. Then, on October 5th, Stevens, tranquilized and on a ventilator, expired. It was now that federal officials decided to go public with the news.

"An isolated case of anthrax infection was confirmed on Thursday in a Florida hospital," announced Tommy Thompson, Secretary of Health and Human Services, at a press conference. When asked whether the patient had contracted it naturally or was the victim of an extremist act, Thompson was unequivocal. "There is no evidence of terrorism," he stated matter-of-factly.[7] He next attempted to explain how Stevens might have come into contact with the bacterium. "The man could have picked up the infection from his clothes," the official said, "and [he] was known to have drunk water from a creek recently," alluding to the family's trip to North Carolina.[8] Such assertions, however, were met with scorn by those in the scientific community who were familiar with the pathogen and its modes of transmission. "Inhalation anthrax—from a stream?" asked Robert Kadlec, a physician at the Pentagon with considerable research experience with the microbe. "You gotta be shittin' me."[9]

As it stood, a handful of scientists already knew Secretary Thompson's remarks were, in effect, off the mark, even as he was making them. This is because the Centers for Disease Control and Prevention (CDC), the previous afternoon, had flown a sample of the bacterium to a laboratory in Arizona, with the latter team wrapping up its analysis before the federal official called his press conference. The lab was headed by microbiologist and evolutionary biologist Paul Keim, who, seven years earlier, had taken part in a classified, CIA-sponsored project at Los Alamos National Laboratory designed to differentiate among the numerous varieties of anthrax bacteria.

Keim's team, which specialized in microbial genetics, worked nonstop to nail down the type that had invaded Stevens' body, and within a matter of hours determined that it was the highly virulent Ames strain. It was a discovery that came as a bolt from the blue, since this was the same laboratory variant that scientists at Fort Detrick had long used in their vaccine studies aimed at protecting American troops from anthrax infection. They had selected it because it was one of the hardiest strains, being able to circumvent existing vaccines and resist nearly all antibiotics. "A vaccine that could protect against Ames," reports the *Washington Post*, "would offer the highest protection for troops exposed to deadly germs on the battlefield."[10] In terms of the number of U.S. labs having access to this distinctive pathogen, it was quite small at the time, no more than a dozen, five of them housed at Fort Detrick itself.[11] And Keim, of course, knew what this meant. "The implications were that this was a bioterrorism event," he said.[12]

Straightaway, the Federal Bureau of Investigation (FBI), together with several other government bodies, formed an investigative unit and set about scrutinizing all aspects of the "Amerithrax" case, as the government now designated it. Because the offices of American Media constituted ground zero, they provided the logical starting point. And, as expected, investigators detected traces of the microorganism in the building, as well as learning about Ernesto Blanco and Stephanie Dailey, the two other employees who suspected they had been exposed to it. A few days later, investigators discovered more anthrax spores in Blanco's van, as well as at a post office where he picked up the mail each day for American Media. Yet this news was withheld from the public. In fact, Florida Governor Jeb Bush, in the coming days, would clamp down on all media coverage of the crisis in his purported determination to protect the state's reputation as a tourist destination and agricultural center. But there was a downside in that the ensuing absence of information would fan Floridians' fears.

"Local media complained—the FBI wasn't talking and neither was the governor's office, and the CDC wasn't returning calls from journalists," says Guillemin.[13] It is doubtful, of course, that the abrupt inaccessibility of information helped tourism and agriculture; rather, it may have raised new questions as to the true extent of the danger in the Sunshine State.

As to the additional pair of American Media employees who tested positive for anthrax exposure, one of them, Stephanie Dailey, remained healthy. Ernesto Blanco, on the other hand, became quite ill and checked into Cedars Medical Center in Miami, where he was diagnosed with pneumonia secondary to inhalation anthrax. What followed would be a protracted, touch-and-go ordeal, although he would ultimately prevail. "Blanco's treatment," writes journalist Adrian Sainz, "was a venture into uncharted waters."[14] In part, this was because the elderly mailroom courier, who had suffered a mild stroke several weeks earlier, displayed symptoms not typically associated with anthrax poisoning, thus making his case unpredictable and its course, irregular.

Regarding the American Media building itself, the publishing company shuttered it once investigators verified the presence of the deadly bacterium. How the pathogen had gotten into the building remained a mystery at this juncture, although those who were knowledgeable about anthrax knew it almost certainly didn't happen by accident. Bolstering their suspicion was the incident's timing; that is, in the wake of the World Trade Center and Pentagon attacks, a moment of profound vulnerability for the American people. Even so, federal officials did not come forward with an acknowledgment that the pathogen's presence in the American Media building was quite likely a terrorist act. They also did not reveal that federally-funded scientists were using the same rare strain of the bacterium for the studies they were carrying out in highly restricted biocontainment laboratories. The only concession was a statement informing the public about the creation of the Amerithrax fact-finding team.

"While officials stress there is no indication the discovery of anthrax in South Florida is linked to any terrorist activity," reads an ABC News report dated October 8, 2001, "the FBI has assumed the lead in the investigation, with the cooperation of law enforcement, local and state health workers, and Centers for Disease Control and Prevention officials."[15] The news report added that Attorney General John Ashcroft continued to insist that the government's investigation was not a criminal inquiry.[16] Thus far, it had been painted either as an epidemiological or an environmental investigation.

Whereas the FBI and other government organizations, presumably to prevent panic, persisted in denying that a bioterror attack was strongly suspected, the CDC adopted a different tack. It, too, strove to discourage undue alarm, but it also furnished a dose of public education in anticipation of additional cases of the infection.

"Anthrax is not contagious," reads the agency's October 4th press release. "The illness is not transmitted person to person."[17] The statement continues,

> Although anthrax starts out with flu-like symptoms, it rapidly progresses to severe illnesses, including pneumonia and meningitis. If anyone has been exposed, antibiotics are the appropriate preventive treatment. CDC has an emergency supply of antibiotics ready for distribution.[18]

To its credit, the CDC issued further press releases in the succeeding days, fact sheets providing information about the bacterium's incubation period and further descriptions of symptoms that might emerge in the infected. To reassure those assumed to be at greatest risk—the workforce of American Media, together with the communities surrounding the publisher's Florida headquarters—the agency announced that the National Pharmaceutical Stockpile had enough antibiotics "to treat several thousand people in Palm Beach County, if needed."[19]

As things stood, thousands of people living in the region, although not experiencing symptoms, were already in a panic. "It's Cipro, Cipro, Cipro," said the manager of a local drugstore, referring to residents' determination to get their hands on the preferred antibiotic, ciprofloxacin, that was used to treat anthrax poisoning at the time.[20] Sales skyrocketed at the pharmacy, essentially tripling in the days after the American Media incident.

Unbeknownst to anyone in Palm Beach County—or to the specialists at the Centers for Disease Control, for that matter—a thirty-eight-year-old woman in New York City had also begun receiving Cipro. Unlike the rattled residents of South Florida, however, she was actually experiencing symptoms of the cutaneous form of anthrax. The most common type of the infection, the one in which the pathogen enters the body through the skin, it comprises over 95 percent of all cases and is less lethal than the inhalation variety.[21] It had been on October 1, 2001, upon developing a dark lesion near her shoulder, that she started taking the antibiotic.

The Manhattan Attacks

Erin O'Connor was her name and she worked in the New York City offices of NBC News, where she was an assistant to newscaster Tom Brokaw. In late September, the studio's incoming correspondence included an envelope addressed to the newsman, one that had been mailed on September 25th from St. Petersburg, Florida, and carried no return address. Removing the letter from the envelope, O'Connor read the inconsistently punctuated, misspelled message:

THE UNTHINKABEL.
SEE WHAT HAPPENS NEXT[22]

For O'Connor, what happened next occurred while she was attaching the letter to its envelope for filing. "As she shot a staple into the porous paper," says Robert Graysmith, "a 'relatively crude' substance spilled onto her."[23] A white powder similar to talc, O'Connor quite naturally was concerned about it, especially in light of the letter's menacing message.

Later that day, she noticed a rash on her neck, with more symptoms soon to emerge. "Over the next several days she developed a softball-size wound on her shoulder," writes David Willman, "along with a low-grade fever."[24] Although her doctor thought he might be looking at an infected spider bite, he didn't rule out the more troubling possibility of cutaneous anthrax, particularly since O'Connor had recently made contact with the mysterious powder.[25] It was during this office visit, then, that he prescribed Cipro and reported the potential exposure to city health officials. But then came a curious turn of events. Because the authorities were aware that previous mailed-in threats had amounted to naught, they were skeptical of the doctor's report. "All these events had been hoaxes," says epidemiologist Joel Ackelsberg, Director of the Emergency Readiness and Response Unit of the New York City Department of Health. "[M]y approach was that they were going to continue to be hoaxes."[26]

Be that as it may, city officials relayed the physician's concerns to the FBI, with the federal agency, in turn, liaising with NBC's security division. At the Bureau's request, the latter scoured the company's mail until the letter was located, after which tests were performed on it. And what came next was precisely what city health officials had expected: negative results. No trace of anthrax spores were detected on the document, meaning it could not have been the source of infection. Rather, it was merely a warning of forthcoming events. For this reason, officials shelved the case. "According to the health department's protocol for suspicious powders, no follow-up of O'Conner's [sic] case was required," says Jeanne Guillemin of MIT's Security Studies Program.[27]

In the ensuing days, Erin O'Connor's lesion progressed in a manner characteristic of cutaneous anthrax—it blackened and hardened—but tests performed on it were inconclusive. Meanwhile, a second opinion backed up that of her doctor, who had concluded by now that the bacterium was indeed the cause of her condition. The second physician went so far as to note that the lesion was "as classic-looking as you can get."[28]

Finally, dermatologist Marc Grossman obtained tissue samples from the patient on October 9th and sent them to the CDC, where infectious disease specialists performed a biopsy. Perhaps because O'Connor had been receiving antibiotics, the results were negative, but further analyses using what the agency's pathologists described as "two novel immunohistochemical assays" did confirm that signs of anthrax infection were present in the samples.[29] "They also did serology, or blood tests," says Grossman, "that showed that she was developing anthrax antibodies."[30] So the question remained: given that the cryptic letter to Tom Brokaw had been a hoax, how did O'Connor contract the infection? And this raised the possibility of a second letter to NBC studios.

It was not a far-fetched notion. After O'Connor was diagnosed with anthrax poisoning, Casey Chamberlain, an NBC intern, reminded her that another letter had arrived around the same time as the first one. This second letter also contained an ominous message, along with a powder—it was a gray, grainy type resembling brown sugar—which the intern dumped into a wastebasket before routing the document on to O'Connor.[31] Chamberlain added that she had also become ill, developed a sore on her leg, and received antibiotics from her doctor. At the time, her condition was thought to be an allergic reaction to a skin medication she had recently been prescribed. Now, however, it appeared otherwise.

Chamberlain would share this crucial information with the FBI on October 12th, soon after the CDC determined that the O'Connor biopsy was positive. Following this confirmation, New York City officials and the FBI sped into action, with the latter interviewing NBC employees for any information that might prove useful in what had advanced to a criminal inquiry. And it was during this process that Chamberlain told investigators about the second letter to NBC, a disclosure that constituted a seminal moment in the investigation. Within hours, health workers began testing the entire NBC workforce for anthrax poisoning, while the company's offices, which occupied the third floor of 30 Rockefeller Center in midtown Manhattan, were shut down and a systematic search of the crime scene initiated. When investigators, after combing through more of the company's discarded documents, succeeded in retrieving the second letter, they found it was postmarked September 18th from Hamilton, New Jersey, where a processing center handles the incoming mail from nearby Princeton. As expected, it tested positive for anthrax. "O'Connor did not remember seeing this letter before," says Leonard Cole, "but she must have come in contact with it between September 19 and 25."[32] At last, the source of her infection had been identified.

Terse, handwritten, and misspelled, the message itself was unnerving. It read,

09-11-01

THIS IS NEXT
TAKE PENACILIN NOW

DEATH TO AMERICA
DEATH TO ISRAEL

ALLAH IS GREAT[33]

It was evident that the letter, which was a photocopy, had either been penned by an anti-American, anti-Semitic Muslim or crafted to appear as such. Also intriguing, the paper had been hand-cut to a precise square, then creased to create a classic pharmaceutical fold so as to hold the spores without spillage. These measures, exacting as they were, betrayed an expertise in such matters. As to the message, it was printed in uppercase, or capital, letters—a peculiar but prevalent practice among those who compose anonymous threats. "Every maniac from Zodiac to the Unabomber favored uppercase printing," says Graysmith. "It was harder to match to an individual and gave a more terrifying aspect to messages, as did purposeful misspellings."[34] Obviously, whoever was behind the document was at the top of their game, be it a lone individual or multiple collaborators masquerading as one.

Proliferation and Panic

On this same Friday, October 12th, Judith Miller, a reporter at the *New York Times*, opened a letter that contained a handwritten threat. Like the letter to NBC, it contained a white powder, one that smelled sweet, and it spilled onto her. Of all the *Times*' staff, the fact that only one such communiqué had arrived and was addressed to Miller seemed significant. What made it especially disconcerting was that Simon & Schuster, just ten days earlier, had published her book, *Germs: Biological Weapons and America's Secret War*, the subject being the U.S. government's classified biowarfare projects and strategies.[35] And there was more. The *Times*, a month before that, had printed Miller's article, *U.S. Germ Warfare Research Pushes Treaty Limits*, which called into question the American biodefense community's compliance with the 1972 global pact banning the development of bioweapons.[36] It was an article that contained a troubling revelation. "Earlier this year," Miller wrote, "the Pentagon drew up plans to engineer genetically a potentially more potent variant of the bacterium that causes anthrax, a deadly disease ideal for germ warfare."[37] Whether or not such publications were the reason Miller became a target would never be established for certain, but the possibility of such a link was undeniable. At the moment, however, the priority was to determine if the letter itself was contaminated, since lives might be at stake and the newspaper's staff was in a collective state of shock. "Whoever did this," Miller said, "had spread panic with only a few anthrax spores, or perhaps only baby powder, and the price of a few stamps."[38]

Securing the *New York Times* building, authorities searched the premises while health officials analyzed the fine, aromatic particles. It was a race against the clock to ascertain if the world's most prominent newspaper had become the target of bioterrorists. And the reassuring news arrived two days later: the building proved to be anthrax-free, and the powder, negative for the pathogen. It had been another deception.

As these nerve-wracking events were unfolding, investigators elsewhere in the city were checking the homes of NBC employees who had been exposed to the anthrax-laced Brokaw letter in case they may have carried the microorganism back to their residences with them. In this way, their family, friends, and neighbors could become infected as well, thus expanding the sphere of damage. And the outcome of the inspections was daunting. "[T]he homes of Erin O'Conner [sic] and Casey Chamberlain," writes Guillemin, "both proved significantly contaminated."[39]

By now, of course, it was neither possible to keep a lid on the story nor advisable to do so, representing, as it did, a public health threat. With the *New York Times* and NBC offices in lockdown and nearly four hundred NBC employees being tested for anthrax poisoning, some of whom were already being administered Cipro as a precaution, it was obviously time for city leaders to address the bio-attack in a public forum.

To this end, Mayor Rudolph (Rudy) Giuliani, on Friday, October 12th, spoke to reporters at a noon press conference in the course of which he acknowledged that a letter sent to NBC News had tested positive for anthrax. Recognizing the panic that might ensue, he then sought to calm the public's fears, just as he would the next day at a similar media appearance. "We had a lot of people go to emergency rooms," Giuliani said. "[T]he fact is that the surveillance system that we have does not yet indicate any [additional] incidents of Anthrax."[40] Rather pointedly, he was inferring that such emergency room visits were probably unnecessary, while imploring the city's residents to remain clear-headed and composed.

It was a message the mayor would reiterate two days later when he and his team, in an afternoon press conference, again addressed the city's unremitting anxiety. "I was informed that since 7:00 a.m. until 1:50 we had 82 calls," he said, referring to the number of potential anthrax exposures that had been reported thus far that day, none of which had been confirmed.[41] Pointing out that disreputable individuals were exploiting the situation, Giuliani added that city officials had examined twenty-four suspicious packages during the same seven-hour period, some of them smelling suspiciously like baby powder and none appearing to be carrying the bacterium. It was, to be sure, a frustrating, fluid situation the mayor and his team were struggling to contain.

Certainly it is true that New York City was in the midst of an emotional storm what with the anthrax-tainted letter and the continuing hoaxes. And worsening matters was the fact that the bio-attack had come on the heels of the World Trade Center tragedy from which the city was still reeling. Bodies were still being dragged from the rubble and thousands more people remained missing as the menacing letters began to arrive. Bioterrorism, then, was achieving its goal. A single confirmed anthrax letter in the city had ballooned into a full-blown crisis, with the citizenry, vulnerable and sensitized owing to the Twin Tower attacks, overestimating the degree of danger.

Reminding New Yorkers that the letter to NBC News dated back to September 25th, Giuliani sought to ease anxieties by making the case that, in the city's favor, over two weeks had passed with no additional infections being verified. "[I]f anyone else was going to be infected, it would have happened by now," he said.[42] His well-intentioned argument, however, contained a flaw in that it was based on the assumption of impeccable diagnostics. As Jeanne Guillemin notes, no new anthrax infections were known to have emerged recently "but it didn't mean they had been correctly diagnosed or reported."[43] The fact is, more cases were indeed occurring but they were not being recognized as such, thereby placing both young and old at grave risk.

On September 28th, for instance, the mother of a seven-month-old boy took him to her office with her. A producer of *World News Tonight*, the woman worked at ABC News in Manhattan. During the visit, her co-workers made substantial contact with the infant, hugging and nestling him, before a babysitter took him back to the family's home.

By the next morning, the boy was symptomatic. "[A] red sore the size of a half-dollar appeared on the back of the infant's left arm," writes Cole.[44] Believing it to be a spider bite, the family's physician placed him on the antihistamine Benadryl and sent him home. When the boy's condition deteriorated, he was admitted to New York University Medical Center, where he was started on antibiotics. But even as the treatment appeared to be producing a visible improvement in the lesion, his red blood cell count went into decline and soon he began exhibiting the early signs of kidney failure.[45] In the ensuing days, the staff performed blood transfusions and continued to administer antibiotics, with the fortunate outcome being that the infant did manage to survive the ordeal. Subsequent to this, his physician, after watching the news coverage of the NBC case, realized the culprit was probably the anthrax bacterium, which the Centers for Disease Control verified. On October 13th, the boy's name was added to the list of anthrax victims, making his infection the second confirmed case in the city and the fifth in the nation.

Next up was Claire Fletcher, a twenty-seven-year-old assistant to CBS news anchor Dan Rather. After a lesion appeared on her face on October 1st, she was placed on an antibiotic regimen that effectively combatted the infection. Only later, as the facts emerged about the NBC incident, would her condition be understood for what it was, cutaneous anthrax. But officials were on top of the situation by this point. Even as Fletcher was alerting the authorities to her apparent exposure, an inspection of CBS News headquarters was underway and turning up spores on various surfaces, among them Dan Rather's desk. No one at the network could recall opening an ominous, powder-laced letter, however, with the dispiriting conclusion being that Fletcher must have become contaminated in another fashion. Soon, a series of infections at the *New York Post* would bear out this concern.

It was an occurrence that began with thirty-one-year-old Joanna Huden, an entertainment writer at the newspaper, who noticed an unusual blister on her finger, assumed it was a type of bite, and obtained treatment for it.[46] However, Huden, upon learning more about the anthrax-laced letter sent to NBC News, realized the lesion may have been caused by the pathogen and notified the authorities. In short order, investigators descended on the *Post*'s headquarters, where potentially contaminated materials had already been collected and were awaiting analysis. It seems the newspaper's management, soon after the mayor revealed the NBC case, had taken measures to collect such materials.

"A *Post* memo a week earlier had asked employees to hunt through their work areas for any suspicious mail and place it in a bin for later testing," says

Graysmith.[47] A seemingly responsible course of action, it was not a wise move in actual practice. Rather than making the office environment safer, the procedure may have placed more employees in contact with the pathogen as they rummaged through their files and wastebaskets. As well, the searches may have spread the microorganism to other pieces of mail, with this further increasing the chances of infection. Tellingly, three additional cases of cutaneous anthrax would soon be diagnosed at the *Post*, one of which almost certainly occurred during this misguided retrieval process.[48]

From the thorough-going investigation carried out at the newspaper's headquarters, officials learned that a powder-filled envelope had previously been found and destroyed. More unsettling, they discovered on the premises another envelope, which, like the NBC missive, had been mailed in Princeton, New Jersey, and processed by the Hamilton Processing Center, in nearby Hamilton, New Jersey. Yet there was an important difference: this second envelope, which was damp, had not been opened; rather, the powder it contained appeared to have leaked from it. If so, it could have infected anyone who had come into contact with it, from its place of origin in New Jersey to the offices of the *New York Post*. For that matter, any parcels, correspondence, or mailbags the tainted envelope touched may have become infectious as well. As for the contents of the envelope, when investigators removed the letter they found it contained the same photocopied message as the one mailed to Tom Brokaw at NBC News.

Partly as a result of the *Post* cockup, businesses in the metropolitan area now instructed their employees not to handle suspicious letters or packages, while government investigators looked more closely at the risks faced by employees of the U.S. Postal Service. By the end of the month, well over six thousand postal workers in New York and New Jersey would either be tested for anthrax exposure or placed directly on Cipro as a precaution.[49] And although a handful would be diagnosed with the cutaneous form of the infection, they would all survive.

Unfortunately, it would be a different outcome for two others in the region, both of whom died after inhaling the spores. One was Kathy Nguyen, a Vietnamese immigrant who worked at the Manhattan Eye, Ear, and Throat Hospital, and lived alone in the Bronx. Despite exhaustive efforts, health officials were unable to determine how or where she had encountered the pathogen. The other casualty was ninety-four-year-old Ottilie Lundgren, who lived in the adjacent state of Connecticut and whose local post office tested positive for the bacterium. Not surprisingly, when this news became public, reaction was swift and sweeping. "Uncertain about the risks," says Guillemin, "people all over the country and especially in the Northeast refused to touch their mail."[50] To be sure, bioterrorism was succeeding spectacularly in spreading fear across the nation and restricting the behavior of untold numbers of citizens. It was also inspiring hoaxers far beyond the boundaries of New York City, this simple, cost-effective means of traumatizing others and creating social disruption.

Imitation and Intimidation

"Anthrax hoaxes have gripped America and the rest of the world," wrote Paul Harris in *The Guardian* at this juncture, "causing far more chaos than any of the genuinely poisoned letters."[51] Highlighting the scope of the problem, Harris provided numerous examples of how ersatz anthrax letters were wreaking havoc from Europe to South America. In Paris, for instance, thirty-four people were hospitalized for observation in mid–October as a result of four separate hoaxes in the French capital. Meanwhile, employees at a telecommunications firm in Lima, Peru, were forced to vacate after the company received a hostile letter carrying an odd substance.[52] It was in the United States, however, that the menace was most pervasive.

Erupting after the announcement of the death in Palm Beach County and accelerating after Giuliani's press conference about the first New York City case, a plethora of anthrax threats beset the nation. The targets ranged from public schools and courts of law to the Internal Revenue Service, the Social Security Administration, even the Federal Bureau of Investigation. As for non-institutional targets, they ran the gamut from the *RMS Queen Mary*—the luxury liner permanently moored in Long Beach, California—to a pair of Home Depot stores in Pennsylvania, both of which received packets of powder from a resentful former employee. In the latter episode, the stunt cost the home improvement chain over a million dollars in lost revenue, since the two stores, among other consequences, were forced to close for several days.[53] And some threats hit even closer to home, such as the FedEx worker in Louisiana who slipped a granular substance into a package and delivered it to a woman on his route, and the man who secretly opened his neighbor's mail and placed a dusty carpet deodorizer in it.[54] According to an FBI report, one man went so far as to scribble the word "anthrax" on a powder-filled envelope that contained "a birthday card for his mother."[55] For sheer number, however, the warnings sent to family planning and abortion clinics were by far the most copious.

"During the second week of October," reports *CNN*, "more than 280 threatening letters were mailed to women's reproductive health clinics on the East Coast."[56] More would follow, nearly three hundred of them. Some, but evidently not all, were the work of one man, a prison escapee whom the authorities subsequently captured and in whose possession they found a sheath of pro-life leaflets, a rifle, and a pipe bomb.[57] Obviously, urgent action was needed to control the frenzy of bio-threats.

A decisive step in this direction occurred on October 21st, when the United Kingdom, which was also experiencing a staggering number of bogus claims, updated its existing laws pertaining to terrorist threats. Until now, its legislation had addressed explosive devices only, so the revision expanded it to include biological, chemical, and nuclear threats. The penalty: a maximum of seven years in prison. "It sends the clearest possible signal that we will not tolerate these hoaxes and [the] fear and widespread disruption they cause," declared Prime Minister Tony Blair, referring to the hastily prepared legislation.[58]

The United States government, by comparison, would not be nearly so timely. Within a few weeks, the House of Representatives passed the Anti-Hoax Terrorism Act of 2001 and sent it to the Senate, but it was not enacted. It would not be until three years later that such a measure, the Stop Terrorist and Military Hoaxes Act of 2004, was signed into law, thereby making bio-threats a federal offense. It is surprising, of course, that the federal government straggled in so important a matter, not least because two leading members of Congress were among those targeted by the anthrax operation. But unlike the bacteria-laced letters sent to NBC News and the *New York Post*, those dispatched to the congressmen carried an even more dangerous form of the pathogenic agent, one heretofore unseen by scientists.

A Capitol Offense

The person or persons behind the Amerithrax attacks placed Congress in their sights in early October 2001, mailing anthrax-spiked letters to Senate Majority Leader Tom Daschle of South Dakota and Senator Patrick Leahy of Vermont, Chairman of the Senate Judiciary Committee. Arguably the most influential Democrats in the nation at the time, the two lawmakers were among those working on the USA Patriot Act, an acronym for *Uniting and Strengthening America by Providing Appropriate Tools Required to Intercept and Obstruct Terrorism.* Identical in most respects to the New York City attacks, those targeting Washington, D.C., were set in motion in Princeton, New Jersey.

It was in this city that the Daschle and Leahy letters were mailed, then transferred to Hamilton, New Jersey, where they were sorted by automated, high-speed equipment at the Hamilton Processing Center. It is now believed that the machinery, which postmarks and routes incoming mail, may have crushed the powdery substance in the Leahy letter and caused it to seep through the envelope, become attached to the barcode scanner, and spread to other pieces of mail. Among the consequences, over four hundred postal facilities in New Jersey became cross-contaminated with the pathogen. Not only that, the microbes traveled to other states as well. In the picturesque

New England town of Wallingford, Connecticut, for instance, three million anthrax spores were detected in the local post office; such was the expanse of the dispersal pattern.[59] It is also instructive to note that this development was almost certainly unanticipated and illustrates a major disadvantage of bioterrorism discussed earlier in this book, namely the difficulty its perpetrators face in controlling the microorganisms they unleash.

From New Jersey, the two anthrax-tainted letters traveled to the Brentwood Postal Facility in the nation's capital. Employing thousands of workers and sorting several million letters and parcels each day, the Brentwood center handles the entirety of U.S. mail bound for the twelve government buildings that comprise the Capitol Complex on Capitol Hill. In addition to numerous government agencies, this cluster of buildings also holds the offices of the nation's senators and state representatives.

As it turned out, one of these edifices, the Russell Senate Office Building which housed Patrick Leahy's offices, would be spared from exposure because the letter addressed to the lawmaker was misrouted. And while the Daschle letter would be misrouted as well—the hand-printed envelope is thought to have been the reason—it was still delivered, albeit behind schedule. In fact, it would not be until a week later, on Monday, October 15th, that it arrived at Daschle's suite in the Hart Senate Office Building.

In one respect, the delay was fortunate, since Washington, D.C., was now aware that it was at risk of a biological attack. Only three days earlier, Rudy Giuliani had publicly revealed the existence of the first anthrax-tainted letter in New York City, the upshot being that the capital had been placed on alert, its intervention strategy primed. Tom Daschle was fortunate, too, in that he was attending a meeting in the Capitol Building on this inauspicious morning, a distance away from his offices in the Hart building. Forty members of his staff would not be so lucky, however, being in his suite and on the job when the poisonous missive arrived.

It was a young female intern who examined the envelope, which was sealed with tape and carried the return address of an elementary school in a small town in New Jersey. In itself, this was unremarkable, since lawmakers often receive messages from students as a part of classroom projects. Only later would investigators discover that this particular school did not exist. Believing the letter to be benign, then, the intern proceeded to open it.

"She felt a small bulge at one end, but slit open the flap anyway," writes Graysmith. "A puff of choking gray powder disgorged itself onto the desktop."[60] Like the messages to NBC and the *New York Post*, the letter was a photocopy, as well as being moist and hand-printed on a pharmaceutically-folded sheet of paper. "YOU CAN NOT STOP US," it said.[61] Once again, the English was less than standard. And then came the thrust:

WE HAVE THIS ANTHRAX.
YOU DIE NOW.
ARE YOU AFRAID?

DEATH TO AMERICA.
DEATH TO ISRAEL.
ALLAH IS GREAT.[62]

The intern pitched the letter into the wastebasket and phoned the police, who, in turn, alerted health authorities. Within minutes, officials activated the capital's biohazard protocol, with John Eisold, the chief physician at the Capitol Complex's clinic, dispatching a five-member team to the Daschle suite.[63]

"I grabbed a big bottle of Cipro," says Norman Lee, the doctor who headed up the team. "We also brought doxycycline and enough culturettes for the swabs."[64] While law enforcement, using portable immunoassay devices, screened Daschle's offices and the adjoining suite of Senator Russ Feingold for signs of the bacterium, the medical team set about testing the two lawmakers' staffs. Since the results would not be available for at least forty-eight hours, the team, as a preventive measure, offered Cipro to the seventy workers who were present in the Daschle and Feingold suites and instructed them to report to the Capitol clinic the next day for follow-up care. As it turned out, far more people would require medical attention because the first responders, in the pandemonium of the moment, did not turn off the air conditioning, the result being that it continued circulating air, possibly contaminated with anthrax, throughout the Hart building for another thirty minutes.

Ultimately, the number of people checked for anthrax poisoning would exceed six thousand, twenty-eight of whom would test positive for exposure to the pathogen, meaning its presence could be detected on nasal swabs. None of the exposures would progress to infection, however, due presumably to the prompt cleansing of the victims' nasal cavities and rapid administration of Cipro. Certainly the timely intervention by the Capitol's medical team, together with law enforcement, appears to have effectively contained the incident. Nevertheless, it was imperative that the Hart building be sealed off and decontaminated, a three-month process during which it was necessary to relocate thousands of government workers. Then, too, the damp, leaky anthrax letters—the one mailed to Leahy was subsequently located elsewhere in the city—caused significant cross-contamination to other pieces of mail. Investigators detected traces of anthrax at the White House, Pentagon, Supreme Court, Central Intelligence Agency, State Department, Justice Department, and the Bureau of Alcohol, Tobacco, and Firearms.

Most endangered, though, were the postal facilities in the metropolitan area. At the Brentwood center, for instance, an employee attempted to clean an apparatus on which anthrax spores had accumulated. Not realizing he was dealing with the highly virulent pathogen, the man used a compressed-air device to dislodge the microbes, which caused them to rain down on him. And symptoms soon erupted. "His face and body became swollen, the membrane around his lungs become inflamed, and he grew unsteady on his feet," writes Graysmith. "Within days he could only speak haltingly and had lost his short-term memory."[65] Not surprisingly, more postal workers would become ill before the situation stabilized. And the situation would indeed stabilize, and fairly soon, with a fragile sense of normalcy returning to the battered nation.

It would be on October 26, 2001, that President George W. Bush signed into law the Patriot Act, a radical piece of legislation ostensibly designed to help prevent terrorist attacks on American soil in the future. By this point, the preponderance of damage caused by the Amerithrax offensive had been done. Shell-shocked, the nation had endured the first major bioterror assault in its history and was preparing to grapple with the aftermath, which would include exorbitant costs. Case in point: the decontamination of the Capitol Complex. It would cost the American taxpayer nearly thirty million dollars to make safe the buildings on its campus.[66] Decontamination expenses in Florida and New York City would likewise be considerable. And there was a substantial human cost, with five people perishing in three regions of the country, all due to inhalation anthrax. Seventeen others suffered from anthrax poisoning but survived, the majority of the cases being cutaneous in nature. And still another twelve thousand people were thought to have been exposed to the pathogen and offered antibiotic therapy as a precaution. All told, the Amerithrax attacks proved to be the most widespread, expensive, and successful biological assault in U.S. history. Yet the question remained: who committed this unprecedented crime?

Pursuit of the Bioterrorists

Among the earliest answers to be proposed were those pointing to foreign sources, namely, the Islamic terrorist elements accused of planning and executing the 9/11 attacks in New York City and Washington, D.C. White House officials wasted no time fingering al-Qaeda as the suspect, with President Bush publicly linking the attacks to the organization during an October 23rd photo-op when asked about the perpetrators' identities. "[T]here's no question that anybody who would mail anthrax with the attempt to harm American citizens is a terrorist," Bush said. "And there's no question that al Qaeda is a terrorist organization."[67] Acknowledging he had no evidence to support this hypothesized link between al-Qaeda and the biological offensive, the president's words nevertheless made it clear the White House was pointing toward the extremist organization as the guilty party.

But although the Bush-Cheney administration would continue to promote this narrative for the next two years, the FBI, within weeks, would shift its focus to the possibility of a domestic, lone-wolf terrorist, the type who functions independently of organizations and states. In fact, the prospect of such a solitary figure would eventually become the FBI's only hypothesis, and the agency would pursue it relentlessly for years.

But perhaps the most provocative thesis to emerge was the one detailed in the recent book, *The 2001 Anthrax Deception*, written by Harvard-educated scholar Graeme MacQueen, Founder and Director of the Centre for Peace Studies at Canada's McMaster University.[68] A theory that has existed for many years, one that has been constructed and advanced in large measure by journalists, it claims a surprising number of adherents, among them a handful of prominent political figures. It is both intriguing and audacious, even rather jarring, since it suggests the complicity of the Bush-Cheney White House in the Amerithrax attacks.

False Flag Theory

As its foundational proposition, the thesis holds that elements connected to the executive branch of the U.S. government—specifically, neoconservative collaborators linked to high-level figures in the White House—played a central role in the attack. The objective: to justify the U.S.-led invasion of Iraq.

In terms of particulars, the theory asserts that the architects of the Amerithrax attacks sought to rally support for the incursion into the Middle Eastern nation by creating the illusion that al-Qaeda, based in Afghanistan, was targeting the United States with letters contaminated with pathogens provided by Saddam Hussein's biowarfare program in Iraq. It was purportedly for this reason that the letters were designed to appear as if they were the product of a radical Islamic organization like al-Qaeda.

In reality, the "Muslim" letters, with their conspicuous misspellings and Western-style misstatements—"Allah is Great" rather than *Allahu Akbar* or its English equivalent, "God is Greatest," for example—proved to be rather unconvincing, at least to investigators. "It was as if someone had tried to frame Native Americans for the crime by inserting a note in the letters announcing, 'White man in heap big trouble,'" says MacQueen.[69] But no matter; the documents were allegedly crafted to convince the American public that Islamic terrorists were behind the attacks, a public that had just been traumatized by the horrors of 9/11, was presumably gullible, and could easily be misled by talk of Allah, America, and Death.

The false flag theory further asserts that the United Nations Security

Council was also meant to be duped, if not by the letters' text then by the anthrax contained within them. On February 5, 2003, at a time when the White House was pressing other nations to join its impending war in the Middle East, Secretary of State Colin Powell appeared before the UN and delivered a speech containing false allegations about Iraqi weapons of mass destruction, both biological and nuclear. "The attempt to link Saddam [Hussein] to the anthrax attacks was just as fraudulent—and just as significant—as the attempt to link Saddam to 9/11, Al Qaeda and nuclear weapons," writes journalist Glenn Greenwald.[70] In subsequent years, Powell would concede that his presentation to the august body, which proved to be the nadir of his career, was not based on accurate information and had tarnished his reputation.

In regard to the speech itself, the Secretary of State began his discussion of bioweapons, and anthrax in particular, by underscoring the lethality of the bacterium and its potential to decimate the American population.[71] "Less than a teaspoon of dry anthrax in an envelope shut down the United States Senate in the fall of 2001," declared Powell, brandishing a small vile of faux anthrax in front of the television cameras. "This forced several hundred people to undergo emergency medical treatment and killed two postal workers just from an amount ... about this quantity that was inside of an envelope."[72] He then proceeded to describe the profound threat that Iraq, with its alleged stockpiles of the pathogen, continued to pose to the United States. "Saddam Hussein [may] have produced 25,000 liters ... enough to fill tens upon tens upon tens of thousands of teaspoons."[73] Although the Secretary of State did not declare, categorically, that Iraq had furnished the microbes contained in the Amerithrax letters, his words were exquisitely hewn to insinuate it had done so. "[H]e made sure Iraq, anthrax and the Senate were all mentioned together," says MacQueen.[74] And Powell's verbal gymnastics proved to be persuasive, above all to the public viewing the televised speech. "[H]e had single-handedly convinced many skeptical Americans that the threat posed by Iraqi dictator Saddam Hussein was real," reports the *Washington Post*.[75]

And there is more. While the false flag theory holds that the paramount objective of the bacteria-laced letters was to help build a case for the invasion of Iraq, MacQueen proposes a second, more immediate purpose for the lethal letters as well. It concerned the perpetrators' reason for selecting as their victims the nation's two leading Democratic senators, Tom Daschle and Patrick Leahy. MacQueen contends that it was an exceptionally vicious episode of political strong-arming.

In late September and early October of 2001, Daschle and Leahy, by virtue of their roles in Congress, were deeply involved in the creation of the controversial Patriot Act, both of whom had strong reservations about its constitutionality and its reach. Of even greater consequence for the bill's fate, the two lawmakers were in a position to slow or stall the legislation. As Senate Majority Leader, Daschle could, and did, delay the bill on the grounds that it would hand too much power to President Bush. The congressman insisted that it be amended to reduce the president's authority to wage future wars. Likewise, Senator Leahy, as Chairman of the Senate Judiciary Committee, slowed passage of the Patriot Act, since his committee's job was to assess the impact it might have on civil liberties. He was worried the bill would lead to the erosion of citizens' rights.

So it was that the Bush/Cheney White House, frustrated by such resistance, demanded that Congress put aside its misgivings and approve the Patriot Act by October 5, 2001. Yet the demand from the Oval Office was not heeded, with the deadline passing without the bill being approved. In large part, the delay was the result of objections registered by Daschle and Leahy. Accordingly, between October 6th and October 9th, anthrax-laced letters were mailed to Daschle and Leahy, with the potency of the pathogen, as noted earlier, being far greater than that contained in the letters dispatched to media figures.[76] "[G]iven Democratic control of the Senate and the importance of quickly getting these two senators on board, it is obvious why Daschle and Leahy would be key targets of intimidation for anyone wanting the bill passed," writes MacQueen.[77] In the end, this act of coercion, as the false flag theory depicts it, was effective, since shortly after the bacteria-laden Daschle document arrived on Capitol Hill, Congress, in a state of panic, overwhelmingly approved the Patriot Act in spite of numerous concerns on both sides of the aisle.

Along the same lines, the theory hold that the perpetrators' choice of media figures was significant as well. Case in point: Judith Miller, the *New York Times* reporter who received a letter containing simulated anthrax. Although it did not harbor the actual bacterium, the document itself appears to have been prepared by the same perpetrators who sent the poisoned letters to others. Subsequent to this, Miller joined the ranks of those who zealously supported the Bush administration's campaign to invade Iraq. So intense was her advocacy for the incursion that numerous commentators dismissed her as a shill for the White House, and, on more than one occasion, questioned the precision of her reporting. "[W]ildly inaccurate," is how the *Columbia Journalism Review* described some of her articles about Iraq's alleged weapons of mass destruction.[78] They were charges Miller would address in *The Story: A Reporter's Journey*, her 2015 memoir.[79]

Lastly, the false flag theory offers an explanation as to why the first anthrax victim was Robert Stevens, an employee of American Media, Inc., in Lantana, Florida. As it happens, Lantana was also the temporary home of four of the Middle Eastern men who would subsequently be alleged to have taken part in the ghastly attacks on the World Trade Center and the Pentagon.

Not only did they live in Lantana prior to 9/11, the four men also shared, and were on friendly terms with, the same real estate agent as Robert Stevens, Gloria Irish. And Gloria's husband, Mike Irish, was an editor-in-chief at American Media, Inc., where Stevens was an employee. The false flag theory, then, holds that the true perpetrators of the Amerithrax attacks selected Stevens and American Media to be the first victims in order to associate the Florida bio-attack with the four alleged 9/11 hijackers, and to do it by means of proximity and mutual friendships. As it stands, all other theories, including those put forth by the FBI, dismiss as a coincidence the fact that the initial anthrax victims and the four of the alleged hijackers all happened to reside in the same small, Southern town—its population was only 9,400 at the time—and, moreover, that they also shared the same real estate agent.

As for the Bush administration's story, the false flag theory, as previously outlined, argues that the White House narrative about the anthrax attacks was erroneous from the start; that there was no evidence, direct or circumstantial, of foreign involvement in the Amerithrax offensive. Instead, the attacks were, part and parcel, a domestic undertaking and seemingly with government complicity, if not government provenance. As it happens, this certainty of a home-grown entity masterminding the biological assault marks the principal point of agreement between adherents of the false flag theory and investigators for the FBI, the latter also being were convinced that the attacks were executed by a domestic entity. The difference is that the Bureau believed there was only one perpetrator and no government connivance. In the years that followed the attacks, three suspects would come into the agency's crosshairs.

Case 1: Ayaad Assaad

It was approximately three weeks after the 9/11 attacks on the World Trade Center and the Pentagon that the Federal Bureau of Investigation began exploring its single-actor theory. The event that sparked the inquiry began when an anonymous tipster sent a typewritten letter to military police at the Marine Corps Base—Quantico, thirty-five miles south of Washington, D.C.

The letter was postmarked September 26, 2001, meaning it was mailed one day after the two anthrax-tainted letters were sent to the *New York Post* and Tom Brokaw at NBC News. At this point, no anthrax infections, most notably that of Robert Stevens in Lantana, Florida, had been made public.

Received on October 2nd at the Quantico base, the letter warned that a scientist who hailed from the Middle East, Ayaad Assaad, was a religious fanatic and a potential terrorist, that he had the means to carry out a biological strike on American soil, and that he had expressed the will to do so. The document thereafter furnished details about Assaad's family life, and, even more startling, the particulars of the classified biodefense research he had conducted in the past. Accordingly, the source of the message appeared to be a person, or persons, who had worked in close proximity to the fifty-three-year-old scientist and knew him well.

In reality, Assaad was neither an Islamic zealot nor a terrorist. He was an Egyptian American microbiologist who had worked for ten years in USAMRIID's biodefense program, situated at Ford Detrick, until budget cuts in 1997 forced the institute to terminate a third of its staff. Subsequent to this, he secured a senior-level position in toxicology at the Environmental Protection Agency.

What *is* significant is that Assaad had been among a handful of scientists who had been scorned and harassed by a clique of coworkers while at Fort Detrick, a group that christened itself the "Camel Club."[80] Congregating during the Gulf War years, the group's participants were openly hostile to their Middle Eastern colleagues at the facility. Plausibly, one or more members of this blinkered crew may have sent the poison-pen letter in an attempt to frame Assaad for the actual anthrax-laden letters that were now in the mail, even though the accused was no longer associated with Fort Detrick.

So it was that FBI investigators decided to pursue the allegations, and to this end interrogated the scientist on October 3rd for several hours. To Assaad's relief, they determined he was not a religious zealot or political extremist and absolved him of the accusations. Even so, he remained incensed by the malicious letter, and he did not hesitate to make his feelings known to the press. "My theory is whoever this person is knew in advance what was going to happen [and set me up as the] scapegoat."[81] And the FBI concurred. The fact that someone sent a warning to authorities about a bio-attack merely twenty-four hours after such an attack had been set in motion hardly seemed coincidental, and it prompted the Bureau to turn its attention to those inside the biodefense establishment. But this was not the agency's only reason; two additional issues caused it to look into this community as well. The first involved the strain of the anthrax bacterium and the second, the technology and expertise required to refine it.

Within forty-eight hours of receiving and refuting the accusations against Assaad, the FBI received the results of the initial anthrax analysis from Paul Keim's laboratory in Arizona. The Bureau was notified, on October 5th, that Floridian Robert Stevens had been exposed to the Ames strain, with the same variant being confirmed a few hours later in another South Florida case, that of mail courier Ernesto Blanco. Then, on October 14th, Keim informed federal officials that the letter dispatched to Tom Brokaw in Manhattan contained the Ames variant as well. And forty-eight hours after that, he relayed the same findings to the authorities, except in this case they concerned the variant that had been sent to Senator Daschle at the Capitol

Complex. "Keim's strain identification," writes Guillemin, "forged an undeniable link to the U.S. defense establishment," since this particular variant, as we have noted, was maintained in high-security laboratories under the auspices of the Department of Defense.[82] Keim's findings also linked the government's Ames strain to the attacks in Florida, New York, and Washington, D.C., meaning they were not unrelated incidents or copycat crimes but a coordinated sequence of assaults.

Further bolstering the FBI's suspicions about a biodefense insider or band of accomplices, the Ames strain contained in the Daschle letter had been milled to an extraordinary degree, a degree previously assumed to be unattainable. So advanced was the technological process that refined the microbe, so minuscule were the spores, that they were nearly vaporous when dispersed. By modifying the pathogen to render it so infinitesimally small, its engineers ensured not only that it would float in the air to the point of near-suspension, but also that it would be exceedingly lethal when inhaled due to its ability to lodge deeply in the lungs. At the time, the consensus was that civilian research centers in the United States lacked the technical sophistication to craft an agent of this type, as did most of the military's biodefense labs. As to the very few that might possess the requisite equipment coupled with the know-how, investigators named as possibilities the Battelle Memorial Institute in Ohio and the Dugway Proving Ground in the Utah desert, where biological and chemical weapons are tested ostensibly for defensive purposes.

To be sure, the nation would be in grave danger if the source of the Amerithrax attacks proved to be embedded somewhere among the scores of scientists and technicians holding security clearances. But this is what Paul Keim's findings seemed to suggest. They also appeared to indicate that a profound security breach had not only taken place, but that the brilliant, if devious, creators of the enhanced pathogen possessed "the scientific capacity to launch more attacks," in the words of Jeanne Guillemin.[83] Accordingly, the FBI, gripped by the prospect of further germ warfare assaults, redoubled its efforts to track down the killers—or, as the Bureau would insist until it closed the case, "the killer." And it would be ten months after the Amerithrax attacks that the agency would visibly zero in on a "person of interest."

Case 2: Steven Hatfill

It was the summer of 2002 and the White House was still fostering the notion that al-Qaeda had committed the anthrax attacks using Iraqi-supplied pathogens. In a sharp departure from the Bush-Cheney position, however, the FBI set off in pursuit of an American scientist, Steven Hatfill, an accomplished virologist having a background in biodefense research. An intelligent, creative, and ambitious man, Hatfill was remarkably successful even if he was not always in favor with his colleagues. Partly, this was because he was not one to hold his tongue, although he was certainly no terrorist according to those who knew him well.[84]

Born in 1953 in St. Louis, Missouri, the FBI's "person of interest" grew up in the central Illinois town of Mattoon. In 1978, after graduating from college and completing a stint in the army, he moved to Africa—specifically, Rhodesia (later renamed Zimbabwe) and South Africa—where he may or may not have been associated with those nation's militaries. Suffice it to say, more than a little mystery remains attached to this period of his life, with there existing wildly divergent accounts of his deeds. As to the reports of his foreign military involvement, author and editor Leonard Cole writes that Hatfill professed to having participated in covert activities while in Africa. "An intelligence analyst who knows Hatfill," says Cole, "confirmed to me that Hatfill had been involved with special operations there of an unspecified nature."[85]

What is known with certainty is that Hatfill was awarded a medical degree in 1984 from the University of Zimbabwe (Godfrey Huggins School of Medicine), after which he completed a residency in hematological pathology, acquired two masters degrees, and pursued doctoral studies in molecular biology at Rhodes University in South Africa. Along the way, he also led a scientific expedition to Antarctica.

In 1994, the United Kingdom became his home for a stretch, with Hatfill relocating to England for a year to carry out biomedical research at an Oxford-affiliated site. Next up was a fellowship at the National Institutes of Health in Bethesda, Maryland, followed by a National Research Council fellowship at Fort Detrick (USAMRIID). While at the latter post, the Department of Defense granted Hatfill a basic security clearance, one that would allow him access to the USAMRIID biosafety labs which housed the Ebola and Marburg viruses. This was necessary because his research centered on viral biowarfare agents, his objective being to help protect the population from such pathogens.

After completing his fellowship at USAMRIID, Hatfill continued his research, this time in the private sector, by accepting a position at the Science Applications International Corporation (SAIC) in Maclean, Virginia. A company that performs biodefense work for the U.S. government, he apparently had no complications at the firm until August 2001, when he was required to undergo a routine, CIA-conducted polygraph test. Because the results turned out to be inconclusive, the Department of Defense, while it reviewed the matter, suspended his existing security clearance as well as withholding a top-secret authorization that was necessary for certain SAIC projects. It was a turn of events that was likely the reason SAIC felt it necessary to terminate him in March 2002, even though it retained him as a consultant; this, in conjunction with a reporter's phone call to the company that same month informing it that Hatfill was among those scientists considered a "person of interest" in the Amerithrax attacks. So it was that he left SAIC at this juncture, although he landed another position without difficulty, his suspended security clearance notwithstanding.

Louisiana State University, in short order, tapped the virologist to serve as Associate Director of its National Center of Biomedical Research and Training. "Hatfill had been highly recommended to Louisiana State University by two very highly placed friends" from the National Institutes of Health, writes Graysmith, an auspicious development both for LSU and for Hatfill himself in view of his circumstances.[86] Unfortunately, the job would not last long due to the FBI's ever more invasive presence in his life.

As the Bureau had done with other scientists known to have had access to biodefense labs prior to the Amerithrax attacks, it approached Hatfill in January 2002, and questioned him about his activities. It also administered a polygraph test, which he passed.[87] Yet he was not off the hook. Six months later, the FBI contacted him again with further questions, with the virologist meeting with investigators once more and offering them the opportunity to inspect his home, car, and off-site storage unit without a search warrant, which they did. And like the initial interview and polygraph test, neither this second interview nor the ensuing searches unearthed any incriminating material. Then, without warning in August, the FBI turned up yet again on his doorstep, this time with a court order in hand and surrounded by television crews. As news helicopters circled overhead, investigators swarmed the premises and made off with his computer and other items. It was a spectacle that seemed staged for maximum publicity, even though the Bureau denied it had tipped off the media in this high-profile case that had yet to yield a meaningful lead in eight months.

"[T]he FBI had been under pressure to solve the case, both from Congress and the general public," writes Marilyn Thompson, the former Assistant Managing Editor of Investigations at the *Washington Post*. "Already under fire for its failure to detect and thwart the plans of the 9/11 hijackers, the bureau badly wanted to prove its competence by making an anthrax arrest."[88]

As it stood, the agency had kept Hatfill under constant surveillance throughout the summer, which had not been the case for the twenty-nine other "persons of interest" it was said to be monitoring. And Hatfill knew the FBI was singling him out. Between the heightened surveillance and the highly publicized raid on his home, he suspected that he was being set up as the fall guy in the Amerithrax case.

Among the reasons the FBI claimed the virologist had come to its attention: he had spent ten days in England in November 2001, a trip that coincided with a hoaxed anthrax threat being mailed to the United States from London.

In reality, Hatfill, shortly after the attacks, had traveled to the town of Swindon, England, to undergo training by Hans Blix, chief of the United Nations' WMD inspection program. The UN had selected Hatfill to join its cadre of bioweapons inspectors, the plan being to dispatch him to Iraq to search for any biowarfare agents that Hussein's military might still possess. Fortunately, a fellow scientist who escorted Hatfill vouched for the fact that the virologist had not traveled into the city of London during the trip and therefore could not have mailed the threat.

Investigators also pointed to an entry on Hatfill's résumé, one from the late 1990s when he was serving as a biodefense consultant to the Metropolitan Medical Strike Force of Washington, D.C. Among his areas of knowledge, he listed the following:

> [F]ormer U.S. and foreign BW programs, wet and dry BW agents, large-scale production of bacterial, rickettsial and viral BW pathogens and toxins, stabilizers and other additives, with former BG [*Bacillus globigii*] simulant production methods, open air testing and vulnerability trials, single and 2 fluid nozzle dissemination, bomblet design, munitions programs, and former Soviet BW programs.[89]

Because Hatfill's description demonstrated an interest and knowledge of biowarfare, and because he had developed training programs designed to prepare emergency responders for bio-attacks, investigators found him suspicious. It was a contorted logic that Hatfill would later liken to Kafka's absurdist novel, *The Trial*.[90] Of particular significance to the FBI, Hatfill had long argued, correctly as it happens, that the United States was ill-equipped to handle a large-scale bioterror assault, with investigators hypothesizing that he may have launched the Amerithrax attacks in order to trigger widespread panic and thereby stimulate the federal government into increasing its funding of the biodefense industry. It was a stretch, without a doubt.

A third source of suspicion centered on another entry on Hatfill's résumé, an entry claiming he had obtained a Ph.D. in molecular biology from Rhodes University. In actuality, he had finished his dissertation and progressed to doctoral candidacy status, but, for whatever reason, did not receive the degree itself. And this was taken as support for the position that he might be a bioterrorist, suggesting, as it did, an impairment in his integrity.

Then there was the draft of a novel investigators retrieved from the computer confiscated from his home. An aspiring writer, Hatfill had penned a thriller in which the bubonic plague is set loose upon a population. It was a tale that had been born years earlier at a dinner party during which the guests—military officials, journalists, and Hatfill himself—spontaneously invented a narrative about a bioterror assault within the United States. Reflecting on the story sometime later, Hatfill decided to weave it into a novel. Upon finishing the text, he registered it with the U.S. Copyright Office and shared it with a friend, a former CNN correspondent, whose help Hatfill solicited in securing a literary agent. While the manuscript, then, was not a secret, and although anthrax did not make an appearance in the story and the U.S. Postal Service was not the route of the fictitious plague's spread, the FBI nevertheless considered the novel a red flag. Shortly after the Bureau downloaded the draft from Hatfill's computer, it appeared in the news media and quickly caused the virologist to appear as something of a madman.

Lastly, *Newsweek* magazine reported that an FBI bloodhound, one that had been exposed to the scent of the Amerithrax letters, identified the same scent at a Denny's restaurant in Louisiana, one the virologist supposedly patronized. It was a curious assertion, since the pristine letters—forensics specialists had been unable to detect fingerprints or DNA evidence on them—had subsequently been decontaminated by irradiation, causing independent experts to express serious doubts about the validity of the FBI's claim.[91] And equally disconcerting was a follow-up to the story itself. "[T]he *Baltimore Sun*," writes journalist David Tell, "phoned the managers of all 12 Denny's restaurants in the state of Louisiana, each of whom insists that no such bloodhound search as is recounted by *Newsweek* has ever been performed on his premises."[92] If the *Baltimore Sun*'s findings are accurate and no tracking dogs were actually used, it raises questions as to who fed the specious information to *Newsweek* and for what purpose.

In the end, the Bureau did not have a strong case against Hatfill. In fact, it had no case at all, with the agency itself conceding a lack of evidence linking the virologist to the Amerithrax attacks. All the same, its agents continued tailing their quarry around the clock, and, by Hatfill's own account, they did so quite conspicuously. But he would have none of it. On August 25, 2002, he called a press conference, one in which he blasted the FBI and its tactics. Complaining of repeated leaks of information calculated to impugn his reputation and railing against the agency's attempts, as he perceived the situation, to stress him to the breaking point, he called into question the motives behind the events.

"This assassination of my character appears to be part of a government-run effort to show the American people that it is proceeding vigorously and successfully with the anthrax investigation," Hatfill declared in the nationally televised appearance on the steps of his attorney's office. "The FBI can be seen to be on the job, the press is hot on the trail and the public is satisfied … that progress in the anthrax letter attacks is being made."[93] He added, "I want to look my fellow Americans directly in the eye and declare to them, 'I am not the anthrax killer.'"[94] At the conclusion of his comments, Hatfill, clearly distraught, tearfully embraced a friend.

Besides proclaiming his innocence before the same media that had been castigating him—a piece in the *New York Times* excoriated the FBI for not pursuing even more aggressively a certain "person of interest" who fit the virologist's description—Hatfill asked that he be permitted to furnish handwriting samples to the Bureau for comparison with the writing in the Amerithrax letters.[95] During the press conference, he wondered openly why he, who was under suspicion, had to suggest this standard procedure to the FBI rather than the Bureau initiating it of its own accord. Additionally, Hatfill volunteered publicly to take a blood test. This is because the press was reporting that he was known to have taken Cipro in the past, which was true, but the news accounts neglected to mention that he had been prescribed the antibiotic by his physician when he was about to undergo sinus surgery. A blood test, Hatfill knew, would confirm that he had not been vaccinated against anthrax in the previous three years and had no antibodies to the pathogen in his system.

Hatfill also produced and distributed to the news media his SAIC timesheets which established that he had been on the job in Maclean, Virginia, on those days and times when the anthrax letters were mailed in New Jersey. Further verifying his presence on those days was the entirety of his former SAIC crew of coworkers. As Hatfill made clear in the press conference, he could not have undertaken the eight-hour roundtrip drive to New Jersey to post the letters while he was at work in another state.

And finally, Hatfill, throughout the investigation itself, made the case over and again that he was a virologist with a specialty in the Ebola virus. He was not a bacteriologist with an expertise in anthrax. Still, the FBI continued to pursue him as the likely terrorist, even to the point of draining a pond near his home the following year at a cost to taxpayers of a quarter of a million dollars.[96] Meanwhile, LSU, which remained highly supportive of him, was forced to suspend him because his government funds were cut off, with the virologist, by 2003, being nearly insolvent.[97] And so he made his last stand.

Determined to clear his name while penalizing those he believed had wronged him, Hatfill, commencing in August of this same year, filed a series of lawsuits against the Federal Bureau of Investigation, the Department of Justice, Attorney General John Ashcroft, the *New York Times*, and *Vanity Fair* magazine, among other parties. His complaints ranged from libel and defamation of character to violations of privacy stemming from the FBI's illegal leaks to media outlets. He then spent the next five years in court, until, at long last, he was vindicated.

Ultimately, the court dismissed his suit against the *New York Times* on the grounds that he was now a public figure and could not prove malicious intent on the part of the newspaper. By comparison, his complaint against *Vanity Fair* was resolved in his favor, with Hatfill settling out of court for an undisclosed sum. Even more significant, the Department of Justice, having within its jurisdiction his principal nemesis, the FBI, was ordered to furnish him with an annuity worth 5.8 million dollars for its having leaked information about the investigation. It was a victorious moment for the exhausted Hatfill. After the judgment was handed down, his attorneys released a statement underscoring the importance of the nation's courts in ensuring that citizens' rights are respected by the government and the media, while also expressing their optimism that the legal and financial ramifications of the case would discourage similar conduct by federal investigators and journalists in the future. "We can only hope that the individuals and institutions involved are sufficiently chastened by this episode to deter similar destruction of private citizens in the future—and that we will all read anonymously sourced news reports with a great deal more skepticism," they said.[98]

Two months later, on August 8, 2008, the FBI issued a statement exonerating Steven Hatfill. It declared there was nothing to tie him to the Amerithrax attacks, nothing to connect him to the flask of spores that had taken the lives of five people. It was the admission Hatfill's attorneys had insisted upon, and it irritated Iowa Senator Charles Grassley who was serving on the Judiciary Committee. "We've had a seven-year investigation and $15 million spent on it and one of the 'people of interest' bought off for $5.8 million over what was obviously an F.B.I. screw-up," he charged.[99]

The senator would be similarly dissatisfied with the next "lone wolf" the Bureau would point to as the terrorist. It was a suspect who would emerge from a renewed search that started in 2006 as the Hatfill case was winding its way through the courts in what was shaping up to be a defeat for the Bureau. At a crossroads, FBI director Robert Mueller replaced the head of the Amerithrax investigation, Richard Lambert, with Vincent Lisi and Edward Montooth. In turn, this pair ordered investigators to re-examine the case in order to discover any clues or suspicious individuals that may have been overlooked. Shortly thereafter, the Bureau had a new prime suspect.

Case 3: Bruce Ivins

As is often the case, the FBI's strategy involved the practice of psychological profiling, a process whereby an expert in human behavior formulates a theoretical portrait of the perpetrator's personality traits. In some cases, the profile turns out to be fairly accurate, while in other instances it proves to be wide of the mark. All the same, the Bureau places considerable stock in the procedure, and its presumed value in the Amerithrax case was no exception.

So it was that an FBI behavioral consultant put together a psychological portrait of the Amerithrax killer, which the Bureau then shared with numerous collaborators, among them Abigail Salyers, President of the American

Society of Microbiology. Investigators hoped Salyers, being well-integrated into the scientific community, might be of help in compiling a list of potential suspects from among her professional acquaintances. To this end, they explained to her that the agency had concluded that the Amerithrax culprit was likely "a loner and a loser, and probably a nerd."[100] Even though this was a narrow characterization, Salyers nevertheless found it too broad to apply to those in her organization. "That describes at least half of our members!" she exclaimed.[101] But one person it did not describe, oddly enough, was the man whom the Bureau would eventually tag as the probable killer.

This new target was neither a recluse nor a flop, but rather an extroverted, witty, and generous man who was considered a "team player" by his colleagues.[102] A highly successful scientist who had toiled at Fort Detrick for twenty-seven years, he also was the 2003 recipient of the Pentagon's highest non-military honor, the "Decoration for Exceptional Civilian Service," which paid tribute to his work in anthrax vaccinology. That said, he was not without his share of emotional difficulties, being prescribed antidepressants and anxiolytics since 1999 and undergoing various forms of therapy. Bruce Ivins was his name, and he was a forty-seven-year-old senior microbiologist at USAMRIID when the Bureau latched onto him as a suspect.

Born in Lebanon, Ohio, in 1946, Ivins earned undergraduate and graduate degrees in microbiology from the University of Cincinnati, following which he was awarded a post-doctoral fellowship in the Department of Bacteriology and Immunology at the University of North Carolina—Chapel Hill. In 1980, he accepted a position at USAMRIID and thereafter devoted himself to helping develop an anthrax vaccine that would protect against multiple strains of the bacterium, as well as those that had been altered by enemy biowarfare scientists. The existing serum, it seems, was ineffective against artificially-manipulated anthrax in which various strains had been combined in order to render it more deadly. And by all accounts, Ivins' efforts, which advanced the field of vaccinology in important respects, were hailed by his peers far and wide. "In the eyes of his colleagues he was one of the world's anthrax experts," writes Guillemin.[103] An FBI report likewise acknowledged Ivins' proficiency, particularly in research making use of the Ames strain of the bacterium. "[C]onsidered an expert in the growth, sporulation, and purification of *Bacillus anthracis*," reads a Bureau summary, "[h]e has personally conducted and supervised Ames anthrax spore productions for over two decades."[104] It was because of this background that he would be among the handful of scientists selected to analyze the pathogens contained in the Amerithrax letters.

Whereas Steven Hatfill, as previously noted, came under the FBI's watchful eye fairly early in the Amerithrax investigation—the accusations against him commenced in mid-2002—Bruce Ivins, by comparison, was regarded as a trusted scientist during this same period. Like most Americans, he, too, had been stunned by the destruction of the World Trade Center and the Pentagon and the staggering loss of life that ensued, with these events triggering in him a mix of intense emotions.

"I am incredibly sad and angry at what happened, now that it has sunk in," Ivins wrote to a friend in mid–September 2001. "Sad for all of the victims, their families, their friends."[105] He added that he was also "angry at those who did this, who support them, who coddle them, and who excuse them."[106] His emotional state was transformed on October 4th, however, when he learned that a Florida man had been diagnosed with inhalation anthrax, a surprising development that ignited the microbiologist's curiosity.[107] Spending the better part of the day at his computer emailing his colleagues, Ivins suggested that the incident may have been a rare instance of natural anthrax poisoning.[108] Unaware that Paul Keim in Arizona was already on the case, he also touched base with the CDC and asked if its specialists had thought to establish the bacterium's strain—and, if so, whether it was one known to South Florida.[109] Obviously, he yearned to be involved in the study of the pathogen, and the following week he would come a step closer when the NBC/Brokaw letter arrived at Rockefeller Center in Manhattan.

On October 12th, shortly after investigators located the tainted Brokaw document at NBC headquarters, the letter and its lethal contents were rushed to the New York City Department of Health laboratories. A few hours later, a sample of the spores was flown to the CDC in Atlanta and another to Keim's microbial genetics lab in Arizona, with the letter itself, along with its envelope and the bulk of the spores, being sent to the Special Pathogens Laboratory (SPL) at USAMRIID. Here, the contaminated materials were hand-delivered to John Ezzell, chief of the SPL and the senior microbiologist who would be in charge of their analysis. It was a fitting facility given that its role was to examine biological and environmental materials for the presence of biothreats. Not surprisingly, the SPL was about to become a hive of activity.

Three days later, to be precise, a postman delivered the next letter on the terrorists' hit list, this one to Senator Daschle's suite in Washington, D.C. Wasting no time, first responders placed it inside of two Ziploc bags and handed it over to officials, who hurried it to the Special Pathogens Laboratory like the Brokaw letter before it. Upon arrival, Ezzell took custody of the document, securing it in a third Ziploc bag and storing it in a refrigerator pending his analysis which was slated to begin later that evening. By this point, the SPL had been designated the principal facility for the study of the Amerithrax agent, with this opening the possibility of additional USAMRIID scientists being assigned to assist with the project. For obvious reasons, they would need to be experienced with the anthrax bacterium, preferably the Ames strain, with this prospect galvanizing Bruce Ivins because it meant he might be ascribed a hands-on role in the historic biomedical event that was unfolding within walking distance of his office.

As things stood, the microbiologist had been feeling exasperated at being kept at arm's length up to this point in the investigation, with one of his colleagues describing him as "an absolute manic basket case" during this period.[110] Certainly Ivins' reaction was understandable in view of his extensive experience with the Ames strain of bacterium, probably more than anyone else at Fort Detrick. But the tide would turn on October 17th, with his mood shifting to exhilaration as he was at last brought into the fold. Because it was necessary to assess the spores' concentration, or density, the task was assigned to him owing to his substantial knowledge of the procedure.

What Ivins lacked knowledge of, or so he claimed, was the type of highly-refined spores he saw under the microscope when he undertook the assessment later that day. It was a sight that rattled him. "[H]e wouldn't, couldn't, stop talking about it," said Gerry Andrews, the head of USAMRIID's Bacteriology Division, referring to the extraordinary milling of the anthrax. "It was like smoke. That's what he said it looked like.... He said it was just hovering in the air."[111] When discussing the experience with another colleague, Ivins described a second emotion he also felt upon observing the microorganism. "It scared the shit out of me," he said. "I've never seen anything with such quality, high grade in all my life.... It's the first time I've ever been scared [at USAMRIID]."[112] In due course, Ivins would become alarmed as well by the FBI's decision to focus on him as the perpetrator of the Amerithrax murders.

To nail down the killer, the Bureau set up shop in the Fort Detrick region, an undertaking that included the participation of the nearby Naval Medical Research Center in what would essentially serve as a temporary forensics lab devoted to the case. And for the next six years, this FBI operation liaised with that of Paul Keim in Arizona. More specifically, the Bureau, beginning in 2002 and continuing for the next few years, required all U.S. institutions that possessed the Ames strain of the bacterium to submit samples of its holdings to this local lab, with identical samples being sent to Keim's microbial genetics outfit in Arizona. The submissions would constitute a national repository, one that would be anonymized for security purposes, meaning the people overseeing the storage operation would not be privy to the samples' places of origin or the identities of the scientists who had submitted them. Once the repository received all of the country's samples, they would be compared to the anthrax extracted from the Daschle letter in the hope of securing a match. Only then would the origin of the submission be revealed.

Of course, the sampling procedure itself was flawed: the person or persons who perpetrated the Amerithrax assault, if they were the ones submitting holdings from their labs, could simply send cultures of anthrax that had not been used in the crime. Nevertheless, the FBI decided it was necessary to proceed in this manner due to the large number of specimens to be collected, over a thousand of them, since a single research institution could conceivably contain several labs in which numerous studies were being conducted using different batches of the Ames variant.

Bruce Ivins, strangely enough, had recommended such a project—a storehouse of anthrax samples to be used for comparison purposes—long before the FBI actually launched it, and he even volunteered to help get it off the ground. Not only that, he offered to contribute specimens from his own laboratory, one of which, ironically, would eventually become central to the investigation and incriminating to Ivins himself. As the microbiologist would explain to the Bureau, it was a sample with a most unusual provenance.

"He mentioned several cultures by name, including a batch made mostly of Ames anthrax that had been grown for him at an Army base in Dugway, Utah," reads a joint investigative report by the *ProPublica* organization, *PBS Frontline*, and the *McClatchy Company*."[113] The Dugway Proving Ground, as noted earlier, conducts experiments with biological and chemical biowarfare agents in what is arguably the most sophisticated biodefense program in the nation. As such, it has the ability to process pathogens and chemical substances in ways far beyond the capacities of other sites, USAMRIID included.

In 2002, Ivins submitted several of his holdings to the FBI and Keim labs, among them a sample of his Dugway batch. Comprised mainly of spores from the Utah facility, he had previously combined them with a reserve of non–Dugway spores he also kept in his lab, the result being a one-of-a-kind mixture. In the anonymization process, RMR-1029 would be its designation. At this juncture, Ivins, unlike Steven Hatfill, was not yet a "person of interest" in the crime.

Once all of the nation's samples had been collected and catalogued in the repository, the next step was to compare them to the anthrax extracted from the Daschle letter. Despite the fact that the researchers were meticulous in measuring the samples' genetic similarities and differences, however, the existent DNA fingerprinting method was unable to draw distinctions among them. At this level of analysis, the cultures were still too homogeneous. A more advanced technique was needed, one that could examine the morphology—the form and structure—of the spores in each submission. If the researchers could detect minor differences in the spores' appearances, and if they could then trace these differences back to the DNA itself, it might provide them with a means of distinguishing among the numerous specimens of Ames-strain anthrax. More to the point, it would permit them to match, at the genetic level, the Amerithrax pathogen to the sample containing the same microbe.

With this aim in mind, a team at The Institute for Genomic Research

(TIGR) in Rockville, Maryland, one consisting of corporate president Claire Fraser and researchers Tim Reid and Jacques Revel, entered the picture. Determined to find the needle in the haystack, these scientists, between 2002 and 2003, devised a technique that enabled them to separate the numerous Ames-strain specimens based on morphology, and to then link the specimens' unique features to their genetic codes. As a result, they were able to isolate a submission that was essentially identical to the anthrax used in the attacks. Labeled RMR-1029, it was the one sent to the repository by Bruce Ivins.

At this point, it is important to note that Ivins was not the only person who had access to the spores in his RMR-1029 sample; scores of USAMRIID researchers made use of anthrax originating from the same lot. "[T]here are dozens, if not hundreds, of scientists, contractors, students, professors, who used that same anthrax, the very anthrax that would have the same genetic components as RMR-1029," writes lawyer Paul Kemp.[114] Unfortunately, USAMRIID did not keep a detailed record of the researchers who extracted shares for their own experiments, therefore investigators could not interview them or inspect their labs.[115] Furthermore, the Fort Detrick facility was not the only research center to possess RMR-1029. "Sixteen domestic government, commercial, and university laboratories ... had virulent RMR-1029 Ames strain *Bacillus anthracis* material in their inventory prior to the attacks," reads an FBI search warrant affidavit.[116] Curiously, the Bureau, although it was in possession of this information, did not release it publicly. Instead, the agency singled out Ivins as the sole custodian of the RMR-1029 spores. "We have a flask that's effectively the murder weapon from which those spores were taken that was controlled by Dr. Ivins," said a lawyer representing the District of Columbia in a 2008 press conference.[117] Once the FBI identified what it considered this telltale link to Ivins' flask, it stopped searching for other matches to the anthrax that was used in the attacks.

It is possible, of course, that the unique spores that comprised RMR-1029 originated from those grown for Ivins at the Dugway Proving Ground, the ones with which he subsequently mixed his own supply of non–Dugway spores. On this point, a subsequent analysis revealed that the former accounted for eighty-five percent of the content in Ivins' RMR-1029 sample, with the latter, the non–Dugway spores, making up the remaining fifteen percent.[118]

The FBI also called attention to a problem that occurred in the microbiologist's collection of the spores prior to their submission. It seems the Bureau rejected Ivins' original RMR-1029 offering from February 2002, because he used a different type of test tube than that stipulated by the agency. Certainly it was not the only submission to be discarded; numerous labs initially provided specimens the Bureau refused to accept. Yet in a procedural oversight, the FBI forwarded a copy of Ivins' excluded sample to Keim's lab, where it was stored, and, in due course, would prove damaging to the microbiologist.

In April 2002, the FBI ordered Ivins to send it another specimen and he obliged, although the willful scientist once again did so in his own fashion rather than comply with the procedures set forth by the Bureau. This is because the sampling method specified by the FBI in the Amerithrax case was different from that which was standard practice in the field of microbiology, with Ivins adhering to his profession's own customs. As it turned out, this would affect the composition of his sample and cast a shadow of suspicion over him.

So now there existed two submissions of Ivins' RMR-1029 spores in Keim's lab in Arizona, with the pair purportedly being drawn from the same flask. Upon examination by researchers, however, it was discovered that Ivins' first sample was different from his second one. Whereas the original submission, the one the Bureau rejected due to the type of test tube used, proved to be morphologically identical to that in the Amerithrax attacks, the subsequent submission was different from the pathogen used in the attacks. Seizing on this discrepancy, the FBI concluded that Ivins had knowingly tampered with the latter sample, the hypothesis being that the microbiologist, between his first and second submissions, found out that the TIGR team had devised a technique that would allow it to more precisely analyze the morphology of anthrax and thus trace the incriminating spores back to him. The Bureau discounted other reasons for the disparities between the two cultures, such as inconsistencies or irregularities in Ivins' collection methods.

Interestingly, a few years later investigative reporters learned that Ivins had, in fact, furnished the FBI with several anthrax specimens from his lab, some of which contained spores that were virtually indistinguishable from those in the Amerithrax attacks. Furthermore, he had done so even after he learned that investigators suspected him of the crime. The Bureau did not make this fact public, however, nor did it share the information with Ivins' attorneys, which could have helped exonerate him.[119] It was a disturbing omission on the part of the investigation's leadership.

The fact is, the Bureau, having decided in 2006 that the microbiologist might be guilty, set about amassing evidence, if only circumstantial, to support its contention while simultaneously downplaying or, it would appear, withholding information that contradicted the case it was attempting to construct. Although it had not yet let Steven Hatfill off the hook, the agency seemed to be aware it would soon be in need of a new prime suspect. Accordingly, it looked to other features of the crime for evidence against Bruce Ivins.

Lab Spill. A few months after the Amerithrax attacks, for instance, a technician assigned to Ivins' lab accidentally splattered a small amount of anthrax. Although Ivins cleaned the spill himself and was prompt in doing so, he did not report the incident to his superiors as prescribed by protocol. In due course, his misdeed came to light, at which point he owned up to it and apologized for having neglected to follow proper procedure even as he assured those in charge that he had made certain the area was free of the pathogen before proceeding with his work. All the same, it was a serious lapse for a scientist conducting germ warfare research in a biocontainment lab, and the FBI pointed to the deed, which had occurred five years earlier, as a demonstration of his lack of professional integrity.

Return Address. Another issue concerned a possible link between the return address on the Daschle and Leahy letters and a past event in the microbiologist's personal life. Pillars of their community in Frederick, Maryland, Bruce Ivins and his wife Diane were highly regarded by their friends and neighbors. Bruce played the keyboards in a Celtic band and at the Catholic church the couple attended, performed juggling acts for local children at Mullinix Park, and volunteered with the Red Cross. Diane, for her part, headed an anti-abortion group, the Frederick Right to Life organization. The pair was also supportive of the Mississippi-based American Family Association (AFA), a Christian fundamentalist outfit having as its mission the return of an errant American society, as the group perceived it, to Biblical tenets. The AFA was vehemently anti-abortion, with this position being to the couple's liking.

In 1999, the AFA filed a lawsuit centering on an incident that had taken place in Wisconsin the previous year, one in which a student at a Christian school was disciplined in such a way as to provoke a state investigation. The legal dispute stemmed from the fact that a pair of government social workers, responding to a tip-off, arrived at the school and interviewed the child without the consent of the school's administration. The AFA lawsuit, then, addressed what the group considered to be a case of state intrusion into a private religious institution.

As it happened, Bruce and Diane Ivins resumed their donations to the AFA, which they had allowed to lapse, a month after the organization filed the lawsuit. This was, part and parcel, the extent of their involvement, with the FBI assuming the couple's renewed contributions were intended to help cover the AFA's legal costs. As to the way in which this incident was related to the Amerithrax case, it had to do with the name of the school. The child in the lawsuit was a fourth grade student at Greendale Baptist Academy near Milwaukee. On the anthrax-tainted letters sent to Senators Daschle and Leahy, the return address, which turned out to be spurious, was listed as, "Fourth Grade, Greendale School, Franklin Park, NJ." Predictably perhaps, the fact that both the litigation and the lethal letters referred to the "fourth grade" at a school called "Glendale" set off alarms at FBI headquarters. It also was not lost on investigators that the American Family Association had long been antagonistic toward Senators Daschle and Leahy, both of whom were pro-choice on the subject of pregnancy termination. The Bureau's hypothesis, then, was that Ivins had addressed the letters in such a way as to slyly reference the AFA lawsuit and his own anti-abortion sentiments.

Work Schedule. Further evidence, again circumstantial, on which the agency placed considerable stock were the USAMRIID time sheets, which revealed that Ivins had returned to his laboratory each night between September 14th and September 16, 2001, despite the fact that his research projects did not necessitate it. Although he arrived and departed at different times on those nights, he remained in his lab for exactly two hours and fifteen minutes on each occasion. It was as if he had allotted himself a precise amount of time per stay. The FBI also noted that he returned to his lab for such visits during the first five nights of October 2001, although on these occasions he remained for periods ranging from twenty minutes to nearly four hours. As to the significance of this pattern, the Bureau pointed out that his two clusters of nightly visits occurred shortly before the first and second sets of Amerithrax letters were mailed to Florida, New York, and Washington, D.C., the suspicion being that he was preparing the anthrax-laced documents for posting on these occasions.[120]

When questioned about such comings-and-goings, the microbiologist explained that "home was not good" during this period and he needed to "escape" to his workplace.[121] Surely it is true that he was enduring a considerable amount of turbulence at this juncture, with the FBI confirming that "Dr. Ivins was undergoing significant stress in both his home and work life."[122] The Bureau did not accept his explanation that he was seeking respite in his laboratory, however, dismissing it as unconvincing.

Yet there was more to the picture. Investigative reporters, upon subsequently inspecting the USAMRIID time sheets, found that Ivins' evening visits to his lab actually began in August 2001, weeks before the 9/11 attacks and the ensuing bioterror incidents designed to appear as a feature of the same radical Islamic plot. And since his "escapes" to his lab, as he described them, pre-dated the 9/11 offensive by a month, it raised the question as to whether his visits in September and October, like those in August are presumed to have been, were for reasons other than refining anthrax and mailing it to media and political figures.

Mental Instability. Lastly, the FBI underscored the fact that Ivins was in individual and group therapy for anxiety and depression at the time of the Amerithrax attacks, had developed a drinking problem as well, and was prescribed psychotropic medication for his symptoms. Investigators argued that these facts, based in part on the agency's examination of a cache of emails he sent to his friends, attested to his emotional instability and behavioral unpredictability during the period in which the bioterror offensive was planned and executed.

It is true, of course, that Ivins was enduring emotional distress before, during, and after the 9/11 and Amerithrax attacks. What the Bureau neglected to mention is that his symptoms did not appear to have affected his outward behavior or his job performance. Other than his pronounced interest in the anthrax attacks, which could be expected of a top anthrax researcher who had devoted his career to the pathogen, Ivins' demeanor was apparently unremarkable. This is not to say his mental state could not have led him to secretly engage in mass murder, but simply to draw attention to the fact that those who had known and worked with him for several years did not consider his mood inappropriate or his behavior to be bizarre, erratic, or otherwise noteworthy. He also was stable enough to retain his security clearance and proceed without interruption in his sensitive position at the biodefense facility.

As to motive, the Bureau hypothesized that Ivins was worried that his vaccine research was about to lose its financial support—a reasonable concern at the time—and that a limited bioterror attack would demonstrate the need for the serum. Thus, the Amerithrax attacks, in this scenario, were part of a scheme to ensure continued funding. Alternatively, the FBI suggested that he may have thought a bio-attack would increase sales of the genetically-engineered anthrax vaccine on which he held two associated patents. The latter premise was a bit shaky, however, since government-employed scientists normally do not collect royalties for their work. Earnings customarily return to the government itself.

So it was that the Bureau, having gathered what it considered to be sufficient circumstantial evidence against Ivins coupled with possible motives, classified him as a "person of interest" in 2006, stepped up its scrutiny in 2007, then pulled out all of the stops in 2008 in much the same way it had done in the Steven Hatfill case. Agents combed through Ivins' personal belongings in his home, as well as confiscated work-related materials such as his research files and equipment.[123] They also conducted conspicuous surveillance. On a twenty-four-hour basis for approximately a year, agents sat in parked cars in front of his house presumably to intimidate the scientist and alienate him from his neighbors. If nothing else, the tactic implied, not so subtly, that he was a treacherous and perhaps dangerous individual. Yet during this same period, Ivins retained his high-level security clearance at Fort Detrick and his access to the lethal microorganisms housed in its biosafety labs. Because of such incongruities in the federal authorities' treatment of him, it was unclear what they truly believed about their new suspect.

On other occasions, investigators confronted the Ivins family in public. One incident occurred at a retail mall, where they waylaid the microbiologist and accused him of mass murder. They assailed his wife, too, who had accompanied him on the shopping trip, asking her point-blank if she was aware that her husband was a killer.[124] Bearing in mind that Ivins was already enduring marital turmoil and undergoing treatment for depression, public condemnations of this sort may have further stressed his marriage and accelerated his emotional decline.

In a similar vein, federal agents paid a visit to the couple's adult daughter, telling her that her father was the bioterrorist behind the Amerithrax murders. "The agents also offered her twin brother the $2.5 million reward for solving the anthrax case—and the sports car of his choice," reports the *Washington Post*.[125] Then, too, they spoke at length with Ivins estranged brother, whose opinion of the microbiologist was what one would expect given that the two had refused to talk to each other for sixteen years. As a result, the FBI, by encroaching on past and present family relationships, may have ramped up the anxiety Ivins' was experiencing.

It therefore did not come as a shock when he was found unconscious in his home on March 19, 2008. Hospitalized briefly at the time, the following month Ivins checked himself into a rehabilitation center where he completed a four-week treatment regimen presumably for alcohol abuse. Unfortunately, his mental condition remained in decline, his inpatient stay notwithstanding, such that it was by now fairly obvious to most everyone who knew him that he was deeply depressed. And while some of his coworkers were unaware that the Federal Bureau of Investigation was pursuing him, and quite aggressively, the observations of those who were aware of this fact were astute. "It would be overstating it to say he looked like a guy who was being led to his execution, but it's not far off," says W. Russell Bryne, an infectious disease specialist and Ivins' former supervisor at USAMRIID.[126] Whereas some thought Ivins was distraught because he was convinced the Bureau was determined to seek an indictment, others emphasized the stress he was enduring due to the ordeal's drain on his financial resources, the microbiologist having hired an attorney to defend himself. "He didn't have any more money to spend on legal fees," recalls a friend.[127] Certainly the onset of financial problems could be expected to contribute to the emotional deterioration of a man who is already despondent.

Then, on June 27, 2008, Ivins' downward spiral reached a new low when the FBI announced its settlement with Steven Hatfill, meaning that Hatfill was no longer a "person of interest." Ivins seemed to have believed, quite rightly as it turned out, that he was the next suspect in line, with the settlement clearing the way for the agency to press for his indictment. Shortly thereafter, on July 10th, he was abruptly hospitalized in a mental health facility, although it remains unclear whether institutionalization was imperative at this point.

It seems that Ivins' group therapist, an entry-level counselor at a clinic in his neighborhood, contacted authorities and claimed he was homicidal. She also took out a restraining order ostensibly to prevent him from attacking her. These were, of course, sensational assertions by any standard, and Ivins' friends and colleagues found them at odds with their own observations of him. Curious, too, was the counselor herself. As the *Washington Post* subsequently reported, the woman, under a different name, had been a member of a motorcycle gang, had a history of heroin, cocaine, and PCP abuse, and had just completed three months of home detention—house arrest—as a punishment for a drunk-driving arrest.[128] On the application for the restraining order, it was also noted that she misspelled her job title, a possible point of relevance in that it suggests she may have lacked the educational or intellectual qualifications for her work as a counselor. When the newspaper published her striking claims about Ivins four weeks later, the counselor no longer worked at the clinic. According to a colleague, Ivins had already suspected she was cooperating with the FBI.[129]

Whatever the case, the microbiologist, on July 10, 2008, was on the job at USAMRIID, where he participated in a top-level conference centering on a vaccine that was in development for the bubonic plague. Although his colleagues at the conference, including the vaccine's developer, did not consider Ivins' behavior to be aberrant at the meeting, authorities arrived and ushered him to an inpatient psychiatric facility based on the counselor's claims about his potential homicidal behavior. The upshot: Ivins' security clearance was revoked, his access to his lab rescinded, and his employment at USAMRIID terminated as of September.

Two weeks later, the fraught scientist discharged himself from the mental health facility and returned home. Telling his daughter a few days later that he was suffering from tension headaches, he asked her to purchase a large bottle of Tylenol for him—not an unusual request in light of his circumstances. But Ivins had other plans for the painkiller. On Sunday, July 27th, police were called to his home, where they found the microbiologist sprawled unconscious of the bathroom floor in what appeared to be an acetaminophen overdose. Rushed to the hospital, he died two days later. Although there was no suicide note at the scene, and while the FBI did not order an autopsy of its prime suspect's body, the death was classified a suicide due to the profound stress Ivins had been enduring, including his impending job loss. In short order, the Bureau announced it had been about to indict him for the anthrax murders and planned to seek the death penalty. Not everyone, however, bought the FBI's story.

"There's nobody easier to convict than a dead man," said Keith Olbermann, an MSNBC political commentator who was quick to challenge the FBI's handling of the investigation.[130] The fact is, the Bureau had nothing but circumstantial evidence to support its allegations, meaning it had a thin case that might not have won a conviction, let alone the death penalty. In fact, some were skeptical the agency was about to charge Ivins at all; rather, doubters suggested that his death had conveniently handed the FBI an opportunity to make this claim and, in so doing, report that it had succeeded in getting its man. "If he was about to be charged," says Byrne, "no one who knew him well was aware of that, and I don't believe it."[131]

Yet this was not the only aspect of the case that met with suspicion. In the months and years after Ivins' demise, as damaging information about him was methodically leaked to the press, a chorus of critics insisted that the Bureau was deliberately painting a distorted picture of the dead man so as to foster an aura of guilt. Among these critics was Richard Lambert, the FBI inspector who, from 2002 to 2006, served as head of the Amerithrax investigation itself. Nine years later, in the spring of 2015, Lambert filed a lawsuit against the Federal Bureau of Investigation to address what he claimed were the agency's retaliatory actions against him, actions that stemmed from concerns he had voiced to his superiors about the investigation while it was still underway and afterwards. In his legal complaint, the former FBI inspector described what he perceived as the Bureau's "efforts to railroad the prosecution of Ivins in the face of daunting exculpatory evidence." The document continues,

> Following the announcement of its circumstantial case against Ivins, Defendants DOJ and FBI crafted an elaborate perception management campaign to bolster their assertion of Ivins' guilt. These efforts included press conferences and highly selective evidentiary presentations which were replete with material omissions.[132]

While the former FBI insider was careful not to state that Ivins was wholly blameless in the Amerithrax attacks, he did assert that there was "a wealth of exculpatory evidence to the contrary which the FBI continues to conceal from Congress and the American people."[133]

Yet putting aside for a moment the exculpatory evidence the agency purportedly withheld, many of the facts that *were* available to Congress and the American people still failed to support Ivins' guilt. First and foremost, the microbiologist passed the FBI's polygraph tests, a finding that was glossed over throughout the course of the investigation. Then, too, his handwriting samples did not match those of the person or persons who addressed the anthrax envelopes and printed the messages in the letters. The Bureau also could not place Ivins in key spots at critical times, most notably in St. Petersburg, Florida, and Princeton, New Jersey, when the letters were known to have been mailed from these locations. Furthermore, investigators found no traces of Ames-strain anthrax in his home or car, nor were they able to detect microscopic fibers from the letters, the envelopes, or the tape used to seal them in his home or office. Even the amount of anthrax deployed in the attacks, as measured by weight, was not missing from the supply of the pathogen the microbiologist kept in his laboratory. To be sure, there were gaping holes in the FBI's case.

It is also worth noting that Ivins' colleagues at USAMRIID, a group that was among his most ardent defenders, pointed out that he worked exclusively with wet anthrax, whereas dry anthrax was used in the Amerithrax offensive, the latter being an altogether different entity and one that none of the researchers at Fort Detrick employed in their research; it was the scientists at the U.S. Army facility in Dugway, Utah, and the Battelle Memorial Institute in Ohio who worked with wet anthrax. Experts also explained that no one person could have pulled off the Amerithrax attacks, from obtaining and processing the spores in such an exceptional fashion to mailing the pathogen-laced letters without leaving any traces of DNA, fingerprints, or other identifying material on the envelopes and mailboxes. A lone-wolf terrorist simply could not have managed the multifaceted operation in its entirety. Rather, the attacks, insisted the FBI's critics in the scientific community, required a plurality of specialists contributing to discrete stages of the process.

Supporting these voices was that of Senator Patrick Leahy, who, as one of the targets of the attacks, had a personal investment in the matter. Although he did not rule out the possibility that Ivins may have been the perpetrator, he put no stock in the lone-wolf theory. "If he is the one who sent the letter," said the congressman, "I do not believe in any way, shape or manner that he is the only person involved in this attack on Congress and the American People."[134]

Senator Leahy made these comments in a packed assembly on September 17, 2008. A tense and at times adversarial meeting, it included not only FBI Director Robert Mueller and members of Congress, but also such interested parties as Robert Hatfill, the Bureau's previous target. Because it was an oversight hearing held by the Senate Judiciary Committee, it was presided over by the committee's chairman, Iowa senator Chuck Grassley, who was himself skeptical of the Amerithrax inquiry's findings. As the congressman was aware, it had been the most expensive investigation in the history of the FBI, and one that, in the end, cost the American taxpayer an estimated one hundred million dollars.

"Congress and the American people deserve a complete accounting of the FBI's evidence, not just a selective release of a few documents and a briefing or two," Grassley said in the hearing. "There are many unanswered questions the FBI must address before the public can have confidence in the outcome of the case."[135]

Grassley then proposed that an independent review of the Bureau's detective work be undertaken, one that would re-examine the nine thousand witness interviews the agency had conducted, along with the eighty searches its agents had performed and the six thousand items they had collected. The FBI, however, opposed the recommendation. Its position was that such an evaluation would interfere with its ongoing Amerithrax investigation, which had not been officially terminated despite the fact that the agency had publicly named the late Bruce Ivins as the likely perpetrator. What was eventually agreed upon, then, was a different sort of review, albeit one that would still be of value. Led by a team of experts from the National Research Council (NRC) under the auspices of the National Academy of Sciences, it would be an appraisal of the scientific quality of the investigation.

So it was that the NRC group, in a 1.1 million dollar study conducted over a two-year period, revisited the science behind the Amerithrax inquiry, and on February 15, 2011, released its findings. A blow to the FBI's reputation, the NRC report was, in effect, a compilation of areas in which the Bureau's research lacked scientific rigor. For instance, the National Research Council revealed that the genetic analyses the FBI used to implicate Ivins and his flask of RMR-1029 were not nearly as convincing as the Bureau had professed. "The scientific data alone do not support the strength of the government's repeated assertions ... as in 'the scientific analysis coordinated by the FBI Laboratory determined that RMR-1029, a spore-batch created and maintained at USAMRIID by Dr. Ivins, was the parent material for the anthrax used in the mailings,'" the report stated.[136]

The NRC team also noted that the FBI failed to consider a process known to biologists as "parallel evolution," in which identical or nearly identical genetic mutations occur independently in two or more batches of spores. In such a case, the distinctive anthrax in Ivins' flask at USAMRIID may have had an unknown twin at another location, namely, in the research lab from which the Amerithrax attacks were perhaps launched. Here, the spores' unique mutations would have arisen randomly or under conditions similar to those in Ivins' laboratory.

Most significantly, though, scientists at the Lawrence Livermore National Laboratory, who also contributed to the NRC study, discovered that the anthrax used in the Amerithrax attacks contained inordinately high levels of silicon, an additive which had been used to ensure that the pathogen floated in the air. What was most telling about this unexpected finding is that the process of aerosolizing, or weaponizing, the microorganism by affixing sizable amounts of silicon requires both highly specialized skills and, even more implicative, an outsized fermenter. And this ruled out USAMRIID. It did not, however, exclude other sites that also conducted biodefense work for the federal government—again, the U.S. Army's Dugway Proving Ground and the Battelle Memorial Institute. A startling revelation by the Livermore scientists, the *Wall Street Journal* published a report on the findings and concluded that they exonerated the troubled microbiologist, at least as a lone-wolf terrorist. "[N]o matter how weird he may have been, [Ivins] had neither the set of skills nor the means to attach silicon to anthrax spores."[137] When questioned by the news media about the critical Livermore discovery, the FBI declined to respond.[138]

Ultimately, the National Research Council's study implicitly reinforced the view held by certain members of Congress and numerous scientists, namely, that the Bureau's all-consuming campaign to assign responsibility to a single bioterrorist may have allowed the real perpetrators to go free. A flawless analysis of a defective federal investigation, the NRC report caused the FBI's leadership to bristle, although the Bureau continued to insist that the Amerithrax investigation had been top-notch and its findings, solid.

In 2010, the FBI formally closed the case. The agency's conclusion was, and remains today, that Bruce Ivins was in all probability the sole perpetrator of the anthrax attacks. Despite repeated requests, the Bureau has not shared the inquiry's case materials with members of Congress or the public.

After the Storm

While questions remain as to the true identity of the person or persons responsible for the Amerithrax assault, what *is* indisputable is that the attacks themselves triggered a jaw-dropping surge in federal funding for biodefense research and preparedness, and they did so with lightning speed. "Since 2001," reports a piece in the *New York Times Magazine*, "senior members of both the Obama and Bush administrations ... have consistently placed biodefense at or near the top of the national-security agenda," with nearly eighty billion dollars having been invested in research and readiness since the ordeal.[139]

The FBI's controversial, eight-year investigation likewise generated substantial change, mainly in the form of scientific progress. Its debatable conclusions notwithstanding, the inquiry brought together such branches of knowledge as genomics, molecular epidemiology, and microbiology, thereby validating and advancing the nascent field of microbial forensics. "[T]he Amerithrax investigation was groundbreaking," writes Paul Keim and his colleagues, in that it "pioneered new approaches to the investigation of microbial-based crimes."[140]

Regarding medical countermeasures, while a newer, safer anthrax vaccine has not yet emerged despite the U.S. government having poured hundreds of millions of dollars into the effort, another post–9/11 program, Project BioShield, substantially increased the nation's stockpile of the existing vaccine. In addition, researchers are making headway in experimental measures aimed at treating anthrax illness in children. A pediatric antibiotic is in development, for instance, that is administered orally and designed to combat anthrax infection, tularemia, and community-acquired bacterial pneumonia. Although *solithromycin*, as it known generically, is also effective in adults, its primary value lies in its comparative safety for a pediatric population.[141]

Noteworthy as well, the Food and Drug Administration, in May of 2015, announced the approval of an infusible drug for the treatment of inhalation anthrax. The medication, under the trade name *Anthrasil*, is made from the plasma of individuals who have been vaccinated against the bacterium and whose serum, as a result, contains antibodies to it.

Still another approach makes use of nanotechnology. Perhaps the most innovative project of this type centers on a microscopic device that, when inserted into the human body, detects lethal bacteria and releases an arsenal of antibiotics in response. The antibiotics, moreover, are programmed to target those parts of the body most affected by the particular pathogens. If successful, the Pentagon hopes to make the device available to military personnel for protection on the battlefield.

And demonstrating considerable merit in the fight against anthrax infection are certain naturally-occurring entities, among them a novel form of marine life recently discovered off the coast of Santa Barbara, California. Itself a form of bacteria known as an actinobacterium, the entity produces *anthracimycin*, an antibiotic that has been found to incapacitate anthrax and may therefore be of value in eliminating the infection in humans. It may also be effective against the difficult-to-treat MRSA, or methicillin-resistant Staphylococcus aureus.[142]

And yet, while these and other advances in the inhibition and eradication of anthrax infection are heartening, it must be kept in mind that they are designed to respond to an exposure, not to prevent one. And this is a serious concern since the chances of a bio-attack may actually be greater today than at the time of the Amerithrax assault owing to the proliferation of biodefense facilities across the United States.

Within the scientific community are researchers like Keith Rhodes, the former Chief Technology Officer of the Government Accountability Office, who warn that the United States now has within its borders far more biosafety level-3 and level-4 laboratories than is necessary or prudent.[143] It is an assertion backed up by data: whereas only a handful of U.S. sites, prior to the Amerithrax offensive, were licensed to experiment with the type of virulent microorganisms used in bio-assaults, this number soared to over four hundred laboratories after the 9/11 attacks, with more than fifteen thousand scientists and technicians having access to the pathogens.[144] This state of affairs, experts argue, sharply increases the chances of an accident or, alternatively, a deliberate strike on the population. "Scientists, security experts and legislators are now pondering various ideas to prevent lab-based terrorism," says

John Dudley Miller in a *Scientific American* report on the proliferation of U.S. biosafety labs.[145]

And there are additional areas of concern. Alongside the heightened risks posed by the steep rise in new biodefense operations are the threats that continue to exist in the nation's long-established facilities. Among such hazards are those stemming from equipment malfunctions, staff negligence, insufficient training, and infiltration by extremists and those who abet them.

An incident at the Dugway Proving Ground illustrates the problem. In May 2015, the germ warfare operation disclosed that it had shipped live anthrax bacteria to other research centers via the commercial carrier Federal Express, anthrax that should have been rendered harmless by irradiation prior to mailing. "[A]nthrax bacteria were shipped out at least 74 times to dozens of labs in the U.S. and at least five foreign countries from January 2005 to May 2015," writes journalist Alison Young, citing the findings of an investigation conducted by the Centers for Disease Control and Prevention.[146] Because of the debacle, it was necessary to treat thirty-one people with antibiotic therapy.

As to the reason for the potentially fatal foul-up, it turns out that the method used by the Dugway facility to sterilize the bacterium was inconsistent in killing the microorganism. Then, too, Dugway technicians, the investigation found, claimed to have performed verification tests to ensure the spores were dead, yet proceeded to disregard the test results and ship live spores anyway.[147] Providentially, Bruce Ivins, before his death, repeatedly alerted his colleagues to the unreliability of certain irradiation practices in incapacitating the anthrax bacterium and called for stricter measures, among them a standardized sterilization protocol for all U.S. laboratories coupled with improved verification procedures.[148]

In terms of bioterrorism as a means of mass exposure, while no further strikes have occurred on U.S. soil since the autumn of 2001, the possibility of a bio-attack will no doubt persist well into the future. Furthermore, of those pathogens likely to be deployed, there are several reasons to assume anthrax will remain the microorganism of choice. Certainly it is known that al-Qaeda, an organization that continues to evolve, was experimenting with the bacterium before and after the events of September 11th, and that it not only formed a committee devoted to biological, chemical, and nuclear attacks but was also constructing a laboratory in Afghanistan specifically to weaponize anthrax.

Al-Qaeda, however, is not the only organization that is a cause for concern; others exist as well. In addition, new extremist groups will no doubt emerge in the coming years, some of which can be expected to explore the use of biological agents against civilian targets. It is imperative, then, that society identify those terrorist organizations that appear to be intent upon doing harm through biological means and thwart their efforts to acquire lethal agents. And this means enhanced vigilance. "Better intelligence and biosecurity measures are essential," write René Pita and Rohan Gunaratna, a common-sense conclusion as well as a challenging one in that it must be accomplished in such a way as to protect the population even as it respects the freedom and privacy of the individual citizen.[149]

Chapter Notes

Introduction

1. Longfellow (2010).

Chapter 1

1. Hammarlund, et al. (January 29, 2007).
2. Goldman, et al. (May, 2007).
3. Bell (2010, p. 104).
4. Blackburn, in Lively (July 13, 2014).
5. Steers (2001), p. 49.
6. Kolata (November 13, 2001).
7. Steers (2001).
8. *Ibid.*, p. 54.
9. Rybicki (January 1990).
10. Federal Bureau of Investigation (Undated).
11. Mayor (2009).
12. *Ibid.*; Bisset (December 1979).
13. Martin, et al. (2007).
14. Dire, et al. (September 3, 2013).
15. Oxford Dictionary.
16. Sawyer (2007).
17. *Ibid.* p. 54.
18. Johnson (Undated).
19. *Ibid.*
20. Romano, et al. (2008).
21. Dire, et al. (September 3, 2013).
22. Romano, et al. (2008).
23. Lorenzi (December 3, 2007).
24. *Ibid.*
25. Trevisanato (2007).
26. Mayor (2009), p. 189.
27. *Ibid.*
28. *Ibid.*
29. Mayor (Autumn 1997), p. 36.
30. *Ibid.*
31. Frischknecht (2003).
32. Johnson (Undated).
33. Martin, et al. (2007).
34. *Ibid.*, p. 2.
35. Mayor (2009), p. 8.
36. Frischknecht (2003), p. S47.
37. *Ibid.*
38. Riedel (2004).
39. *Ibid.*
40. *Ibid.*
41. *Ibid.*
42. Harris (December 1992). p. 21.
43. Williams and Wallace (1989).
44. *Ibid.*, pp. 19–20.
45. Farmer, in Kristof (March 17, 1995).
46. Riedel (2004).
47. Williams and Wallace (1989), p. 35.
48. *Ibid.*
49. Kristof (March 17, 1995).
50. Riedel (2004).
51. Kristof (March 17, 1995).
52. Williams and Wallace (1989).
53. Smith (1999, p. 155).
54. *Ibid.*
55. Williams and Wallace (1989), p. 127.
56. Smith (1999, p. 155).
57. Sanders, in Williams and Wallace (1989), p. 126.
58. Kristof (March 17, 1995).
59. Drayton (May 9, 2005).
60. Williams and Wallace (1989).
61. Kunkle (February 27, 2006).
62. Regis (1999), p. 79.
63. *Ibid.*
64. *Ibid.*, p. 66.
65. *Ibid.*
66. Mahlen (October 2011).
67. Regis (1999).
68. Williams and Wallace (1989).
69. Regis (1999), p. 119.
70. Nixon (November 25, 1969).
71. Banting, in Bliss (1984), pp. 284–285.

72. Mauroni (2003), p. 123.
73. Alibek and Handelman (1999).
74. *Ibid.*, p. 299.
75. *Ibid.*, p. 19.
76. Cole (May 2, 2005), p. 1110.
77. Department of Defense (Nov. 8, 2010), p. 247.
78. Jansen, et al. (June 2014), p. 488.
79. Stern (August 1999).
80. bin Laden interview, TIME Staff (January 11, 1999).
81. Cullison and Higgins (December 31, 2001).
82. *Ibid.*
83. *Ibid.*
84. Pita and Gunaratna (May 15, 2009).
85. *Ibid.*
86. Gellman (March 23, 2003, p. A01.
87. *Ibid.*
88. Stephen (November 22, 2001).
89. Gellman (March 23, 2003).
90. *Ibid.*
91. Mowatt-Larssen (January 2010), p. 6.
92. *Ibid.*
93. Pita and Gunaratna (May, 15, 2009).
94. *Ibid.*
95. Simon (2013), p. 94.
96. Gerstein (2009), p. 10.

Chapter 2

1. Robespierre (2007), p. 115.
2. Post (2000), p. 280.
3. *Ibid.*, p. 281.
4. *Ibid.*
5. *Ibid.*, pp. 282.
6. Stern (2003), p. 141.
7. Stern and Modi (2010), p. 263.
8. Searcely and Santora (November 18, 2015); Mauro (2013).
9. Institute for Economics and Peace (2015), p. 22.
10. Mauro (2013).
11. Stern (2003), p. 141.
12. Beam (1983).
13. Sageman (March/April 2008).
14. *Ibid.*, p. 39.
15. *Ibid.*
16. Michael (2012), p. 101.
17. Army of God (Undated).
18. *Ibid.*
19. Bray, in Clarkson (February 19, 2002).
20. Post (2000), p. 283.
21. Post (2004), p. 82.
22. *Ibid.*
23. Post (2000), p. 284.
24. Dees (1997), p. 42.
25. *Ibid.*
26. Robles (June 20, 2015).
27. Roof, in Robles (June 20, 2015).
28. Kurzman and Schanzer (June 16, 2015).
29. Greenberg (January 8, 2015).
30. Kurzman and Schanzer (June 16, 2015).
31. Watson (February 6, 2002).
32. Paterson (January 22, 2016).
33. In Smith and Moncourt (2009), p. 505.
34. Stammer (May 15, 1987).
35. Alexander, in Bari (1994) pp. 266–267.
36. Foreman, in Bari (1994), p. 268.
37. Stammer (May 15, 1987).
38. Rowell (1996).
39. Stern (1999), p.66.
40. Casagrande (Fall/Winter 2000), p. 100.
41. *Ibid.*
42. Skawińska (2009), p.9.
43. *Ibid.*
44. *Ibid.*, p. 9.
45. IRA Overpowers Crew, Sinks British Ship in Irish Bay (February 8, 1981), p. 6.
46. Skawińska (2009).
47. 9/11 Not as Bad as IRA, Says Doris Lessing (October 24, 2007).
48. Taylor (May 11, 1996).
49. Gerstein (2009), p. 198.
50. *Ibid.*, p. 104.
51. *Ibid.*
52. Gurr and Cole (2002), p. 41.
53. *Ibid.*, p. 43.
54. *Ibid.*
55. *Ibid.*, p. 43.
56. *Ibid.*
57. *Ibid.*, p. 53.
58. *Ibid.*
59. *Ibid.*, p. 66.
60. *Ibid.*
61. Gerstein (2009), p. 108.
62. Markel (September 29, 2014).
63. Olson, in Taylor (May 11, 1996), p. 33.
64. Schutzer, et al. (2005), p. 1242.
65. *Ibid.*
66. *Ibid.*, p. 1243.
67. Kristof (July 12, 2002).
68. Bailey (October 14, 2001).
69. United Nations Security Council (May 29, 2007).
70. *Ibid.*

Chapter 3

1. Whalen (May 12, 2009).
2. Schmidt (2008).
3. *Ibid.*, p. 1.
4. *Ibid.*, p. 3.
5. Hannemyr (1999).
6. *Ibid.*
7. *Ibid.*
8. *Ibid.*
9. Levy (2010).
10. *Ibid.*, p. 30.
11. *Ibid.*
12. *Ibid.* (2010), p. 10.
13. *Ibid.*, p. 55.
14. *Ibid.*
15. Himanen (2001).
16. Gates (February 2, 1976).
17. *Ibid.*
18. Delfanti (2013), p. 111.
19. *Ibid.*
20. *Ibid.*
21. St. John, in Schrage (January 31, 1988).
22. *Ibid.*
23. Grushkin, et al. (November, 2013).
24. Zimmer (March 5, 2012).
25. International Human Genome Sequencing Consortium (February 15, 2001).
26. Collins (1999).
27. Delfanti (2013), p. 111.
28. Wohlsen (2011), p. 19.
29. *Ibid.*, p. 52.
30. Patterson, in McKenna (January 7, 2009).
31. Grushkin, et al. (November, 2013).
32. *Ibid.*
33. *Ibid.*
34. Belew, in Newitz (February 26, 2002).
35. Liptak (June 13, 2013).
36. Philipkoski (April 20, 2001).
37. Holloway (March 1, 2013).
38. Bollier (2002), p. 77.
39. Jefferson (July 5, 2013).
40. Holloway (March 1, 2013).
41. Clinton, in Garrett (December 15, 2011).
42. Herfst, et al. (2012); Imai, et al. (June 21, 2012).
43. Elbright, in Enserink (November 23, 2011).
44. Taubenberger, et al. (October 6, 2005).
45. van Aken (2006), p. 10.
46. Jorgensen, in Zimmer (March 5, 2012).
47. Endy, in Guthrie (2009).
48. Gates, in Levy (April 15, 2010).
49. *Ibid.*
50. Bobe, in Charisius, et al. (January 24, 2013).
51. Biohackers of the World, Unite (Sept. 6, 2014).
52. Wohlsen (2011/2012), p. 182.
53. Vahid (October 18, 2011).
54. *Ibid.*
55. Charisius, et al. (January 24, 2013).
56. Bobe, Jason (June 3, 2013A); Bobe, Jason (June 3, 2013B).

Chapter 4

1. Editorial: Microbiology by Numbers (September 2011), p. 628.
2. Bioterrorism Overview (February 12, 2007).
3. Vaccines, Blood & Biologics: Anthrax (June 17, 2015).
4. *Ibid.*
5. *Ibid.*
6. Injection Anthrax (July 21, 2014).
7. Grunow, et al. (December 2012).
8. *Ibid.*
9. Biederbick (2012), p. 115.
10. Botulism (August 2013).
11. *Ibid.*; Passaro, et al. (March 1998).
12. Botulism (August 2013); Dembek, et al. (2007).
13. Chalk, et al. (February 20, 2014).
14. *Ibid.*
15. *Ibid.*
16. Biederbick (2012).
17. Frequently Asked Questions on Ebola Virus Disease (January 2016).
18. Osterholm, et al. (February 19, 2015).
19. *Ibid.*
20. Frequently Asked Questions on Ebola Virus Disease (January 2016); Marburg Haemorrhagic Fever (November 2012).
21. Frequently Asked Questions on Ebola Virus Disease (January 2016).
22. Geggel (October 7, 2014).
23. Rasmussen, in Geggel (October 7, 2014).
24. Frequently Asked Questions on Ebola Virus Disease (January 2016).
25. Maron (September 25, 2014).
26. Tu and Croddy, in Croddy, et al. (2005).
27. *Ibid.*
28. Ibeji (February 17, 2011).
29. *Ibid.*
30. Brower (2005), p. 262.

31. Smallpox Fact Sheet: Smallpox Disease Overview (February 6, 2007).
32. Boseley and Borger (May 16, 5005).
33. Henderson, in Boseley and Borger (May 16, 2005).
34. Kraft, in Croddy, et al. (2005), pp. 288–289.
35. Raoult (1990).
36. Brenner, et al. (July 2000).
37. Jean, et al. (December 2007), p. 1920.
38. Mayo Clinic Staff (December 16, 2015).
39. Croddy, in Croddy, et al. (2005), p. 315.
40. *Ibid.*
41. Dong, et al. (May 2008).
42. *Ibid.*, p. 186.
43. Hussain (August 5, 2014).
44. *Ibid.*, p. 162.
45. Multidrug-Resistant Tuberculosis (MDR TB) (August 1, 2012).
46. Gurr and Cole (2002), p. 52.
47. Mauroni (2006), p. 110.
48. *Ibid.*
49. Puskoor and Zubay (2005), p. 247.
50. *Ibid.*
51. *Ibid.*
52. Ward and Garrido (2005), p. 193.
53. *Ibid.*, p. 193.
54. Vector-Borne Diseases: Overview (February 2016).
55. *Ibid.*
56. Touma (2005), p. 23.
57. *Ibid.*
58. Morser, et al. (2005)
59. *Ibid.*; Tucker, in Tucker (2000).
60. *Ibid.*, p. 7.
61. Morser, et al. (2005), p. 289.
62. Hickman (1999).
63. Burrows and Renner (December 1999), p. 982.
64. Hickman (1999).
65. Calomiris, in American Society for Microbiology (February 26, 2006).
66. Hickman (1999).
67. Anthrax Hard to Remove from Drinking Water Systems (April 15, 2006).
68. Hickman (1999).
69. Burrows and Renner (December 1999, p. 982).
70. Hickman (1999).
71. *Ibid.*
72. Tucker, in Tucker (2000), p. 8.
73. Gurr and Cole (2002), pp. 52–53.
74. *Ibid.*
75. Mauroni (2006), pp. 115–116.
76. *Ibid.*
77. *Ibid.*
78. Mehta, in Zubay, et al. (2005), p. 156.
79. *Ibid.*, p. 157.
80. Puskoor and Zubay (2005).
81. *Ibid.*, p. 74.
82. *Ibid.*

Chapter 5

1. Opportunity May Be More Important Than Profession in Serial Homicide (editorial) (April 21, 2001), p. 993.
2. Iverson (2003), pp. 8–9.
3. *Ibid.*
4. Opportunity May Be More Important Than Profession in Serial Homicide (editorial) (April 21, 2001).
5. Iverson (2003).
6. Suzuki, in Iverson (2003), p. 302.
7. Iverson (2003), p. 300.
8. Suzuki, in Iverson (2003) p. 302.
9. Iverson (2003).
10. *Ibid.*
11. Franz (July 31, 2002).
12. Zilinskas (2011).
13. *Ibid.*, p. 1.
14. Associated Press (September 11, 1998).
15. *Ibid.*
16. Cutler, in Associated Press (September 11, 1998), p. 2.
17. Thomas (December 4, 1998).
18. Carus and Center for Counterproliferation Research (2003).
19. Braun, in CNN (April 24, 1998).
20. Thomas (December 4, 1998).
21. *Ibid.*
22. Stewart, in Thomas (December 4, 1998).
23. *Ibid.*
24. *Ibid.*
25. Associated Press (June 5, 2009).
26. Brryan Stewart, in Maysh (February 13, 2011).
27. Cundiff, in Associated Press (January 9, 1999).
28. Harson, in Associated Press (July 25, 1996).
29. *Louisiana v Schmidt* (July 26, 2000).
30. *Ibid.*
31. Wong, in *Louisiana v Schmidt* (July 26, 2000).
32. *Ibid.*
33. Dye (January 7, 2006).
34. Oxford Dictionary.
35. Metzker, et al. (October 29, 2002), p. 14296.

36. Hillis, in *Louisiana v Schmidt* (July 26, 2000).
37. Burgess (June 11, 2015).
38. Metzker (October 29, 2002), p. 14297.
39. *Ibid.*
40. Richardson, in Rule (1997), p. 258.
41. Rule (1997).
42. Farrar, in Rule (1997), p. 24.
43. *Ibid.*, p. 28.
44. Colleague, in Rule (1997), pp. 34.
45. Rule (1997).
46. *Ibid.*, p. 44.
47. Farrar, in Rule (1997), p. 67.
48. Supreme Court of Kansas (March 23, 2007).
49. Rule (1997), p. 61.
50. Green, in Rule (1997), p. 74.
51. Rule (1997), pp. 75–76.
52. Supreme Court of Kansas (March 23, 2007).
53. *Ibid.*
54. Rule (1997).
55. Associated Press (November 24, 1995).
56. Supreme Court of Kansas (March 23, 2007).
57. *Ibid.*
58. Rule (1997).
59. Supreme Court of Kansas (March 23, 2007).
60. Rule (1997).
61. Supreme Court of Kansas (March 23, 2007).
62. *Ibid.*
63. Rule (1997), p. 134.
64. *Ibid.*
65. Shea and Gottron (April 17, 2013).
66. Rizzo (January 22, 2015).
67. Carus (2002), p. 46.

Chapter 6

1. Bureau of the Census (December 1981).
2. Latkin (1992).
3. Milne (1986), p. 292.
4. *Ibid.*, p. 292.
5. Weaver (February 29, 1985).
6. In McCormack (1987), p. 21.
7. Zaitz (April 14, 2011—B).
8. Zaitz (April 14, 2011—A).
9. *Ibid.*
10. Abbott (Winter 2015), p. 417.
11. Bhagwan Shree Rajneesh Biography—Religious Figure, Criminal (1931–1990) (2016).
12. Rajneesh (1988), p. 52.
13. *Ibid.*
14. Gordon (1987), p. 13.
15. Sannyasin, in Gordon (1987), p. 14.
16. Gordon (1987), p. 14.
17. Hitchens (2007).
18. *Ibid.*, p. 197.
19. *Ibid.*
20. Price, in Milne (1986), p. 141.
21. Gordon (1987).
22. Milne (1986), p. 186.
23. Bhagwan Shree Rajneesh Biography—Religious Figure, Criminal (1931–1990) (2016).
24. *Ibid.*
25. Geist (September 16, 1981).
26. Prasinos and Jackson (October 1, 2015).
27. Gosnell, et al. (2011), p. 186.
28. *Ibid.*
29. *Ibid.*
30. Carter (August 31, 1990), p. 203.
31. Cody (March 8, 2005).
32. Shay (October 23, 2010).
33. Margaret Hill, in Oregon Experience: Rajneeshpuram (transcript) (November 19, 2012).
34. Shay (October 23, 2010).
35. Carus (2002), p. 51.
36. *Ibid.*
37. Abbott (Winter 2015), p. 420.
38. Incorporation of Rajneeshpuram Opens Door to Development (Part 9 of 20) (July 8, 1985).
39. *Ibid.*
40. Land Use Database: Rajneeshpuram Site (Undated).
41. Goldman (2011), p. 309.
42. Cody (March 8, 2005).
43. *Ibid.*
44. Sheela (2014), p. 235.
45. Karuna, in Guru's Disciples Taking Over in Oregon Town (December 19, 1982).
46. Latkin (1992), p. 265.
47. *Ibid.*
48. Cody (March 8, 2005).
49. *Ibid.*
50. United Press International (November 10, 1983).
51. Bhagwan Shree Rajneesh Biography—Religious Figure, Criminal (1931–1990) (2016).
52. Cody (March 8, 2005).
53. Carus (2002); Incorporation of Rajneeshpuram Opens Door to Development (Part 9 of 20) (July 8, 1985).
54. Carus (2002).

55. Franklin (1992), p. 137.
56. Keyes (June 10, 2014).
57. Flynn (October 7, 2009).
58. Carus (2002).
59. *Ibid.*
60. Carus (2000), p. 134.
61. Testimony of Alma Peralta (May 21, 1990), p. 24.
62. Ma Anand Ava, in Carus (2002), p. 57.
63. Weaver (February 29, 1985).
64. Elmer-DeWitt (September 30, 2001).
65. Thuras (January 9, 2014).
66. Keyes (June 10, 2014).
67. Zaitz (April 14, 2011- B)
68. *Ibid.*
69. Centers for Disease Control and Prevention (April 13, 2012); Weaver (February 29, 1985).
70. Carus (2000), p. 131.
71. Centers for Disease Control and Prevention (April 13, 2012).
72. *Ibid.*, p. 3.
73. *Ibid.*
74. Weaver (February 29, 1985).
75. *Ibid.*
76. Weaver, in Flynn (October 7, 2009).
77. Weaver (February 29, 1985).
78. Weaver (April 14, 2001).
79. Miller, et al. (2001/ 2002), p. 23.
80. Gordon (1987), p. 158.
81. Fitzgerald (1986), p. 360.
82. *Ibid.*, p. 361.
83. Witness, in Miller, et al. (2001/2002), p. 30.
84. *Ibid.*
85. *Ibid.*
86. *Ibid.*
87. *Ibid.*; Gordon (1987).
88. Miller, et al. (2001/2002).
89. *Ibid.*, p. 25.
90. Rajneesh and Company Pull Up Stakes from Oregon As Guru's Vision in Desert Becomes a Mirage (December 30, 1985).
91. Rajneesh, in Rajneesh and Company Pull Up Stakes from Oregon As Guru's Vision in Desert Becomes a Mirage (December 30, 1985).
92. Waslekar (January 20, 1987).
93. Associated Press (July 22, 1986).
94. Miller, et al. (2001/2002), p. 32.

Chapter 7

1. Pollack (March 29, 1995).
2. Reader (2000).
3. Hayashi Ikuo, in Reader (2000), p. 40.
4. Classmate, in Kaplan and Marshall (1996), p. 8; Danzig, et al. (July 20, 2011).
5. Kaplan and Marshall (1996), p. 8.
6. Reader (2000).
7. Kaplan and Marshall (1996), p. 9.
8. Danzig, et al. (July 20, 2011).
9. Reader (2000), p. 44.
10. *Ibid.*, p. 48.
11. *Ibid.*
12. Danzig, et al. (2011), p. 6.
13. Wessinger (2000).
14. Kaplan and Marshall (1996), p. 21.
15. Repp (2011), p. 148.
16. Reader (2000).
17. *Ibid.*, p. 148.
18. Kristof (March 14, 1996).
19. Reader (2000), p. 154.
20. Broad (May 26, 1998).
21. *Ibid.*
22. *Ibid.*
23. *Ibid.*
24. *Ibid.*
25. Takahashi, et al. (January 2004).
26. Keim, in Onion (October 5, 2001).
27. Takahashi, et al. (January 2004), p. 117.
28. Murphy (June 21, 2014).
29. *Ibid.*
30. Hamblin (May 6, 2013).
31. Kaplan and Marshall (1996), p. 235.
32. *Ibid.*
33. *Ibid.*, p. 236.
34. Lifton (1999), p. 172.
35. *Ibid.*, p. 210.
36. *Ibid.*
37. *Ibid.*, p. 184.
38. Murakami (2000), p. 10.
39. *Ibid.*, p. 11.
40. RamaRao, et al. (April 4, 2014).
41. Lifton (1999).
42. Murakami (2000).
43. *Ibid.*, p. 10.
44. Lifton (1999), p. 146.
45. Murakami (2000), p. 118.
46. *Ibid.*
47. Hirose, in Murakami (2000), p. 59.
48. *Ibid.*, p. 59.
49. Ikuo Hiyashi, in Marukami (2000), p. 11.
50. Pletcher (2016).
51. Kenji Ohashi, in Murakami (2000), p. 67.
52. *Ibid.*, p. 67–68.
53. *Ibid.*, p. 72.
54. Kaplan and Marshall (1996), p. 247.
55. *Ibid.*

56. Aya Kazaguchi, in Murakami (2000), p. 51.
57. *Ibid.*, p. 52.
58. In Kaplan and Marshall (1996), p. 260.
59. Pletcher (2016).
60. Ohbu, et al. (July 1997).
61. *Ibid.*, p. 588.
62. *Ibid.*
63. Taneda (May 22, 2009), p. 288.
64. Ohbu, et al. (July 1997), p. 590.
65. *Ibid.*
66. Saito (Autumn 2010), p. 20.
67. Pangi (February 2002), p. 30.
68. *Ibid.*
69. WuDunn (March 22, 1995).
70. *Ibid.*
71. Lifton (1999), p. 41.
72. *Ibid.*
73. Goozner (March 30, 1995).
74. Goozner (May 16, 1995).
75. Hongo (November 11, 2011).
76. *Ibid.*
77. Hongo and Wijers-Hasegawa (September 16, 2006).
78. Amnesty International (September 19, 2016).
79. Kamikuishiki's Gulliver Park Falls. (November 13, 2001).
80. BBC (April 6, 2016).
81. *Ibid.*
82. Saito (Autumn 2010).

Chapter 8

1. Willman (2011), p. 83.
2. Stolberg (September 30, 2001).
3. George, in Stolberg (September 30, 2001).
4. Willman (2011).
5. *Ibid.*
6. Friedlander (1997), p. 467.
7. Thompson, in Green and NATO Staff (2007), p. 187.
8. *Ibid.*
9. Kadlec, in Willman (2011), p. 86.
10. Fainaru and Warrick (November 30, 2001), p. A.01.
11. *Ibid.*
12. Keim, in Willman (2011), p. 95.
13. Guillemin (2011), p. 46.
14. Sainz (February 25, 2002).
15. ABC News (October 8, 2001).
16. *Ibid.*
17. Centers for Disease Control and Prevention (October 4, 2001).
18. *Ibid.*
19. Centers for Disease Control and Prevention (October 11, 2001).
20. Manager, in Cole (2009), p. 48.
21. Friedlander (1997).
22. Anonymous, in Foster (October, 2003), p. 190.
23. Graysmith (2003), p. 54.
24. Willman (2011), p. 96.
25. Cole (2009).
26. Ackelsberg, in Cole (2009), p. 51.
27. Guillemin (2011), pp. 49–50.
28. Unidentified physician, in Altman (October 18, 2001).
29. Shieh, et al. (November, 2003), p. 1901.
30. Grossman, in Ferraro (October 19, 2001).
31. Guillemin (2011).
32. Cole (2009), p. 52.
33. Anonymous, in U.S. Department of Justice (February 9, 2010), p. 2.
34. Graysmith (2003), p. 54.
35. Miller, et al. (2001/2002).
36. Miller, et al. (September 4, 2001).
37. *Ibid.*
38. Miller (October 14, 2001).
39. Guillemin (2011), p. 62.
40. Giuliani, in CNN Transcripts (October 13, 2001).
41. Giuliani, in CNN Transcripts (October 15, 2001).
42. Giuliani, in CNN Transcripts (October 12, 2001).
43. Guillemin (2011), p. 57.
44. Cole (2009), p. 53.
45. *Ibid.*
46. Graysmith (2003).
47. *Ibid.*, p. 74.
48. *Ibid.*
49. *Ibid.*
50. Guillemin (2011), p. 122.
51. Harris (October 20, 2001).
52. *Ibid.*
53. *Ibid.*
54. Federal Bureau of Investigation (December 21, 2001).
55. *Ibid.*, p. 1.
56. CNN (November 30, 2001).
57. *Ibid.*
58. Blair, in BBC (October 21, 2001).
59. Graysmith (2003).
60. *Ibid.*, p. 104.
61. Anonymous, in U.S. Department of Justice (February 9, 2010), p. 2.
62. *Ibid.*

63. Cole (2009).
64. Lee, in Cole (2009), p. 55.
65. Graysmith (2003), p. 99.
66. Simon (2013).
67. Bush (October 23, 2001).
68. MacQueen (2014).
69. *Ibid.*, p. 81.
70. Greenwald (April 9, 2007).
71. DeYoung (October 1, 2006).
72. Powell (October 5, 2003).
73. *Ibid.*
74. MacQueen (2014), p. 167.
75. DeYoung (October 1, 2006).
76. Guillemin (2011).
77. MacQueen (2014), p. 50.
78. Klein (April 22, 2015).
79. Miller (2015).
80. Guillemin (2011).
81. Assaad, in Graysmith (2003), p. 89.
82. Guillemin (2011), p. 63.
83. *Ibid.*
84. *Ibid.*
85. Cole (2009), p. 194.
86. Graysmith (2003), p. 311.
87. *Ibid.*
88. Thompson (2003), p. 192.
89. *Ibid.*, p. 194.
90. Hatfill (August 25, 2002).
91. Tell (September 16, 2002).
92. *Ibid.*
93. Hatfill (August 25, 2002).
94. *Ibid.*
95. Cole (2009).
96. *Ibid.*
97. *Ibid.*
98. Attorneys, in Shane and Lichtblau (June 28, 2008).
99. Grassley, in Lichtblau (August 9, 2008).
100. Salyers, in Thompson (2003), p. 201.
101. *Ibid.*
102. Guillemin (2011).
103. *Ibid.*, p. 4.
104. U.S. District Court for the District of Columbia (October 31, 2007), p. 7.
105. *Ibid.*, p. 13.
106. *Ibid.*
107. Guillemin (2011).
108. *Ibid.*
109. *Ibid.*
110. U.S. District Court for the District of Columbia (October 31, 2007); p. 14.
111. Ivins, in Willman (2011), p. 111.
112. *Ibid.*
113. Engelberg, et al. (October 10, 2011).
114. Kemp, in MacQueen (2014), p. 91.
115. Shane and Lichtblau (August 6, 2008).
116. U.S. District Court for the District of Columbia (October 31, 2007), p. 6.
117. Lawyer, in Shane and Lichtblau (August 6, 2008).
118. Shane and Lichtblau (August 5, 2008).
119. Engelberg, et al. (October 10, 2011).
120. U.S. District Court for the District of Columbia (October 31, 2007).
121. *Ibid.*, p. 9.
122. *Ibid.*, p. 11.
123. Warrick, et al. (August 2, 2008).
124. Goldstein, et al. (August 6, 2008).
125. *Ibid.*
126. Byrne, in Warrick, et al. (August 2, 2008).
127. Friend, in Willman (August 1, 2008).
128. Goldstein, et al. (August 6, 2008).
129. *Ibid.*
130. Olbermann, in *Countdown with Keith Olbermann* (August 4, 2008).
131. Bryne, in Associated Press (August 1, 2008).
132. Lambert v. (1) Holder (2) Mueller (3) Kelly (4) DOJ Unknown Employees (5) FBI Unknown Employees (6) U.S. Dept. of Justice (7) FBI (Filed April 2, 2015), pp. 25–26.
133. *Ibid.*, p. 28.
134. Leahy, in Shane (September 18, 2008).
135. Grassley (February 15, 2011).
136. National Research Council, Board of Life Sciences, Division on Earth and Life Studies, Technology and Law Committee on Science, Policy and Global Affairs Division, Committee on Review of the Scientific Approaches Used During the FBI's Investigation of the 2001 Bacillus Anthracis Mailings (June 1, 2011), p.119.
137. Epstein (January 24, 2010).
138. *Ibid.*
139. Hylton (October 26, 2011).
140. Keim, et al. (2010), p.23.
141. *HHS Funds Drug Development for Bioterror Infections, Pneumonia* (press release) (May 24, 2013).
142. Jang, et al. (July 22, 2013).
143. Miller (October 3, 2008).
144. *Ibid.*
145. *Ibid.*
146. Young (June 18, 2015).
147. *Ibid.*
148. *Ibid.*
149. Pita and Gunaratna (May 15, 2009).

Bibliography

Abbott, Carl (Winter 2015). Revisiting Rajneeshpuram: Oregon's Largest Utopian Community as Western History. *Oregon Historical Quarterly.* Vol. 116, No. 4, pp. 414–447.

ABC News (October 8, 2001). Anthrax Scare Shuts Down National Enquirer. *ABC News.* Retrieved April 12, 2015, http://abcnews.go.com/Entertainment/story?id=102020.

Alibek, Ken, with Stephen Handelman (1999). *Biohazard: The Chilling True Story of the Largest Covert Biological Weapons Program in the World—Told from the Inside by the Man Who Ran It.* New York: Delta (Dell Publishing).

Altman, Lawrence K. (October 18, 2001). A NATION CHALLENGED: NBC; Doctor in City Reported Anthrax Case Before Florida. *New York Times.* Retrieved April 15, 2015, http://www.nytimes.com/2001/10/18/nyregion/a-nation-challenged-nbc-doctor-in-city-reported-anthrax-case-before-florida.html.

American Society for Microbiology (February 26, 2006). Anthrax Spores May Survive Water Treatment. *Science Daily.* Retrieved June 1, 2016, http://www.sciencedaily.com/releases/2006/02/060226115234.htm.

Amnesty International (September 19, 2016). Urgent Action: Thirteen Men at Risk of Execution in Japan. *Amnesty International.* Retrieved September 30, 2016, https://www.amnesty.org/en/documents/asa22/4856/2016/en/.

Anthrax Hard to Remove from Drinking Water Systems. (April 15, 2006). *Clinical Infectious Disease.* Vol. 42, No. 8, pp. iii–iv.

Army of God (Undated). *Army of God Manual.* Army of God (website). Retrieved February 14, 2016, http://www.armyofgod.com/AOGsel3.html.

Associated Press (July 22, 1986). Ma Sheela Gets 4 1/2 Years for Fraud, Wiretapping. *Los Angeles Times.* Retrieved August 3, 2016, http://articles.latimes.com/1986–07-22/news/mn-30934_1_ma-sheela.

Associated Press (November 24, 1995). Kansas Doctor Is Accused in Fire That Killed 2 of Her Children. *New York Times.* Retrieved December 19, 2015, http://www.nytimes.com/1995/11/24/us/kansas-doctor-is-accused-in-fire-that-killed-2-of-her-children.html.

Associated Press (July 25, 1996). Doctor Accused of Injecting His Girlfriend with HIV. *Los Angeles Times.* Retrieved January 13, 2016, http://articles.latimes.com/1996–07-25/news/mn-27692_1_vitamin-injections.

Associated Press (September 11, 1998). Hospital Employee Sentenced to 20 Years for Poisoning Co-Workers. *The Paris News* (Paris, Texas), p. 2.

Associated Press (January 9, 1999). Man Who Infected Son with HIV Gets Life. *Los Angeles Times.* Retrieved January 19, 2016, http://articles.latimes.com/1999/jan/09/news/mn-61920.

Associated Press (August 1, 2008). Two Portraits of Anthrax Suspect. *NBC News.* Retrieved July 3, 2015, http://www.nbcnews.com/id/25972123/ns/us_news-security/t/two-portraits-anthrax-suspect/.

Associated Press (June 5, 2009). Injected with HIV by Dad as Baby, Teen Inspires. *NBC News*. Retrieved January, October 2016, http://www.nbcnews.com/id/31129553/ns/health-aids/t/injected-hiv-dad-baby-teen-inspires/.

Bailey, Ronald (October 14, 2001). Bioterrorism: Searching for Breathing Room/Anthrax Threats: Don't Panic, Be Cool. *SF Gate (San Francisco Chronicle)*. Retrieved April 2, 2016, http://www.sfgate.com/opinion/article/BIOTERRORISM-Searching-for-Breathing-Room-2869900.php.

Bari, Judi (1994). *Timber Wars*. Monroe, Maine: Common Courage Press.

BBC (October 21, 2001). New Law Targets Anthrax Hoaxers. *British Broadcasting Corporation*. Retrieved May 7, 2015, http://news.bbc.co.uk/2/hi/uk_news/1611170.stm.

BBC (April 6, 2016). Aum Shinrikyo: The Japanese Cult Surfacing in Europe. *British Broadcasting Corporation*. Retrieved September 30, 2016, http://www.bbc.com/news/world-asia-35975069.

Beam, Louis (1983). Leaderless Resistance. *The Seditionist*. Retrieved February 12, 2016, http://www.armyofgod.com/LeaderlessResistance.htm.

Bell, Andrew McIlwaine (2010). *Mosquito Soldiers: Malaria, Yellow Fever, and the Course of the American Civil War*. Baton Rouge: Louisiana State University Press.

Bhagwan Shree Rajneesh Biography—Religious Figure, Criminal (1931–1990). (2016). *Bio (A&E Television Networks)*. Retrieved June 27, 2016, http://www.biography.com/people/bhagwan-shree-rajneesh-20900613#spiritual-leadership-.

Biederbick, Walter. (2009) "Terrorism and Potential Biological Warfare Agents" (Chapter Six). In *Unconventional Weapons and International Terrorism: Challenges and New Approaches*, edited by Ranstorp, Magnus, and Magnus Normark. Oxford: Routledge.

Biohackers of the World, Unite. (Sept. 6, 2014). *The Economist*. Retrieved October 17, 2015, http://www.economist.com/news/technology-quarterly/21615064-following-example-maker-communities-worldwide-hobbyists-keen-biology-have.

Bioterrorism Overview. (February 12, 2007). *Centers for Disease and Prevention*. Retrieved May 12, 2016, http://emergency.cdc.gov/agent/smallpox/overview/disease-facts.asp.

Bisset, N. G. (December 1979). Arrow Poisons in China. Part I. *Journal of Ethnopharmacology*, Vol. 1, No. 4, pp. 325–84.

Bliss, Michael (1984). *Banting: A Biography*. Toronto: McClelland and Stewart.

Bobe, Jason (June 3, 2013A). Draft DIYbio Code of Ethics from European Congress. *DIYbio*. Retrieved November 14, 2015, http://diybio.org//?s=ethics.

Bobe, Jason (June 3, 2013B). Draft DIYbio Code of Ethics from North American Congress. *DIYbio*. Retrieved November 14, 2015, http://diybio.org//?s=ethics.

Bollier, David (2002). *Silent Theft: The Private Plunder of Our Common Wealth*. London: Routledge.

Boseley, Sarah, and Julian Borger (May 16, 2005). U.S. Scientists Push for Go-Ahead to Genetically Modify Smallpox Virus. *The Guardian*. Retrieved April 27, 2016, http://www.theguardian.com/society/2005/may/16/research.medicineandhealth.

Botulism. (August 2013). *World Health Organization*. Retrieved April 22, 2016, http://www.who.int/mediacentre/factsheets/fs270/en/.

Brenner, F. W., R. G. Villar, F. J. Angulo, R. Tauxe, and B. Swaminathan (July 2000). *Salmonella* Nomenclature. *Journal of Clinical Microbiology*. Vol. 38, No. 7, pp. 2465–2467.

Broad, William J. (May 26, 1998). Sowing Death: A Special Report; How Japan Germ Terror Alerted World. *New York Times*. Retrieved August 5/12, http://www.nytimes.com/1998/05/26/world/sowing-death-a-special-report-how-japan-germ-terror-alerted-world.html?pagewanted=all.

Brower, Jennifer (2005). "Smallpox" (entry). In *Weapons of Mass Destruction: An Encyclopedia of Worldwide Policy, Technology, and History*, edited by Croddy, Eric A., James J. Wirtz, and Jeffrey A. Larsen. Santa Barbara: ABC-CLIO, Inc.

Bureau of the Census (December 1981). "Number of Inhabitants, Part 39—Oregon" (Chapter A). In *1980 Census of Population, Volume 1: Characteristics of the Population*, U.S. Department of Commerce. Washington, D.C.: U.S. Government Printing Office.

Burgess, Richard (June 11, 2015). State Board Denies Parole to Former Lafayette Doctor Convicted of Injecting Mistress with AIDS, Hepatitis C. *The Advocate*. Retrieved January, December 2016, http://theadvocate.com/news/12622392-123/state-board-denies-parole-to.

Burrows, W. Dickinson, and Sara E. Renner (December 1999). Biological Warfare Agents as Threats to Potable Water. *Environmental Health Perspectives*. Vol. 107, No. 12, pp. 975–984.

Bush, George W. (October 23, 2001). President Says Terrorists Won't Change American Way of Life (press release). *White House Archives*. Retrieved May 25, 2015, http://georgewbush-whitehouse.archives.gov/news/releases/2001/10/print/20011023-33.html.

Carter, Lewis F. (August 31, 1990). *Charisma and Control in Rajneeshpuram: A Community Without Shared Values* (American Sociological Association Rose Monographs). Cambridge: Cambridge University Press.

Carus, W. Seth (2000). "The Rajneeshees (1984)" (Chapter 8). In *Toxic Terror*, edited by Jonathan B. Tucker. Cambridge, Massachusetts: MIT Press.

Carus, W. Seth, Center for Counterproliferation Research: National Defense University (2002). *Bioterrorism and Biocrimes: The Illicit Use of Biological Agents Since 1900*. Amsterdam: Fredonia Books.

Casagrande, Rocco (Fall/Winter 2000). Biological Terrorism Targeted at Agriculture: The Threat to U.S. National Security. *The Nonproliferation Review*, pp. 92–105.

Centers for Disease Control and Prevention (October 4, 2001). Public Health Message Regarding Anthrax Case (press release). *Centers for Disease Control and Prevention*. Retrieved April 13, 2015, http://www.cdc.gov/media/pressrel/r011004.htm.

Centers for Disease Control and Prevention (October 11, 2001). Public Health Message Regarding Anthrax Case (press release—update). *Centers for Disease Control and Prevention*. Retrieved April 13, 2015, http://www.cdc.gov/media/pressrel/r011011a.htm.

Centers for Disease Control and Prevention (April 13, 2012). Case Study III—Salmonella in Oregon. *Forensic Epidemiology Original CDC Scenarios (CDC Public Health Law Program)*. Retrieved July 22, 2016, http://www.cdc.gov/phlp/docs/fe14.pdf.

Chalk, Colin H., Tim J. Benstead, and Mark Keezer (February 20, 2014). Medical Treatment for Botulism. *Cochrane Database of Systematic Reviews*. Retrieved April 22, 2016, http://onlinelibrary.wiley.com/doi/10.1002/14651858.CD008123.pub3/full.

Charisius, Hanno, Richard Friebe, and Sascha Karberg (January 24, 2013). Becoming Biohackers: The Long Arm of the Law. *British Broadcasting Corporation*. Retrieved October 1, 2015, http://www.bbc.com/future/story/20130124-biohacking-fear-and-the-fbi.

Clarkson, Frederick (February 19, 2002). Brand New War for the Army of God? *Salon*. Retrieved February 15, 2016, http://www.salon.com/2002/02/19/gays_10/.

CNN (April 24, 1998). Mother in HIV Case Pleads for Privacy. *CNN*. Retrieved January, July 2016, http://edition.cnn.com/US/9804/24/aids.injection/index.html.

CNN (November 30, 2001). Abortion Foe Suspected in Anthrax Hoax Letters. *CNN*. Retrieved May 6, 2015, http://edition.cnn.com/2001/US/11/29/anthrax.hoaxes/index.html.

CNN Transcripts (October 12, 2001). Giuliani Gives Press Conference on Woman Infected with Anthrax. *CNN*. Retrieved April 25, 2015, http://www.cnn.com/TRANSCRIPTS/0110/12/se.16.html.

CNN Transcripts (October 13, 2001). Anthrax Scare: Giuliana Holds Press Conference. *CNN*. Retrieved April 22, 2015, http://www.cnn.com/TRANSCRIPTS/0110/13/se.04.html.

CNN Transcripts (October 15, 2001). Rudolph Giuliani Holds Press Conference. *CNN*. Retrieved April 22, 2015, http://www.cnn.com/TRANSCRIPTS/0110/15/se.17.html.

Cody, Rachel Graham (March 8, 2005). The Saffron Swami. *The Willamette Week*. Retrieved July 5, 2016, http://www.wweek.com/portland/article-4182-1983.html.

Cole, Leonard A. (May 2, 2005). The Problem of Biological Weapons. *Journal of Clinical Investigation*, Vol. 115, No. 5, pp. 1110.

Cole, Leonard A. (2009). *The Anthrax Letters: A Bioterrorism Expert Investigates the Attacks That Shocked America*. New York: Skyhorse Publishing.

Collins, Francis S. (1999). The Human Genome Project and the Future of Medicine. *Annals of the New York Academy of the Sciences*, Vol. 882, pp. 42–55. Retrieved October 16, 2015, http://onlinelibrary.wiley.com/doi/10.1111/j.1749–6632.1999.tb08532.x/abstract.

Countdown with Keith Olbermann (August 4, 2008). MSNBC (broadcast).

Croddy, Eric A. (2005). "Vector" (entry). In *Weapons of Mass Destruction: An Encyclopedia of Worldwide Policy, Technology, and History*, edited by Croddy, Eric A., James J. Wirtz, and Jeffrey A. Larsen. Santa Barbara: ABC-CLIO.

Cullison, Alan, and Andrew Higgins (December 31, 2001). Forgotten Computer Reveals Thinking Behind Four Years of Al Qaeda Doings. *Wall Street Journal*. Retrieved February 28, 2015, http://www.wsj.com/articles/SB100975171479902000.

Danzig, Richard, Marc Sageman, Terrance Leighton, Lloyd Hough, Hidemi Yuki, Rui Kotani, and Zachary M. Hosford (2011). *Aum Shinrikyo: Insights into How Terrorist Develop Biological and Chemical Weapons*. Washington, D.C.: Center for a New American Security.

Dees, Morris (1997). *Gathering Storm: America's Militia Threat*. New York: Harper-Perennial.

Delfanti, Alessandro (2013). *Biohackers: The Politics of Open Science*. London: Pluto Press.

Dembek Z. F., L. A. Smith, and J. M. Rusnak (2007). Botulism: Cause, Effects, Diagnosis, Clinical and Laboratory Identification, and Treatment Modalities. *Disaster Medicine and Public Health Preparedness*. Vol. 1, No. 2, pp. 122–134.

Department of Defense (Nov. 8, 2010). *DOD Dictionary of Military and Associated Terms* (Joint Publication 1–02). Retrieved February 24, 2015, http://www.dtic.mil/doctrine/dod_dictionary/data/t/7591.html.

DeYoung, Karen (October 1, 2006). falling On His Sword. *Washington Post*. Retrieved May 23, 2015, http://www.washingtonpost.com/wp-dyn/content/article/2006/09/27/AR2006092700106.html.

Dire, Daniel J., Robert G. Darling, Michael J. Burns, John D. Halamka, Edmond A. Hooker II, and John T. VanDeVoort (September 3, 2013). *CBRNE—Biological Warfare Agents: Historical Aspects of Biological Warfare Agents*. WebMD (Medscape References: Drugs, Diseases & Procedures). Retrieved October 8, 2014, http://emedicine.medscape.com/article/829613-overview#showall.

Dong, Jianli, and J. P. Olano, J. W. McBride, and D. H. Walker (May 2008). Emerging Pathogens: Challenges and Successes of Molecular Diagnostics. *Journal of Molecular Diagnostics*. Vol. 10, No. 3, pp. 185–197.

Drayton, Richard (May 9, 2005). An Ethical Blank Cheque. *The Guardian* (Editorial). Retrieved March 29, 2015, http://www.theguardian.com/politics/2005/may/10/foreignpolicy.usa.

Dye, Lee (January 7, 2006). Scientists Use Virus to Trace Assault Suspect. *ABC News*. Retrieved January 21, 2016, http://abcnews.go.com/Technology/story?id=97856&page=1.

Editorial: Microbiology by Numbers. (September 2011). *Nature Reviews Microbiology*. Vol. 9, No. 9, p. 628.

Elmer-DeWitt, Philip (September 30, 2001). America's First Bioterrorism Attack. *TIME Magazine*. Retrieved July 18, 2016, http://content.time.com/time/magazine/article/0,9171,176937,00.html.

Engelberg, Stephen, Greg Gordon, Jim Gilmore, and Mike Wiser (October 10, 2011). New Evidence Adds Doubt to FBI's Case Against Anthrax Suspect. *ProPublica, PBS Frontline, and McClatchy* (joint project). Retrieved July 3, 2015, http://www.propublica.org/article/new-evidence-disputes-case-against-bruce-e-ivins.

Enserink, Martin (November 23, 2011). Scientists Brace for Media Storm Around Controversial Flu Studies. *Science Insider/American Association for the Advancement of Science Magazine*. Retrieved November 3, 2015, http://news.sciencemag.org/2011/11/scientists-brace-media-storm-around-controversial-flu-studies.

Epstein, Edward Jay (January 24, 2010). The Anthrax Attacks Remain Unsolved. *Wall Street Journal*. Retrieved July 25, 2015, http://www.wsj.com/articles/SB10001424052748704541004575011421223515284.

Fainaru, Steve, and Job Warrick (November 30, 2001). Ames Strain of Anthrax Limited to Few Labs. *Washington Post*, p. A.01.

Federal Bureau of Investigation (Undated). Weapons of Mass Destruction. *Federal Bureau of Investigation (FBI)*. Retrieved October 14, 2014, http://www.fbi.gov/about-us/investigate/terrorism/wmd/wmd_faqs.

Federal Bureau of Investigation (December 21, 2001). *FBI Warns Against Anthrax Hoaxes* (National Press Release). Washington, D.C.: FBI Press Office.

Ferraro, Susan (October 19, 2001). Diagnosis Often Hard to Unlock: Testing Is Complex. *New York Daily News*. Retrieved April 15, 2015, http://www.nydailynews.com/archives/news/diagnosis-hard-unlock-testing-complex-article-1.927400.

Fitzgerald, Frances (1986). *Cities on a Hill: A Journey Through Contemporary American Cultures*. New York: Simon & Schuster.

Flynn, Dan (October 7, 2009). Salmonella Bioterrorism: 25 Years Later. *Food Safety News*. Retrieved July 18, 2016, http://www.foodsafetynews.com/2009/10/for-the-first-12/.

Foster, Don (October, 2003). The Message in the Anthrax. *Vanity Fair*, pp. 180–200.

Franklin, Satya Bharti (1992). *Promise of Paradise*. Barrytown, New York: Barrytown/Station Hill Press.

Franz, D. R. (July 31, 2002). *Potential for Biological Terrorism Using the Food Supply Chain*. Presentation, 39th Florida Pesticide Residue Workshop Joint Technical Session. St. Petersburg, Florida.

Frequently Asked Questions on Ebola Virus Disease. (January 2016). *World Health Organization*. Retrieved April 25, 2016, http://www.who.int/csr/disease/ebola/faq-ebola/en/.

Friedlander, Arthur M. (1997). "Anthrax" (Chapter 22). In *Medical Aspects of Chemical and Biological Warfare*, edited by Frederick R. Sidell, Ernest T. Takafuji, and David R. Franz. Washington, D.C.: Office of the Surgeon General.

Frischknecht, Friedrich (2003). The History of Biological Warfare. *European Molecular Biology Organization (EMBO) Reports*. Vol. 4, Special Issue, pp. S47-S52.

Garrett, Laurie (December 15, 2011). The Bioterrorist Next Door. *Foreign Policy Magazine*. Retrieved September 1, 2015, http://foreignpolicy.com/2011/12/15/the-bioterrorist-next-door/.

Gates, Bill (Feb. 2, 1976). An Open Letter to Hobbyists. *Microsoft Archives*. Retrieved October 10, 2015, http://www.microsoft.com/about/companyinformation/timeline/timeline/docs/di_hobbyists.doc.

Geggel, Laura (October 7, 2014). Doctors Puzzled Why Only Some Ebola Patients Bleed. *Live Science*. Retrieved April 25, 2016, http://www.livescience.com/48182-why-ebola-causes-bleeding.html.

Geist, William E. (September 16, 1981). Cult in Castle Troubling Montclair. *New York Times*. Retrieved July 2, 2016, http://www.nytimes.com/1981/09/16/nyregion/cult-in-castle-troubling-montclair.html.

Gellman, Barton (March 23, 2003). Al Qaeda Near Biological, Chemical Arms Production. *Washington Post*, p. A01.

Gerstein, Daniel M. (2009). *Bioterror in the 21st Century: Emerging Threats in a New Global Environment*. Annapolis, Maryland: Naval Institute Press.

Goldman, Armond S., and Frank C. Schmalstieg (May 2007). Abraham Lincoln's Gettysburg Illness. *Journal of Medical Biography*. Vol. 15, No. 2, pp. 104–110.

Goldman, Marion (2011). "Cultural Capital, Social Networks, and Collective Violence at Rajneeshpuram" (Chapter 15). In *Violence and New Religious Movements*, edited by James R. Lewis. New York: Oxford University Press.

Goldstein, Amy, Nelson Hernandez, and Anne Hull (August 6, 2008). Tales of Addiction, Anxiety, Ranting. *Washington Post*. Retrieved July 1, 2015, http://www.washingtonpost.com/wp-dyn/content/article/2008/08/05/AR2008080503747.html.

Goozner, Merrill (March 30, 1995). Japan's Top Cop Survives Assassin's Bid. *Chicago Tribune*. Retrieved September 28, 2016, http://articles.chicagotribune.com/1995–03-30/news/9503300193_1_takaji-kunimatsu-supreme-truth-deadly-nerve-gas-sarin.

Goozner, Merrill (May 16, 1995). Sweep of 130 Sites Culminates 2-Month Probe of Gas Attack. *Chicago Tribune*. Retrieved September 28, 2016, http://articles.chicagotribune.com/1995-05-16/news/9505160272_1_gas-attack-cult-aum-shinri-kyo.

Gordon, James S. (1987). *The Golden Guru: The Strange Journey of Bhagwan Shree Rajneesh*. Lexington, Massachusetts: The Stephen Greene Press.

Gosnell, Hannah, Jeffrey D. Kline, Garrett Chrostek, and James Duncana (2011). Is Oregon's Land Use Planning Program Conserving Forest and Farm Land? A Review of the Evidence. *Land Use Policy*, Vol. 28, No. 1, pp. 185–192.

Grassley, Chuck (February 15, 2011). *Grassley Response to National Academy of Sciences Amerithrax Report* (memorandum). Office of Senator Chuck Grassley. Retrieved July 5, 2015, http://www.grassley.senate.gov/news/news-releases/grassley-response-national-academy-sciences-amerithrax-report.

Graysmith, Robert (2003). *Amerithrax: The Hunt for the Anthrax Killer*. New York: Berkley Books.

Green, Manfred S., and NATO Staff (2007). *Risk Assessment and Risk Communication Strategies in Bioterrorism Preparedness*. Dordrecht, The Netherlands: Springer.

Greenberg, Jon (January 8, 2015). Kohn: Since September 11, Right-Wing Extremists Killed More Americans Than Islamic Extremists. *PunditFact*. Retrieved February 22, 2016, http://www.politifact.com/punditfact/statements/2015/jan/08/sally-kohn/kohn-911-right-wing-extremists-killed-more-america/.

Greenwald, Glenn (April 9, 2007). The Unresolved Story of ABC News' False Saddam-Anthrax Reports. *Salon*. Retrieved May 23, 2015, http://www.salon.com/2007/04/09/abc_anthrax/.

Grunow, Roland, Luzie Verbeek, Daniela Jacob, Thomas Holzmann, Gabriele Birkenfeld, Daniel Wiens, Leonie von Eichel-Streiber, Gregor Grass, Udo Reischl (December 2012). *Deutsches Ärzteblatt International*. Vol. 109, No. 49, pp. 843–848.

Grushkin, Daniel, Todd Kuiken, and Piers Millet. (November, 2013). Seven Myths & Realities About Do-It-Yourself Biology. *Woodrow Wilson Center International Center for Scholars, Synthetic Biology Project* (SYNBIO). Retrieved October 16, 2015, http://www.synbioproject.org/process/assets/files/6673/_draft/7_myths_final.pdf.

Guillemin, Jeanne (2011). *American Anthrax: Fear, Crime, and the Investigation of the Nation's Deadliest Bioterror Attack*. New York: Times Books / Henry Holt.

Gurr, Nadine, and Benjamin Cole (2002). *The New Face of Terrorism: Threats from Weapons of Mass Destruction*. London: I.B. Tauris.

Guru's Disciples Taking Over in Oregon Town. (December 19, 1982). *New York Times*. Retrieved July 11, 2016, http://www.nytimes.com/1982/12/19/us/guru-s-disciples-taking-over-in-oregon-town.html.

Guthrie, Julian (December 20, 2009). Do-It-Yourself Biology Grows with Technology. *San Francisco Chronicle*. Retrieved November 12, 2015, http://www.sfgate.com/science/article/Do-it-yourself-biology-grows-with-technology-3277834.php.

Hamblin, James (May 6, 2013). What Does Sarin Do to People? *The Atlantic*. Retrieved September 11, 2016, http://www.theatlantic.com/health/archive/2013/05/what-does-sarin-do-to-people/275577/.

Hammarlund, Marc, Erik M. Jorgensen and Michael J. Bastiani (January 29, 2007). Axons Break in Animals Lacking β-spectrin. *Journal of Cell Biology*. Vol. 176, No. 3., pp. 269–275.

Hannemyr, Gisle (February 1999). Technology and Pleasure: Considering Hacking Constructive. *First Monday* (journal). Retrieved October 5, 2015, http://journals.uic.edu/ojs/index.php/fm/article/view/647/562.

Harris, Paul (October 20, 2001). Anthrax Hoax Chaos. *The Guardian*. Retrieved May 6, 2015, http://www.theguardian.com/world/2001/oct/21/anthrax.terrorism.

Harris, Sheldon (December 1992). Japanese Biological Warfare Research on Humans: A Case Study of Microbiology and Ethics. *Annals of the New York Academy of Sciences*. Vol. 666, pp. 21–52.

Hatfill, Steven (August 25, 2002). Anthrax Investigation (press conference). *C-SPAN*. Retrieved June 7, 2015, http://www.c-span.org/video/?172241-1/anthrax-investigation.

Herfst, S., E. J. A. Schrauwen, M. Linster, S. Chutinimitkul, E. De Wit, V. J. Munster, E. M. Sorrell, T. M. Bestebroer, D. F. Burke, D. J. Smith, G. F. Rimmelzwaan, A. D. M. E. Osterhaus, and R. A. M. Fouchier (2012). Airborne Transmission of Influenza A/H5N1 Virus Between Ferrets. *Science*. Vol. 336, No. 6088, pp. 1534–1541.

HHS Funds Drug Development for Bioterror Infections, Pneumonia (press release). (May 24, 2013). Washington, D.C.: U.S. Department of Health and Human Services News Division.

Hickman, Donald C. (1999). A Chemical and Biological Warfare Threat: USAF Water Systems at Risk (Counterproliferation Paper No. 3). *Air War College, Air University, United States Air Force*. Retrieved May 24, 2016, http://www.au.af.mil/au/awc/awcgate/cpc-pubs/hickman.htm.

Himanen, Pekka (2001). *The Hacker Ethic: A Radical Approach to the Philosophy of Business*. New York: Random House.

Hitchens, Christopher (2007). *God Is Not Great: How Religion Poisons Everything*. New York/Boston: Twelve (Hachette Book Group).

Holloway, Dustin T. (March 1, 2013). Regulating Amateurs. *The Scientist*. Retrieved October 28, 2015, http://www.the-scientist.com/?articles.view/articleNo/34444/title/Regulating-Amateurs/.

Hongo, Jun (November 22, 2011). Last Trial Brings Dark Aum Era to End. *Japan Times*. Retrieved September 30, 2016, http://www.japantimes.co.jp/news/2011/11/22/reference/last-trial-brings-dark-aum-era-to-end/.

Hongo, Jun, and Yumi Wijers-Hasegawa (September 16, 2006). Asahara's Execution Finalized. *Japan Times*. Retrieved September 30, 2016, http://www.japantimes.co.jp/news/2006/09/16/national/asaharas-execution-finalized/.

Hussain, Hamid Y. (August 5, 2014). Incidence and Mortality Rate of "Middle East Respiratory Syndrome"—Corona Virus (MERS-CoV), Threats and Opportunities. *Journal of Mycobacterial Diseases*. Vol. 4, No. 4, p. 162.

Hylton, Wil S. (October 26, 2011). How Ready Are We for Bioterroism? *New York Times Magazine*. Retrieved July 28, 2015, http://www.nytimes.com/2011/10/30/magazine/how-ready-are-we-for-bioterrorism.html?_r=0.

Ibeji, Mike (February 17, 2011). Black Death: The Disease. *British Broadcasting Corporation*. Retrieved April 23, 2016, http://www.bbc.co.uk/history/british/middle_ages/black disease_01.shtml.

Imai, Masaki, Tokiko Watanabe, Masato Hatta, Subash C. Das, Makoto Ozawa, Kyoko Shinya, Gongxun Zhong, Anthony Hanson, Hiroaki Katsura, Shinji Watanabe, Chengjun Li, Eiryo Kawakami, Shinya Yamada, Maki Kiso, Yasuo Suzuki, Eileen A. Maher, Gabriele Neumann, and Yoshihiro Kawaok (June 21, 2012). Experimental Adaptation of an Influenza H5 HA Confers Respiratory Droplet Transmission to a Reassortant H5 HA/H1N1 Virus in Ferrets. *Nature*, Vol. 486, pp. 420–428.

Incorporation of Rajneeshpuram Opens Door to Development (Part 9 of 20). (July 8, 1985). *The Oregonian/Oregon Live*. Retrieved July 6, 2016, http://www.oregonlive.com/rajneesh/index.ssf/1985/07/incorporation_of_city_opens_do.html.

Injection Anthrax. (July 21, 2014). *Centers for Disease Control and Prevention*. Retrieved April 21, 2016, http://www.cdc.gov/anthrax/basics/types/injection.html.

Institute for Economics and Peace (2015). *Global Terrorism Index*. Sydney: Institute for Economics and Peace.

International Human Genome Sequencing Consortium (February 15, 2001). Initial Sequencing and Analysis of the Human Genome. *Nature*. Vol. 409, pp. 860–921.

IRA Overpowers Crew, Sinks British Ship In Irish Bay. (February 8, 1981). *Chicago Tribune*, Section 3, p. 6.

Iverson, Kenneth (2003). *Demon Doctors: Physicians as Serial Killers*. Tucson: Galen Press.

Jang, Kyoung Hwa, Sang-Jip Nam, Jeffrey B. Locke, Christopher A. Kauffman, Deanna S.

Beatty, Lauren A. Paul, and William Fenical (July 22, 2013). Anthracimycin, a Potent Anthrax Antibiotic from a Marine-Derived Actinomycete. *Angewandte Chemie International Edition*, Vol. 52, Issue 30, pp. 7822–7824.

Jansen, H. J., and F. J. Breeveld, C. Stijnis, and M. P. Grobusch (June 2014). Biological Warfare, Bioterrorism, and Biocrime. *Clinical Microbiology and Infection*. Vol. 20, No. 6, pp. 488–496.

Jean, C. M., and S. Honarmand, J. K. Louie, and C. A. Glaser (December 2007). Risk Factors for West Nile Virus Neuroinvasive Disease, California, 2005. *Emerging Infectious Diseases*. Vol. 13, No. 12, pp. 1918–1920.

Jefferson, Catherine (July 5, 2013). The Role of Codes of Conduct in the Amateur Biology Community. *Royal Society of Biology*. Retrieved November 2, 2015, https://blog.rsb.org.uk/codes-of-conduct-in-the-amateur-biology-community/.

Johnson, Thomas J. (Undated). *A History of Biological Warfare from 300 B.C.E. to the Present*. American Association of Respiratory Care. Retrieved October 14, 2014, http://c.aarc.org/resources/biological/history.asp.

Kamikuishiki's Gulliver Park Falls. (November 13, 2001). *Japan Times*. Retrieved August 27, 2016, http://www.japantimes.co.jp/news/2001/11/13/national/kamikuishikis-gulliver-park-falls/#.V8IAcoCZGbU.

Kaplan, David E., and Andrew Marshall (1996). *The Cult at the End of the World: The Terrifying Story of the Aum Doomsday Cult, from the Subways of Tokyo to the Nuclear Arsenals of Russia*. New York: Random House.

Keim, Paul S., Bruce Budowle, and Jacques Ravel (2010). "Microbial Forensic Investigation of the Anthrax-Letter Attacks" (Chapter 2). In *Microbial Forensics* (Second Edition), edited by Budowle, Bruce, Steven E. Schutzer, Robert G. Breeze, Paul S. Keim, and Stephen A. Morse. San Diego: Academic Press.

Keyes, Scott (June 10, 2014). A Strange but True Tale of Voter Fraud and Bioterrorism. *The Atlantic*. Retrieved July 19, 2016, http://www.theatlantic.com/politics/archive/2014/06/a-strange-but-true-tale-of-voter-fraud-and-bioterrorism/372445/.

Klein, Julia M. (April 22, 2015). Judith Miller Tells Her Side of "The Story." *Columbia Journalism Review*. Retrieved May 26, 2015, http://www.cjr.org/analysis/miller_review.php.

Kolata, Gina (November 13, 2001). New York Was Bioterrorism Target, in 1864. *New York Times*. Retrieved on August 28, 2014, http://www.nytimes.com/2001/11/13/health/new-york-was-bioterrorism-target-in-1864.html.

Kraft, Amy E. (2005). "Tularemia" (entry). In *Weapons of Mass Destruction: An Encyclopedia of Worldwide Policy, Technology, and History*, edited by Croddy, Eric A., James J. Wirtz, and Jeffrey A. Larsen. Santa Barbara: ABC-CLIO, Inc.

Kristof, Nicholas D. (March 14, 1996). Japan Sect's Role in Murder Case Emerges, Prompting Outcry. *New York Times*. Retrieved August 8, 2016, http://www.nytimes.com/1996/03/14/world/japan-sect-s-role-in-murder-case-emerges-prompting-outcry.html.

Kristof, Nicholas D. (March 17, 1996). Unmasking Horror—A Special Report: Japan Confronting Gruesome War Atrocity. *New York Times*. Retrieved December 24/ 14, http://www.nytimes.com/1995/03/17/world/unmasking-horror-a-special-report-japan-confronting-gruesome-war-atrocity.html.

Kristof, Nicholas D. (July 12, 2002). The Anthrax Files. *New York Times*. Retrieved October 15, 2016, http://www.nytimes.com/2002/07/12/opinion/the-anthrax-files.html.

Kunkle, Fredrick (February 27, 2006). Fort Detrick Neighbors Jittery Over Expansion. *Washington Post*. Retrieved January 17, 2015, http://www.washingtonpost.com/wp-dyn/content/article/2006/02/26/AR2006022601423.html.

Kurzman, Charles, and David Schanzer (June 16, 2015). The Growing Right-Wing Terror Threat. *New York Times*. Retrieved February 22, 2016, http://www.nytimes.com/2015/06/16/opinion/the-other-terror-threat.html.

Lambert v. (1) Holder (2) Mueller (3) Kelly (4) DOJ Unknown Employees (5) FBI Unknown Employees (6) U.S. Dept. of Justice (7) FBI. Case 3:15-cv-00147-PLR-HBG. Filed April 2/15.

Land Use Database: Rajneeshpuram Site. (undated). *The Center for Land Use Interpretation*. Retrieved July 6, 2016, http://clui.org/ludb/site/rajneeshpuram-site.

Latkin, Carl A. (1992). Seeing Red: A Social-Psychological Analysis. *Sociological Analysis*. Vol. 53, No. 3, pp. 257–271.

Levy, Steven (April 15, 2010). Geek Power: Steven Levy Revisits Tech Titans, Hackers, Idealists. *Wired*. Retrieved October 17, 2015, http://www.wired.com/2010/04/ff_hackers/.

Levy, Steven (2010). *Hackers: Heroes of the Computer Revolution*. Sebastopol, California: O'Reilly Media, Inc.

Lichtblau, Eric (August 9, 2008). Scientist Officially Exonerated in Anthrax Attacks. *New York Times*. Retrieved June 11, 2015, http://www.nytimes.com/2008/08/09/washington/09anthrax.html?_r=0http://www.nytimes.com/2008/08/09/washington/09anthrax.html?_r=0.

Lifton, Robert Jay (1999). *Destroying the World to Save It: Aum Shinrikyo, Apocalyptic Violence, and the New Global Terrorism*. New York: Metropolitan Books (Henry Holt).

Liptak, Adam (June 13, 2013). Justices, 9–0, Bar Patenting Human Genes. *New York Times*. Retrieved October 20, 2015, http://www.nytimes.com/2013/06/14/us/supreme-court-rules-human-genes-may-not-be-patented.html?_r=0.

Lively, Mathew W. (July 13, 2014). Yellow Fever Plot of 1864 Targeted Lincoln, U.S. Cities. *Civil War Profiles*. Retrieved August 29, 2014, http://www.civilwarprofiles.com/yellow-fever-plot-of-1864-targeted-lincoln-u-s-cities/.

Longfellow, Henry Wadsworth (2010). *The Golden Legend*. Charleston, South Carolina: Nabu Press, p. 34.

Lorenzi, Rossella (December 3, 2007). *Killer Donkeys Were First Bioweapons*. Australian Broadcasting Corporation. Retrieved October 18, 2014, http://www.abc.net.au/science/articles/2007/12/03/2108080.htm.

Louisiana v Schmidt (July 26, 2000). Court of Appeal of Louisiana, Third Circuit, No. 99–1412.

MacQueen, Graeme (2014). *The 2001 Anthrax Deception: The Case for a Domestic Conspiracy*. Atlanta: Clarity Press.

Mahlen, Steven D. (October 2011). Serratia Infections: From Military Experiments to Current Practice. *Clinical Microbiology Reviews*, Vol. 24. No. 4, pp. 755–791.

Marburg Haemorrhagic Fever. (November 2012). *World Health Organization*. Retrieved April 25, 2016, http://www.who.int/mediacentre/factsheets/fs_marburg/en/.

Markel, Howard (September 29, 2014). How the Tylenol Murders of 1982 Changed the Way We Consume Medication. *PBS News Hour*. Retrieved March 25, 2016, http://www.pbs.org/newshour/updates/tylenol-murders-1982/.

Maron, Dina Fine (September 25, 2014). Weaponized Ebola: Is It Really a Bioterror Threat? *Scientific American*. Retrieved April 25, 2016, http://www.scientificamerican.com/article/weaponized-ebola-is-it-really-a-bioterror-threat/.

Martin, James W., George W. Christopher, and Edward M. Eitzen (2007). "History of Biological Weapons: From Poisoned Darts to Intentional Epidemics (Chapter One)." In *Medical Aspects of Biological Warfare*, edited by Zygmunt F. Dembek. Washington, D.C.: Borden Institute/Office of the Surgeon General.

Mauro, Ryan (2013). Boko Haram: Nigeria's Islamist Group. *The Clarion Project*. Retrieved February 10, 2016, http://www.clarionproject.org/factsheet/boko-haram-nigerias-islamist-group.

Mauroni, Albert J. (2006). *Chemical and Biological Warfare: A Reference Handbook*. Santa Barbara: ABC- CLIO.

Mayo Clinic Staff (December 16, 2015). West Nile Virus: Symptoms and Causes. *Mayo Clinic*. Retrieved May 7, 2016, http://www.mayoclinic.org/diseases-conditions/west-nile-virus/symptoms-causes/dxc-20166291.

Mayor, Adrienne (Autumn 1997). Dirty Tricks in Ancient Warfare. *MHQ: The Quarterly Journal of Military History*. Vol. 10, No. 1, pp. 32–37.

Mayor, Adrienne (2009). *Greek Fire, Poison Arrows, and Scorpion Bombs: Biological and Chemical Warfare in the Ancient World*. New York: Duckworth Overlook.

Maysh, Jeff (February 13, 2011). Miracle Love Story of Man Whose Father Injected Him with Aids as a Baby to Avoid Child Payments. *Daily Mail*. Retrieved January, October 2016, http://www.dailymail.co.uk/news/article-1356465/Miracle-love-story-Brryan-Jackson-whos-father-injected-Aids-baby-avoid-child-payments.html.

McCormack, Win (1987). *The Rajneesh Chronicles: The True Story of the Cult That Unleashed the First Act of Bioterrorism on U.S. Soil*. Portland, Oregon: Tin House Books.

McKenna, Phil (January 7, 2009). Rise of the Garage Genome Hackers. *New Scientist*. Retrieved October 15, 2015, https://www.newscientist.com/article/mg20126881–400-rise-of-the-garage-genome-hackers/.

Mehta, Anuj (2005). "Anthrax (*Bacillus anthracis)*" (Chapter Eight). In *Agents of Bioterrorism: Pathogens and Their Weaponization*, edited by Zubay, Geoffrey, et al. New York: Columbia University Press.

Metzker, Michael L., David P. Mindell, Xiao-Mei Liu, Robert G. Ptak, Richard A. Gibbs, and David H. Hillis (October 29, 2002). Molecular Evidence of HIV-1 Transmission in a Criminal Case. *Proceedings of the National Academy of Sciences*, Vol. 99, No. 22, pp. 14292–14297.

Michael, George (2012). *Lone Wolf Terror and the Rise of Leaderless Resistance*. Nashville: Vanderbilt University Press.

Miller, John Dudley (October 3, 2008). Postal Anthrax Aftermath: Has Biodefense Spending Made Us Safer? *Scientific American*. Retrieved July 31, 2015, http://www.scientificamerican.com/article/postal-anthrax-aftermath/.

Miller, Judith (October 14, 2001). Fear Hits Newsroom in a Cloud of Powder. *New York Times*. Retrieved May 5, 2015, http://www.nytimes.com/2001/10/14/national/14LETT.html?pagewanted=1.

Miller, Judith, Stephen Engelberg, and William J. Broad (September 4, 2001). U.S. Germ Warfare Research Pushes Treaty Limits. *New York Times*. Retrieved April 22, 2015, http://www.nytimes.com/2001/09/04/world/us-germ-warfare-research-pushes-treaty-limits.html.

Miller, Judith, Stephen Engelberg, and William Broad (2001/2002). *Germs: Biological Weapons and America's Secret War*. New York: Touchstone (Simon & Schuster).

Milne, Hugh (1986). *Bhagwan: The God That Failed*. New York: St. Martin's Press.

Morser, Kira, Rohit Puskoor, and Geoffrey Zubay (2005). "Salmonella" (Chapter Thirteen). In *Agents of Bioterrorism: Pathogens and Their Weaponization*, edited by Zubay, Geoffrey, et al. New York: Columbia University Press.

Mowatt-Larssen, Rolf (January 2010). *Al Qaeda Weapons of Mass Destruction Threat: Hype or Reality?* Cambridge, Massachusetts: Belfer Center for Science and International Affairs, Harvard Kennedy School.

Multidrug-Resistant Tuberculosis (MDR TB). (August 1, 2012). *Centers for Disease Control and Prevention*. Retrieved May 9, 2016, http://www.cdc.gov/tb/publications/factsheets/drtb/mdrtb.htm.

Murakami, Haruki (2000). *Underground: The Tokyo Gas Attack and the Japanese Psyche*. New York: Vintage International/Vintage Books (Random House).

Murphy, Paul (June 21, 2014). Matsumoto: Aum's Sarin Guinea Pig. *Japan Times*. Retrieved September 11, 2016, http://www.japantimes.co.jp/news/2014/06/21/national/history/matsumoto-aums-sarin-guinea-pig/.

National Research Council, Board of Life Sciences, Division on Earth and Life Studies, Technology and Law Committee on Science, Policy and Global Affairs Division, Committee on Review of the Scientific Approaches Used During the FBI's Investigation of the 2001 Bacillus Anthracis Mailings (June 1, 2011). *Review of the Scientific Approaches Used During the FBI's Investigation of the 2001 Anthrax Letters*. Washington, D.C.: National Academies Press.

Newitz, Annalee (February 26, 2002). Genome Liberation. *Salon*. Retrieved October 18, 2015, http://www.salon.com/2002/02/26/biopunk/.

9/11 Not as Bad as IRA, Says Doris Lessing. (October 24, 2007). *The Telegraph*. Retrieved

March 7, 2016, http://www.telegraph.co.uk/news/uknews/1567144/911-not-as-bad-as-IRA-says-Doris-Lessing.html.

Nixon, Richard (November 25, 1969). *Remarks Announcing Decisions on Chemical and Biological Defense Policies and Programs*. The American Presidency Project. Retrieved March 29, 2015, http://www.presidency.ucsb.edu/ws/?pid=2344.

Ohbu, S., A. Yahashina, T. Yamaguchi, T. Murai, K. Nakano, Y. Matsui, R. Mikami, K. Sakurai, and S. Hinohara (July 1997). Sarin Poisoning on Tokyo Subway. *Southern Medical Journal*. Vol. 90, No. 6, pp. 587–93.

Onion, Amanda (October 5, 2001). Lessons from Failed 1993 Biological Attack. *ABC News*. Retrieved August 10, 2016, http://abcnews.go.com/Technology/story?id=98249&page=1.

Opportunity May Be More Important Than Profession in Serial Homicide (editorial). (April 21, 2001). *British Medical Journal*. Vol. 322, No. 7292, p. 993.

Oregon Experience: Rajneeshpuram (transcript). (November 19, 2012). *Oregon Public Broadcasting*. Retrieved July 6, 2016, http://www.opb.org/television/programs/oregonexperience/segment/rajneeshpuram/.

Osterholm, Michael T., Kristine A. Moore, Nicholas S. Kelley, Lisa M. Brosseau, Gary Wong, Frederick A. Murphy, Clarence J. Peters, James W. LeDuc, Phillip K. Russell, Michel Van Herp, Jimmy Kapetshi, Jean-Jacques T. Muyembe, Benoit Kebela Ilunga,James E. Strong, Allen Grolla, Anja Wolz, Brima Kargbo, David K. Kargbo, Pierre Formenty, David Avram Sanders, and Gary P. Kobinger (February 19, 2015). Transmission of Ebola Viruses: What We Know and What We Do Not Know. *mBio (American Society for Microbiology)*. Retrieved May 30, 2016, http://mbio.asm.org/content/6/2/e00137-15.full.

Oxford English Dictionary, s.v. "phylogenetics."

Oxford English Dictionary, s.v. "(toxic) toxikon."

Pangi, Robyn (February 2002). Consequence Management in the 1995 Sarin Attacks on the Japanese Subway System (Discussion Paper 2002–4). *Belfer Center for Science and International Affairs*. Retrieved September 26, 2016, http://belfercenter.ksg.harvard.edu/files/consequence_management_in_the_1995_sarin_attacks_on_the_japanese_subway_system.pdf.

Passaro, D. J., S. B. Werner, J. McGee, W. R. MacKenzie, and D. J. Vugia (March 18, 1998). Wound Botulism Associated with Black Tar Heroin Among Injecting Drug Users." *Journal of the American Medical Association (JAMA)*. Vol. 279, No. 11, pp. 859–863.

Paterson, Tony (January 22, 2016). Return of Germany's Baader-Meinhof Terrorist Gang—They're Desperate for Cash and Have a Dog. *The Independent*. Retrieved February 1, 2016, http://www.independent.co.uk/news/world/europe/return-of-germanys-baader-meinhof-terrorist-gang-they-re-desperate-for-cash-and-have-a-dog-a6828366.html.

Philipkoski, Kristen (April 20, 2001). Biology Yearns to Be Free. *Wired*. http://archive.wired.com/medtech/health/news/2001/04/43151?currentPage=all.

Pita, René, and Rohan Gunaratna (May 15, 2009). Revisiting al-Qa`ida's Anthrax Program. *CTC Sentinel (Combating Terrorism Center at West Point)*, Vol. 2, No. 5, Article 4. Retrieved February 28, 2015, https://www.ctc.usma.edu/posts/revisiting-al-qaida%E2%80%99s-anthrax-program.

Pletcher, Kenneth (2016). Tokyo Subway Attack of 1995. *Encyclopaedia Britannica Online*. Retrieved September 19, 2016, https://www.britannica.com/event/Tokyo-subway-attack-of-1995.

Pollack, Andrew (March 29, 1995). Japanese Police Say They Found Germ-War Material at Cult Site. *New York Times*. Retrieved August 24, 2016, http://www.nytimes.com/1995/03/29/world/japanese-police-say-they-found-germ-war-material-at-cult-site.html?rref=collection%2Ftimestopic%2FAum%20Shinrikyo.

Post, Jerrold M. (2000). "Psychological and Motivational Factors in Terrorist Decision-Making: Implications for CBW Terrorism" (Appendix). In *Toxic Terror: Assessing Terrorist Use of Chemical and Biological Weapons*, edited by Jonathan B. Tucker. Cambridge, Massachusetts: Belfer Center for Science and International Affairs, Harvard Kennedy School.

Post, Jerrold M. (2004). "Prospects for Chemical/Biological Terrorism: Psychological Incentives and Restraints" (Chapter 6). In *Bioterrorism: Psychological and Public Health Interventions*, edited by Ursano, Robert J., Ann E. Norwood, and Carol S. Fullerton. Cambridge, England: Cambridge University Press.

Powell, Colin L. (October 5, 2003) Remarks to the United Nations Security Council. *U.S. Department of State*. Retrieved May 23, 2015, https://web.archive.org/web/20070109235502/http://www.state.gov/secretary/former/powell/remarks/2003/17300.htm.

Prasinos, Chloe, and Steven Jackson (October 1, 2015). Episode 194: Rajneeshpuram. *99% Invisible* (KALW Public Radio). Retrieved July 2, 2016, http://99percentinvisible.org/episode/rajneeshpuram/.

Puskoor, Rohit, and Geoffrey Zubay (2005). "Smallpox (Variola Virus)" (Chapter Eleven). In *Agents of Bioterrorism: Pathogens and Their Weaponization*, edited by Zubay, Geoffrey, et al. New York: Columbia University Press.

Rajneesh, Bhagwan Shree (1988). "Guilt Is Inverted Revenge" (Chapter 6). In *Hari Om Tat Sat: The Divine Sound—That Is the Truth* (English Discourse series). Retrieved June 29, 2016, http://www.oshorajneesh.com/download/osho-books/Tantra/Hari_Om_Tat_Sat.pdf.

Rajneesh and Company Pull Up Stakes from Oregon as Guru's Vision in Desert Becomes a Mirage. (December 30, 1985). *The Oregonian/OregonLive*. Retrieved August 1, 2016, http://www.oregonlive.com/rajneesh/index.ssf/1985/12/rajneesh_and_company_pull_up_s.html.

RamaRao, Golime, Prachiti Afley, Jyothiranjan Acharya, and Bijoy Krishna Bhattacharya (April 4, 2014). Efficacy of Antidotes (Midazolam, Atropine and HI-6) on Nerve Agent Induced Molecular and Neuropathological Changes. *BMC Neuroscience*. Vol. 15, No. 47. Retrieved September 1, 2016, http://bmcneurosci.biomedcentral.com/articles/10.1186/1471-2202-15-47.

Raoult, D. (1990). Host Factors in the Severity of Q Fever. *Annals of the New York Academy of Sciences*. Vol. 590, pp. 33–38.

Reader, Ian (2000). *Religious Violence in Contemporary Japan: The Case of Aum Shinrikyo*. Honolulu: University of Hawaii Press.

Regis, Ed (1999). *The Biology of Doom: The History of America's Secret Germ Warfare Project*. New York: Henry Holt.

Repp, Martin (2011). "Religion and Violence in Japan: The Case of Aum Shinrikyo" (Chapter 7). In *Violence and New Religious movements*, edited by James R. Lewis. Oxford, England: Oxford University Press.

Riedel, Stefan (2004). Biological Warfare and Bioterrorism: A Historical Review. *Baylor University Medical Center Proceedings*. Vol. 17, No. 4, pp. 400–406.

Rizzo, Tony (January 22, 2015). Debora Green, Imprisoned for the 1995 Arson Murder of Two Children, Is Denied New Sentencing. *Kansas City Star*. Retrieved December 28, 2015, http://www.kansascity.com/news/local/crime/article7939620.html.

Robespierre, Maximilien (2007). *Virtue and Terror*. London: Verso.

Robles, Frances (June 20, 2015). Dylann Roof Photos and a Manifesto Are Posted on Website. *New York Times*. Retrieved on February 20, 2016, http://www.nytimes.com/2015/06/21/us/dylann-storm-roof-photos-website-charleston-church-shooting.html.

Romano, James A., Brian J. Lukey, and Harry Salem (2008). *Chemical Warfare Agents: Chemistry, Pharmacology, Toxicology, and Therapeutics* (Second Edition). Boca Raton, Florida: CRC Press (Taylor & Francis).

Rowell, Andrew (1996). *Green Backlash: Global Subversion of the Environment Movement*. London: Routledge.

Rule, Ann (1997). *Bitter Harvest*. New York: Simon & Schuster.

Rybicki, Edward (January 1990). The Classification of Organisms at the Edge of Life, or Problems with Virus Systematics. *South African Journal of Science*, Vol. 86, pp. 182–186.

Sageman, Marc (March/April 2008). The Next Generation of Terror. *Foreign Policy*, No. 165, pp. 37–42.

Sainz, Adrian (February 25, 2002). Ernesto Blanco Getting Used to Fame as Anthrax Survivor. *The Ledger*. Retrieved April 12, 2015, http://www.theledger.com/article/20020225/NEWSCHIEF/302259984?p=4&tc=pg.

Saito, Tomoya (Autumn 2010). Tokyo Drift? *CBRNe World*. Retrieved September 26, 2016, http://www.cbrneworld.com/_uploads/download_magazines/CBRNe_world_autumn_2010_Tokyo_drift.pdf.

Sawyer, Richard D. (2007). *The Tao of Deception: Unorthodox Warfare In Historic and Modern China*. New York: Basic Books.

Schmidt, Markus (2008). Diffusion of Synthetic Biology: A Challenge to Biosafety. *Systems and Synthetic Biology*. Vol. 2, pp. 1–6.

Schrage, Michael (January 31, 1988). Playing God in Your Basement. *Washington Post*. Retrieved on October 15, 2015, https://www.washingtonpost.com/archive/opinions/1988/01/31/playing-god-in-your-basement/618f174d-fc11-47b3-a8db-fae1b8340c67/.

Schutzer, S. E., Budowle, B., and Atlas, R. M. (2005). Biocrimes, Microbial Forensics, and the Physician. *PLoS Medicine*. Vol. 2, No. 12: e337, pp. 1242 -1247.

Searcely, Dionne, and Marc Santora (November 18, 2015). Boko Haram Ranked Ahead of ISIS for Deadliest Terror Group. *New York Times*. Retrieved February 10, 2016, http://www.nytimes.com/2015/11/19/world/africa/boko-haram-ranked-ahead-of-isis-for-deadliest-terror-group.html.

Shane, Scott (September 18, 2008). Senator, Target of Anthrax Letter, Challenges F.B.I. Finding. *New York Times*. Retrieved July 4, 2015, http://www.nytimes.com/2008/09/18/washington/18anthrax.html.

Shane, Scott, and Eric Lichtblau (June 28, 2008). Scientist Is Paid Millions by U.S. in Anthrax Suit. *New York Times*. Retrieved May 8, 2015, http://www.nytimes.com/2008/06/28/washington/28hatfill.html.

Shane, Scott, and Eric Lichtblau (August 5, 2008). Pressure Grows for F.B.I.'s Anthrax Evidence. *New York Times*. Retrieved June 25, 2015, http://www.nytimes.com/2008/08/05/washington/05anthrax.html?pagewanted=all.

Shane, Scott, and Eric Lichtblau (August 6, 2008). F.B.I. Presents Anthrax Case, Saying Scientist Acted Alone. *New York Times*. Retrieved June 28, 2015, http://www.nytimes.com/2008/08/07/washington/07anthrax.html?ref=washington.

Shay, Roshani (October 23, 2010). Rajneeshpuram Residents Profile. *OshoNews/Online Magazine*. Retrieved July 5, 2016, http://www.oshonews.com/2010/10/23/rajneeshpuram-residents-profile/.

Shea, Dana A., and Frank Gottran (April 17, 2013). *Ricin: Technical Background and Potential Role in Terrorism* (RS21383). Washington, D.C.: Congressional Research Service.

Sheela, Ma Anand (2014). *Don't Kill Him! The Story of My Life with Bhagwan Rajneesh*. New Dehli, India: Fingerprint! (Prakash Books India, Ltd.).

Shieh, Wun-Ju, Jeannette Guarner, Christopher Paddock, Patricia Greer, Kathleen Tatti, Marc Fischer, Marci Layton, Michael Philips, Eddy Bresnitz, Conrad P. Quinn, Tanja Popovic, Bradley A. Perkins, Sherif R. Zaki, and the Anthrax Bioterrorism Investigation Team (November 2003). The Critical Role of Pathology in the Investigation of Bioterrorism-Related Cutaneous Anthrax. *American Journal of Pathology*, Vol. 163, No. 5, pp. 1901–1910.

Simon, Jeffrey D. (2013). *Lone Wolf Terrorism: Understanding the Growing Threat*. Amherst, New York: Prometheus Books.

Skawińska, Mirosława (2009). Bioterrorism and Ecoterrorism—Contemporary Dangers for the International Safety. *Studia Medyczne*. Vol. 16, pp. 7–11.

Smallpox Fact Sheet: Smallpox Disease Overview. (February 6, 2007). *Centers for Disease Control and Prevention*. Retrieved April 28, 2016, http://emergency.cdc.gov/agent/smallpox/overview/disease-facts.asp.

Smith, J., and Andre Moncourt (2009). *The Red Army Faction, a Documentary History: Volume 1: Projectiles for the People*. Oakland: PM Press, Inc.

Smith, Jeffery Alan (1999). *War and Press Freedom: The Problem of Prerogative Power*. Oxford, England: Oxford University Press.

Stammer, Larry B. (May 15, 1987). Environment Radicals Target of Probe into Lumber Mill Accident. *Los Angeles Times*. Retrieved February 27, 2016, http://articles.latimes.com/1987-05-15/news/mn-5213_1_louisiana-pacific.

Steers, Edward, Jr. (2001). *Blood on the Moon: The Assassination of Abraham Lincoln*. Lexington: University Press of Kentucky.

Stephen, Chris (November 22, 2001). Kabul House of Anthrax Secrets. *London Evening Standard*. Retrieved March 2, 2015, http://www.standard.co.uk/news/kabul-house-of-anthrax-secrets-6312737.html.

Stern, Jessica (August 1999). The Prospect of Domestic Bioterrorism. *Emerging Infectious Disease*. Retrieved March 11, 2015, http://wwwnc.cdc.gov/eid/article/5/4/99-0410_article.

Stern, Jessica (1999). *The Ultimate Terrorists*. Boston: Harvard University Press.

Stern, Jessica (2003). *Terror in the Name of God: Why Religious Militants Kill*. New York: Ecco (HarperCollins).

Stern, Jessica, and Amit Modi (2010). "Producing Terror: Organizational Dynamics of Survival" (Chapter 11). In *Terrorism, Security, and the Power of Informal Networks*, edited by Jones, David Martin, Ann Lane, and Paul Schulte. Northampton, Massachusetts: Edward Elgar Publishing.

Stolberg, Sheryl Gay (September 30, 2001). A Nation Challenged: The Biological Threat; Some Experts Say U.S. Is Vulnerable to a Germ Attack. *New York Times*. Retrieved April 11, 2015, http://www.nytimes.com/2001/09/30/us/nation-challenged-biological-threat-some-experts-say-us-vulnerable-germ-attack.html.

Supreme Court of Kansas (March 23, 2007). State of Kansas, Appellee, v. Debora J. Green, Appellant. No. 94,162. Retrieved December 17, 2015, http://caselaw.findlaw.com/ks-supreme-court/1285263.html.

Takahashi, Hiroshi, Paul Keim, Arnold F. Kaufmann, Christine Keyst, Kimothy L. Smith, Kiyosu Taniguchi, Sakae Inouye, and Takeshi Kurata (January 2004). *Bacillus anthracis* Bioterrorism Incident, Kameido, Tokyo, 1993. *Emerging Infectious Diseases* Vol. 10, No. 1, pp. 117–120.

Taneda, Kenichiro (May 22, 2009). "Tokyo: Terror in the Subway" (Chapter 16). In *Medical Disaster Response: A Survival Guide for Hospitals in Mass Casualty Events 1st Edition*, edited by David Goldschmitt and Robert Bonvino. Boca Raton, Florida: CRC Press (Taylor & Francis).

Taubenberger, Jeffery K., Ann H. Reid, Raina M. Lourens, Ruixue Wang, Guozhong Jin, and Thomas G. Fanning (October 6, 2005). Characterization of the 1918 Influenza Virus Polymerase Genes. *Nature*, Vol. 437, pp. 889–893.

Taylor, Robert (May 11, 1996). All Fall Down. *New Scientist* Vol. 150, No. 2029, pp. 32–37.

Tell, David (September 16, 2002). The Hunting of Steven J. Hatfill. *The Weekly Standard*. Retrieved June 4, 2015, http://www.weeklystandard.com/Content/Public/Articles/000/000/001/623rbipi.asp.

Testimony of Alma Peralta (May 21, 1990). Grand Jury Testimony Attached to Hagan Motion to Reconsider Motion to Dismiss Indictment, *U.S. v. Susan Hogan and Sally Croft*, CR-146-MA #588, U.S. District Court Files, p. 24.

Thomas, Jo (December 4, 1998). Man Accused of Injecting H.I.V. in Son. *New York Times*. Retrieved January, July 2016, http://www.nytimes.com/1998/12/04/us/man-accused-of-injecting-hiv-in-son.html.

Thompson, Marilyn W. (2003). *The Killer Strain: Anthrax and a Government Exposed*. New York: HarperCollins.

Thuras, Dylan (January 9, 2014). The Secret's in the Sauce: Bioterror at the Salsa Bar. *Slate*. Retrieved July 18, 2016, http://www.slate.com/blogs/quora/2016/07/19/how_should_you_answer_an_interviewer_s_question_about_adding_value_to_the.html.

TIME Staff (January 11, 1999). Wrath of God. *TIME* magazine. Retrieved February 27, 2015, http://content.time.com/time/world/article/0,8599,2054517,00.html.

Touma, Salwa (2005). "Viral Encephalitis (Flaviviruses)" (Chapter Two). In *Agents of Bioter-*

rorism: Pathogens and Their Weaponization, edited by Zubay, Geoffrey, et al. New York: Columbia University Press.

Trevisanato, Siro (2007). The Biblical Plague of the Philistines Now Has a Name, Tularemia. *Medical Hypotheses*. Vol. 69, No. 5, pp. 1144–1146.

Tu, Anthony, and Eric A. Croddy (2005). "Plague" (entry). In *Weapons of Mass Destruction: An Encyclopedia of Worldwide Policy, Technology, and History*, edited by Croddy, Eric A., James J. Wirtz, and Jeffrey A. Larsen. Santa Barbara: ABC-CLIO, Inc.

Tucker, Jonathan B. (2000). "Introduction" (Chapter One). In *Toxic Terror: Assessing Terrorist Use of Chemical and Biological Weapons*, edited by Jonathan B. Tucker. Cambridge, Massachusetts: Belfer Center for Science and International Affairs, Harvard Kennedy School.

United Nations Security Council (May 29, 2007). Twenty-Ninth Quarterly Report on the Activities of the United Nations Monitoring, Verification and Inspection Commission in Accordance with Paragraph 12 of Security Council Resolution 1284 (1999). *United Nations*. Retrieved April 1, 2016, http://www.un.org/Depts/unmovic/new/documents/quarterly_reports/s-2007-314.pdf.

United Press International (November 10, 1983). L.A. Resident Gets 20 Years for '83 Bombing of Hotel Rajneesh. *Los Angeles Times*. Retrieved July 11, 2016, http://articles.latimes.com/1985-11-10/news/mn-3387_1_three-pipe-bombs.

U.S. Department of Justice (February 19, 2010). Amerithrax Investigative Summary. *U.S. Department of Justice*. Retrieved April 15, 2015, http://www.justice.gov/archive/amerithrax/docs/amx-investigative-summary.pdf.

U.S. District Court for the District of Columbia (October 31, 2007). Search Warrant Affidavit 07-514-M-01.

Vaccines, Blood & Biologics: Anthrax. (June 17, 2015). *U.S. Food and Drug Administration (U.S. Dept. of Health and Human Services)*. Retrieved April 21, 2016, http://www.fda.gov/BiologicsBloodVaccines/Vaccines/ucm061751.htm.

Vahid, Majidi (October 18, 2011). Ten Years After September 11 and the Anthrax Attacks: Protecting Against Biological Threats. *Federal Bureau of Investigation (FBI)*. Retrieved November 10, 2015, https://www.fbi.gov/news/testimony/ten-years-after-9-11-and-the-anthrax-attacks-protecting-against-biological-threats.

van Aken, Jan (2006). When Risks Outweigh Benefits. *European Molecular Biology Organization (EMBO) Reports*. Vol. 7, pp 10–13.

Vector-Borne Diseases: Overview. (February 2016). *World Health Organization*. Retrieved May 21, 2016, http://www.who.int/mediacentre/factsheets/fs387/en/.

Ward, Joseph Patrick, and Maria E. Garrido (2005). "Severe Acute Respiratory Syndrome (SARS)" (Chapter Nine). In *Agents of Bioterrorism: Pathogens and Their Weaponization*, edited by Zubay, Geoffrey, et al. New York: Columbia University Press.

Warrick, Joby, Marilyn W. Thompson, and Nelson Hernandez (August 2, 2008). A Scientist's Quiet Life Took a Darker Turn. *Washington Post*. Retrieved July 1, 2015, http://www.washingtonpost.com/wp-dyn/content/article/2008/08/01/AR2008080102326.html.

Waslekar, Sundeep (January 20, 1987). India's Gurus Fall Victim to West's Shrinking Market. *Pacific News Service*. Retrieved August 3, 2016, http://articles.latimes.com/1987-01-20/local/me-5756_1_western-india/.

Watson, Dale L. (February 6, 2002). The Terrorist Threat Confronting the United States (Testimony). *Federal Bureau of Investigation (FBI)*. Retrieved February 24, 2016, https://www.fbi.gov/news/testimony/the-terrorist-threat-confronting-the-united-states.

Weaver, James (April 14, 2001). Slow Medical Sleuthing. *New York Times*. Retrieved July 23, 2016, http://www.nytimes.com/2001/04/24/science/l-slow-medical-sleuthing-003751.html.

Weaver, James (February 29, 1985). The Town That Was Poisoned. *U.S. Congress. Congressional Record (Procedures & Debates), 99th Congress, 1st Session*, V. 131, 3–4, pp. 4185–4189.

Wessinger, Catherine (2000). *How the Millennium Comes Violently: From Jonestown to Heaven's Gate*. Chatham, New Jersey: Chatham House Publishers.

Whalen, Jeanne (May 12, 2009). In Attics and Closets, "Biohackers" Discover Their Inner Frankenstein. *Wall Street Journal*. Retrieved October 4, 2015, http://www.wsj.com/articles/SB124207326903607931.

Williams, Peter, and David Wallace (1989). *Unit 731: The Japanese Army's Secret of Secrets*. London: Hodder & Stoughton.

Willman, David (August 1, 2008). Apparent Suicide in Anthrax Case. *Los Angeles Times*. Retrieved July 2, 2015, http://articles.latimes.com/2008/aug/01/nation/na-anthrax1.

Willman, David (2011). *The Mirage Man: Bruce Ivins, the Anthrax Attacks, and America's Rush to War*. New York: Bantam Books.

Wohlsen, Marcus (2011/2012). *Biopunk: DIY Scientists Hack the Software of Life*. New York: Current (Penguin).

WuDunn, Sheryl (March 22, 1995). Terror in Tokyo: The Cult: Sect Says Government Staged the Gas Attack. *New York Times*. Retrieved September 27, 2016, http://www.nytimes.com/1995/03/22/world/terror-in-tokyo-the-cult-sect-says-government-staged-the-gas-attack.html.

Young, Alison (June 18, 2015). Army Lab Lacked Effective Anthrax-Killing Procedures for 10 Years. *USA Today*. Retrieved August 1, 2015, http://www.usatoday.com/story/news/2015/06/17/anthrax-shipments-bruce-ivins-emails/28883603/.

Zaitz, Les (April 14, 2011A). Rajneeshees in Oregon: The Untold Story—25 Years After Rajneeshee Commune Collapsed, Truth Spills Out—Part 1. *The Oregonian/OregonLive*. Retrieved June 20, 2016, http://www.oregonlive.com/rajneesh/index.ssf/2011/04/part_one_it_was_worse_than_we.html.

Zaitz, Les (April 14, 2011B). Rajneeshees in Oregon: The Untold Story—Rajneeshee Leaders Take Revenge on The Dalles' With Poison, Homeless—Part 3. *The Oregonian/OregonLive*. Retrieved June 20, 2016, http://www.oregonlive.com/rajneesh/index.ssf/2011/04/part_three_mystery_sickness_su.html.

Zilinskas, Raymond A. (2011). "Diane Thompson: A Case Study" (Entry). In *Encyclopedia of Bioterrorism Defense, 2nd Edition*, edited by Rebecca Katz and Raymond A. Zilinskas. New York: John Wiley & Sons, Inc.

Zimmer, Carl (March 5, 2012). Amateurs Are New Fear in Creating Mutant Virus. *New York Times*. Retrieved October 16, 2015, http://www.nytimes.com/2012/03/06/health/amateur-biologists-are-new-fear-in-making-a-mutant-flu-virus.html.

Index